Selected Topics in Biomedical Circuits and Systems

EDITORS

Minkyu Je
Korea Advanced Institute of Science and Technology, Republic of Korea

Myung Hoon Sunwoo
Ajou University, Republic of Korea

Tutorials in Circuits and Systems

For a list of other books in this series, visit www.riverpublishers.com

Series Editors

Amara Amara
IEEE CASS President

Yen-Kuang Chen
VP - Technical Activities, IEEE CASS

Yoshifumi Nishio
VP - Regional Activities and Membership, IEEE CASS

Published, sold and distributed by:

River Publishers
Alsbjergvej 10
9260 Gistrup Denmark

www.riverpublishers.com

Availability: February 2021

ISBN: Print: 9788770221481
E-book: 9788770221474

Published by River Publishers.

Library of Congress Cataloging-in-Publication Data: February 2021
Editors: Minkyu Je
Myung Hoon Sunwoo
Title: Selected Topics in Biomedical Circuits and Systems

Table of contents

Introduction

The integrated circuits and microsystems play a vital role in a variety of biomedical applications such as life-saving/changing miniature medical devices, surgical procedures with less invasiveness and morbidity, low-cost preventive healthcare solutions in daily life, effective chronic disease management, point-of-care diagnosis for early disease detection, high-throughput bio sequencing and drug screening, and groundbreaking brain-machine interface from deep understanding of human intelligence. In response to such strong demands on biomedical circuits and systems, a considerable amount of effort has been devoted to the research and development in this area, both by industry and academia.

This book, which belongs to the series of "Tutorials in Circuits and Systems," aims to provide readers with overview of trends and directions in the field of biomedical circuits and systems, basic knowledge and understanding of system-level and circuit-level requirements, operation principles, key factors of considerations, and design/implementation techniques, as well as introduction of recent advances in the art for the integrated circuits and microsystems for emerging biomedical applications.

Technical topics covered in this book include:

- Biomedical Microsystem Integration;
- Biomedical Sensor Interface Circuits;
- Neural Stimulation Circuits;
- Wireless Power Transfer Circuits for Biomedical Microsystems;
- Artificial Intelligence Processors for Biomedical Circuits and Systems;
- Neuro-Inspired Computing and Neuromorphic Processors for Biomedical Circuits and Systems.

This tutorial book, entitled *Selected Topics in Biomedical Circuits and Systems*, is ideal for personnel in medical devices and biomedical engineering industries as well as academic staff and postgraduate/research students in biomedical circuits and systems.

CHAPTER 01

Overview of Biomedical Microsystems and Integration Challenges

Minkyu Je

KAIST, Korea

Many factors, such as extended average lifespan, prevailing obesity, and globally aging population, are increasing the healthcare cost dramatically. Recent advances in semiconductor technologies, as well as innovations in IC design techniques, have led to microsystems with sensing and processing capabilities that can supplement, improve, or even entirely replace traditional biomedical diagnostic and therapeutic procedures. Integrated biomedical solutions based on IC technologies can offer remarkably effective ways of timely diagnosis, treatment, and management of diseases at a very low cost never seen before, by providing a seamless interface to various sensors and actuators, high-efficiency operation with various energy sources, high-level integration and miniaturization, embedded intelligence, and connectivity.

This chapter begins with introducing an overview of biomedical ICs and microsystems, which is followed by the discussion of challenges faced when integrating the biomedical microsystems as well as important factors to consider when designing key functional blocks such as sensor and actuator interface circuits, digital control and signal processing circuits, communication circuits, and energy supply and management circuits. Especially, the application dependence of biomedical microsystem designs and implementation approaches are investigated, along with several examples of microsystem design and implementation across different biomedical applications introduced. While we find the optimally crafted system design and implementation strategies can draw the maximum out of currently available technologies, on the one hand, the study, on the other hand, reveals the limitations, challenges, and bottlenecks of the technologies to overcome for a leap to the future generations of biomedical microsystems.

1 Outline

- Overview of Biomedical Microsystems
  - Needs of Biomedical ICs / Microsystems and Current Landscape
  - Typical Format of Biomedical Microsystems
  - Trend, Future Projection, and Vision
- Design Considerations and Integration Challenges
  - Microsystem
  - Sensor and Actuator Interface
  - Digital Control and Signal Processing
  - Communication
  - Energy Supply and Management
- Conclusion

OVERVIEW OF BIOMEDICAL MICROSYSTEMS

2 Why Biomedical ICs and Microsystems? (I)

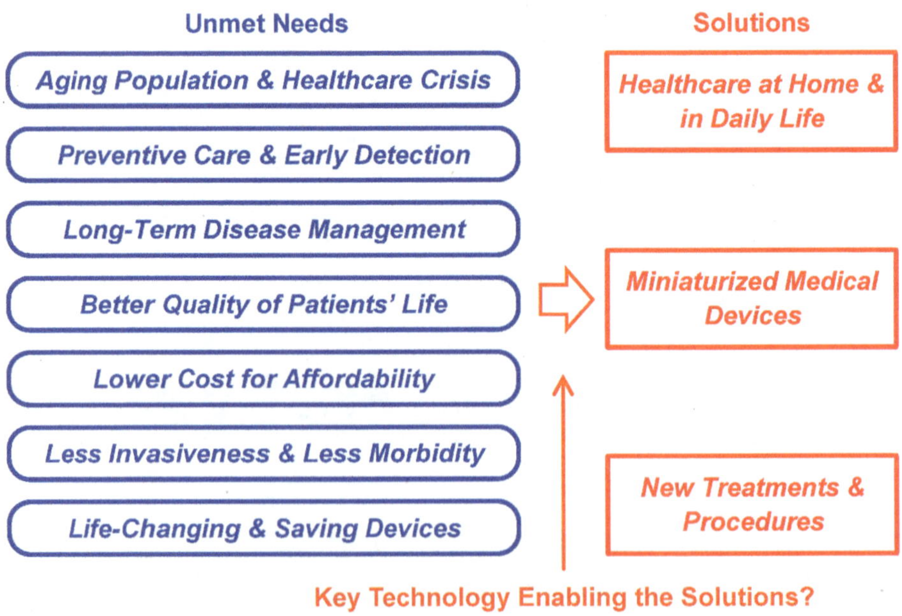

2 Why Biomedical ICs and Microsystems? (I)

Why is IC and microsystem technology important for emerging biomedical applications?

It's because the biomedical IC and microsystem are key technologies enabling the solutions for many critical unmet needs faced by current healthcare and clinical practices.

Unmet needs are many - healthcare crisis caused by the aging population, preventive care and early detection of diseases, long-term management of prevailing chronic diseases, better quality of life for suffering patients, more affordable diagnosis and therapeutics, less invasive procedures leading to less morbidity, and last but not least, life-changing and saving devices.

To address these needs, researchers are working hard to develop relevant solutions such as healthcare at home and in daily life, miniaturized medical devices, and innovative clinical treatment and procedures.

At the center of this development, there is biomedical IC and microsystem technology, as one of the key enablers.

3 Why Biomedical ICs and Microsystems? (II)

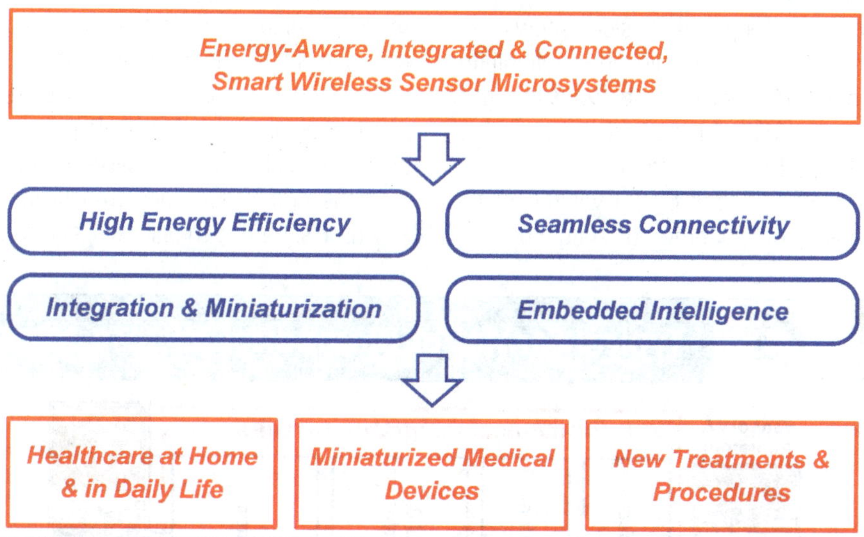

How does the IC and microsystem technology enable emerging next-generation biomedical solutions? What are the great values they add and the key roles they play?

First, the energy-aware IC and microsystem design provides high energy efficiency, which is extremely important in implantable medical devices and portable/wearable continuous health monitoring solutions.

Second, the IC and microsystem technology enables very-high-level integration and miniaturization of medical devices, especially for in implanted diagnostic and therapeutic devices, and surgical procedures with minimal invasiveness.

Third, the wireless function supported by the IC and microsystem technology provides medical devices with seamless connectivity and changes the healthcare paradigm.

Last but not least, the embedded intelligence enabled by smart ICs and microsystems bring the medical devices to the new horizon of closed-loop disease management consisting of sensing, processing, and therapy delivery.

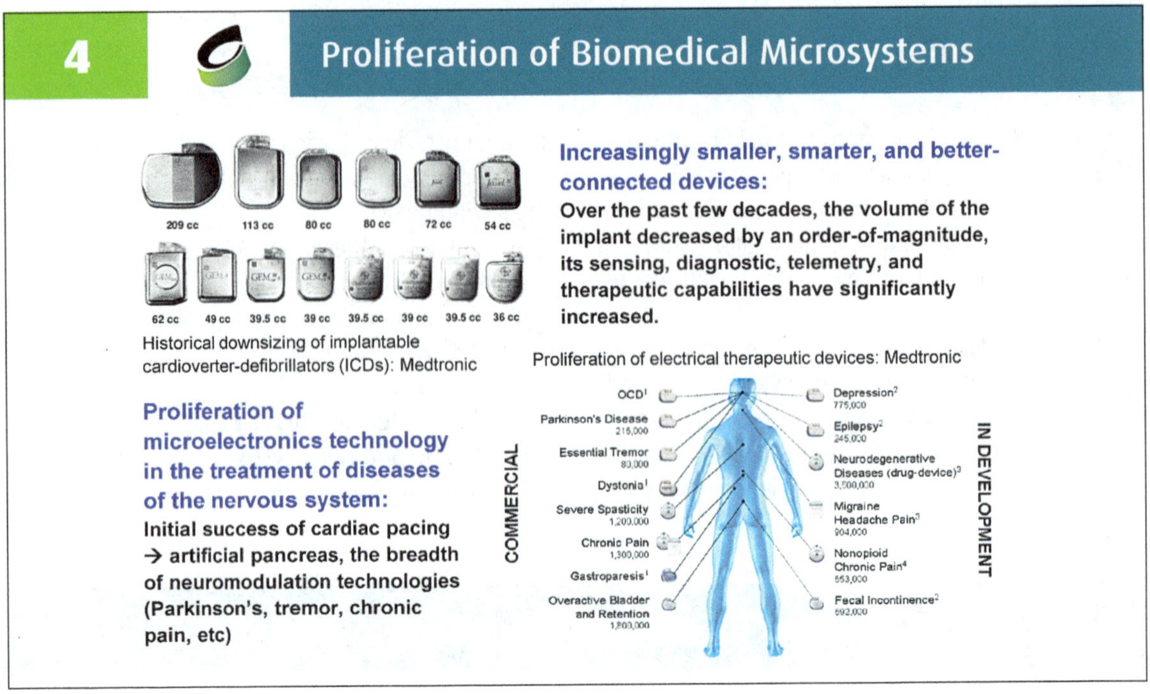

The figure on the upper-left side shows the Medtronic's implantable cardioverter-defibrillator devices from old models used in the 1980s to recent models. It presents a dramatic downsizing of medical devices over the past few decades.

More importantly, in spite of the volume of the implant decreased by an order of magnitude, its sensing, diagnostic, telemetry, and therapeutic capabilities have significantly improved over time.

This pioneering success of cardiac implants led to a proliferation of electrical technology in the treatment for the diseases of the nervous system, such as the artificial pancreas and various neuromodulation implants for Parkinson's, tremor, chronic pain, and so on.

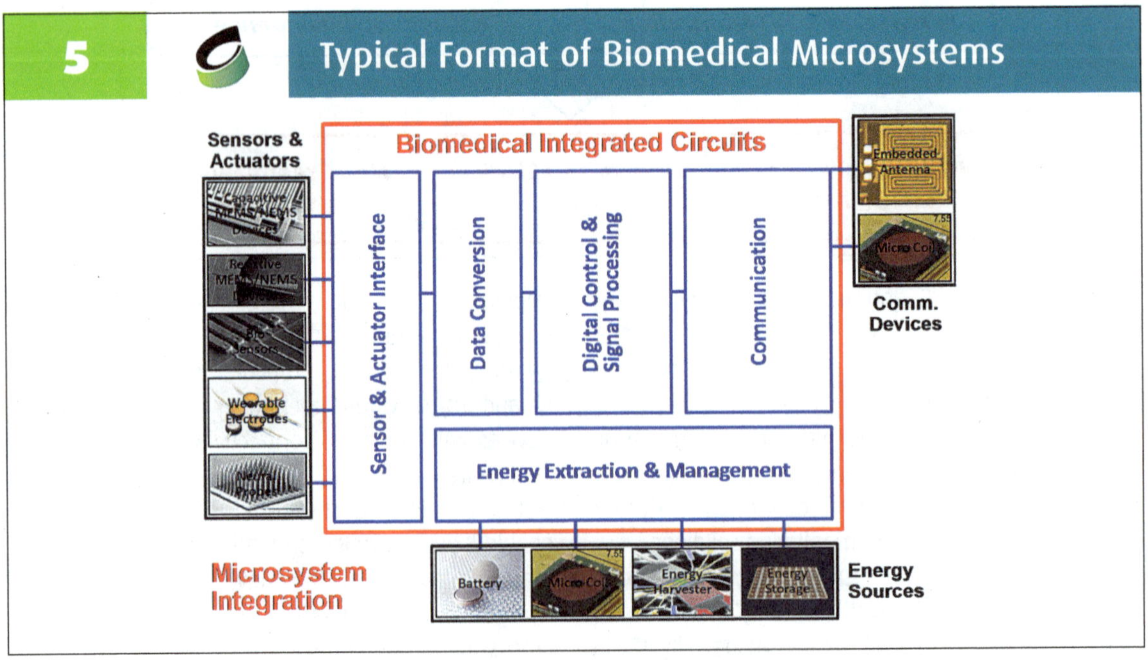

5 Typical Format of Biomedical Microsystems

This slide shows the typical format of the biomedical microsystem, which enables various biomedical solutions, including implantable medical devices, minimally invasive surgical operations, wireless health monitoring, and point-of-care diagnosis.

The whole microsystem platform involves sensors and actuators, energy sources, and passive components for communications while having an ultra-low-power smart biomedical IC technology placed at the center of the platform.

The biomedical IC consists of several subsystems such as sensor interface, data conversion, digital signal processing, energy delivery and management, and communication.

Note that the biomedical IC has a very interface-oriented nature. Except the data conversion and following digital signal processing which are to implement embedded intelligence of the microsystem, all the rest are dealing with interfaces, interactions, and communications, with sensors, actuators, energy sources, and outer world.

6 More-than-Moore Evolution and Biomedical Microsystems

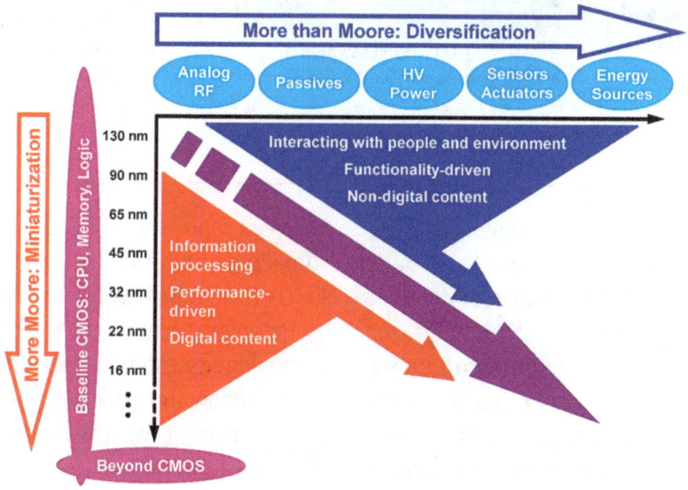

Now, let's try to understand the development of the biomedical IC and microsystem in the context of the semiconductor technology development trend and roadmap.

In addition to continuing tremendous efforts exercised by the semiconductor industry and research societies to push the traditional Moore's law forward, recently a rather new approach of More-than-Moore evolution of semiconductor technology is getting rapidly growing attentions.

More-than-Moore approach explores a new area of microelectronics and nanoelectronics, which reaches beyond the boundaries of conventional semiconductor technologies and applications. It integrates diverse functionalities in the form of system-in-package (SiP) or 3D IC to create innovative solutions for non-traditional applications, rather than focusing on technology scaling to improve performance and reduce power and cost as pursued in More-Moore approach.

The point here is that this More-than-Moore development is greatly relevant to biomedical ICs and microsystems where we integrate different functionalities of integrated circuits, sensors, actuators, passives, and energy sources in highly miniaturized forms using heterogeneous integration technology.

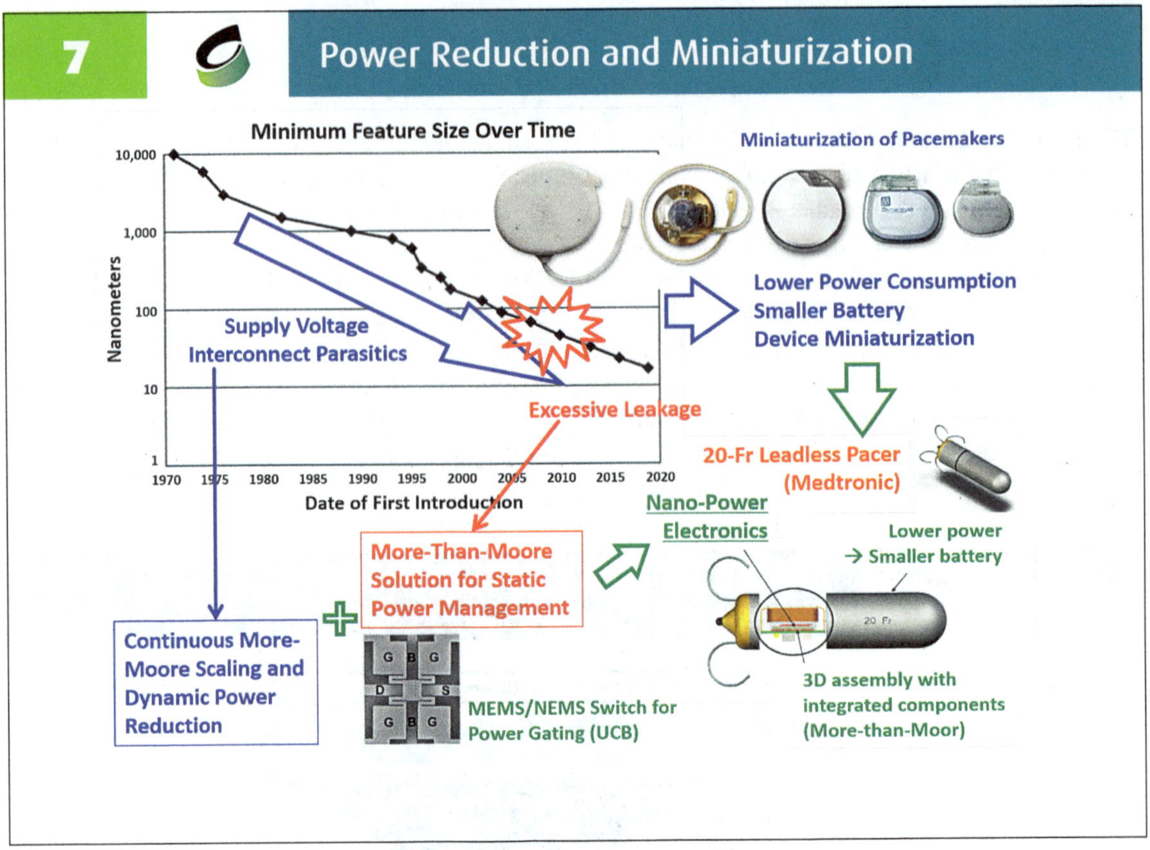

As the scaling of CMOS technology, which is driven by Moore's law continues, the power consumption of integrated circuits and microsystems reduces because of the lower supply voltage and smaller interconnect parasitics. This reduced power consumption allows the use of low-capacity energy sources with smaller size and results in the further miniaturization of medical devices. Such development trends have been well demonstrated by the historical downsizing of implantable pacemakers, for example.

To shrink the size of future medical devices even further, the power consumption should keep reducing so that the devices can be powered with even smaller energy sources. Considering that the count of transistors in the integrated circuits and microsystems keep increasing to implement more complex and sophisticated functions and embed more powerful intelligence, using the most scaled CMOS technology would become mandatory.

However, when the extremely scaled CMOS technology is used for lowering the dynamic power consumption, the static power consumption due to the leakage current becomes more and more significant. To solve this issue, innovative More-than-Moore approaches such as employing MEMS/NEMS switches for ultra-low-leakage power gating may be required.

At the same time, to minimize the form factor of medical devices, more advanced assembly and integration approaches such as 2.5D/3D integration based on through-silicon interposer (TSI) and through-silicon-via (TSV) technologies would need to be introduced.

8 Large-Scale Integration

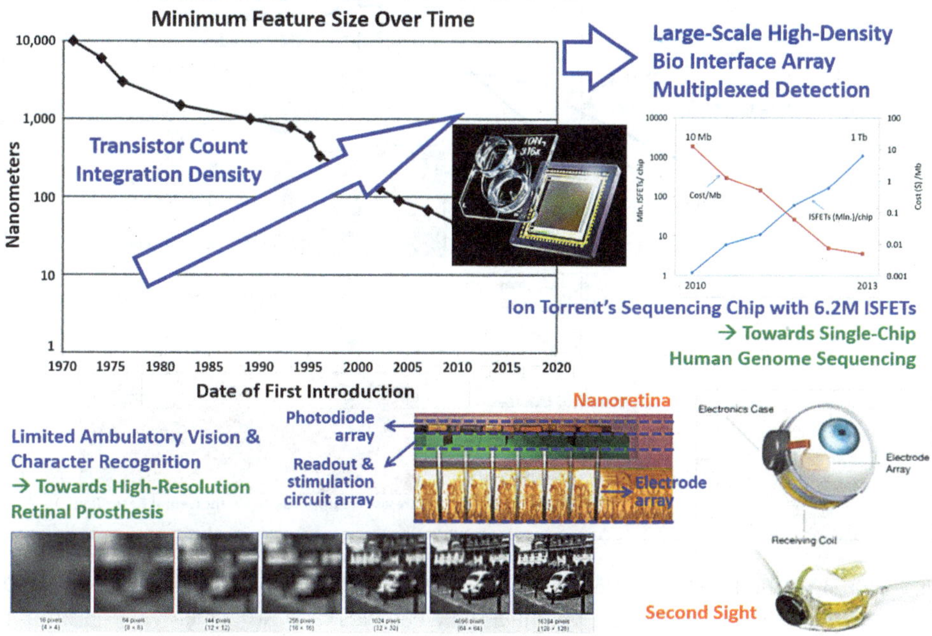

The advancement of CMOS technology also helps to scale up the integration level of medical devices. The scaled CMOS process has enabled large-scale high-density bio interface arrays with a capability of multiplexed operation.

For example, the Ion Torrent's DNA sequencing chip has 6.2 million ion-sensitive field-effect-transistors (ISFETs) integrated inside, and researchers in this field are working for realizing the vision of single-chip human genome sequencing.

The retinal prosthesis is another example. Although the current version of retinal prosthetic devices provides an only limited capability for ambulatory vision and character recognition, there are significant efforts exerted to implement high-resolution retinal prosthetic devices by taking advantages of the scaled CMOS technology.

The channel count as well as density of neural interface arrays for neurological devices and brain-machine interfaces have also been scaled up with the help of advanced CMOS processes.

9 Seamless Connectivity

The operation speed and radio-frequency (RF) performance of CMOS transistors have been improved drastically over the past decades. As a result, highly integrated low-power CMOS radio has been enabled.

By using such CMOS RF transceiver ICs, the capsule endoscopy could be realized. Once the patient swallows the endoscopy capsule, it travels through the gastrointestinal (GI) tract with taking internal images and transmitting the taken images to the external receiver module placed outside the patient body.

The wireless neuroprobe array for neural recording is also implemented using the CMOS RF transceiver IC

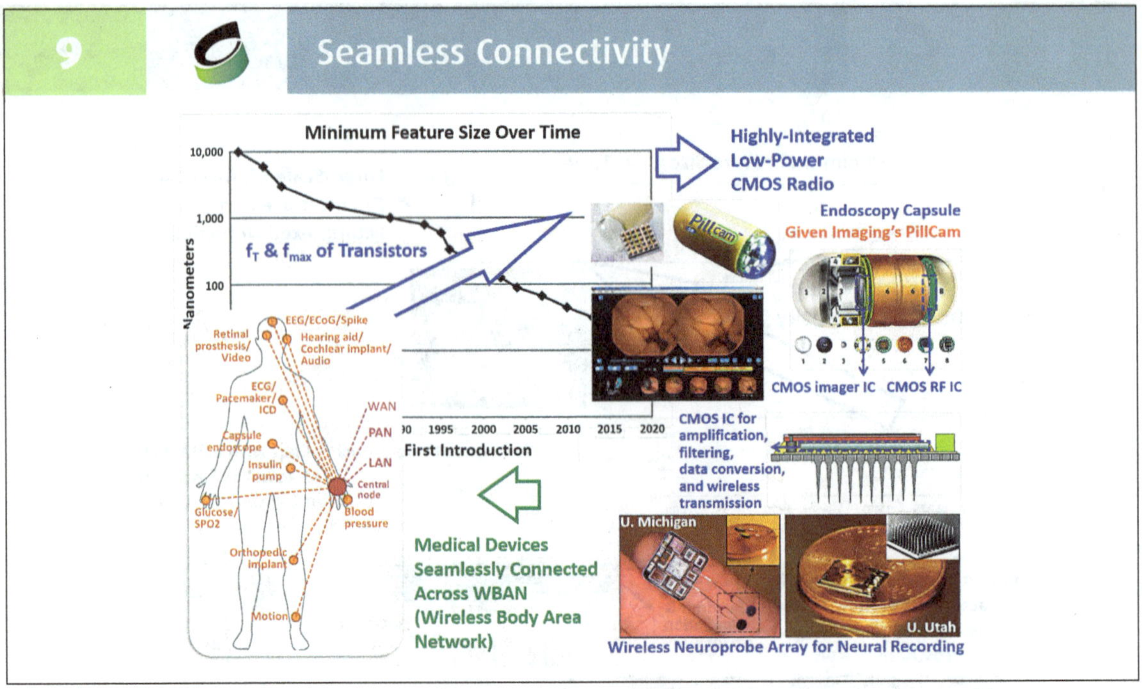

to transmit a large amount of neural signal data such as local field potentials (LFPs) and action potentials (APs) which are recorded from a large number of microelectrodes in the neuroprobe array.

As the medical device technologies advance further, various medical devices in the form of wearables and implants will become present in the area of the human body, and all of them will be connected by wireless communication based on CMOS RF transceivers.

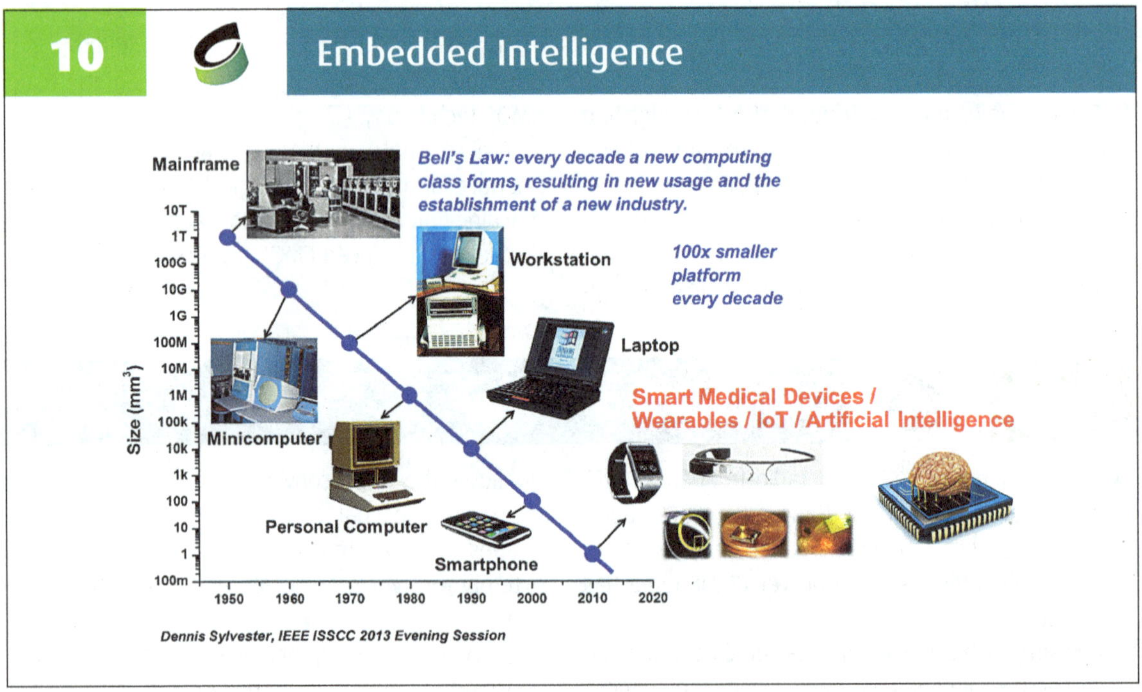

10 Embedded Intelligence

Bell's law states that every decade a new computing class forms, resulting in new usage and the establishment of a new industry. This statement well describes the development history of computers which started with mainframes in the 1950s and has progressed by a series of emergences of new computing classes such as minicomputers in the 1960s, workstations in the 1970s, personal computers in the 1980s, laptops in the 1990s, and finally, smartphones in the 2000s.

From the development history of computers, we can also observe that the physical size of computing platforms has been reduced by 100 times every decade.

Extrapolating this trend of development, a new computing platform with 100 times smaller size than the smartphone is expected to appear. As repeated throughout the development history of computers, this new platform would provide much stronger computing performance than its predecessor even with such a tremendously reduced form factor.

This new computing class is expected to be in the form of smart miniaturized devices such as IoT sensor nodes, wearables, and implants. The next-generation medical devices will be equipped with powerful embedded intelligence, which is based on this extremely miniaturized low-power high-performance computing class even mimicking human intelligence.

11 Trend: Evolution and Diversification

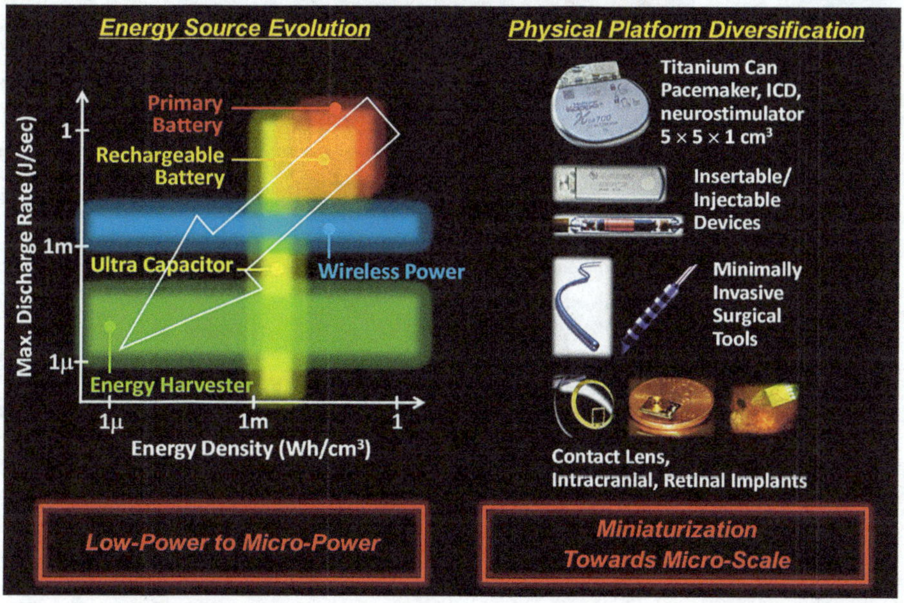

The medical devices are evolving in terms of their energy sources and diversifying in their physical platforms.

As the energy source evolves from primary to secondary batteries, to wireless power, and further to harvested energy, the available amount of energy to operate devices becomes smaller and smaller. To support this trend, biomedical ICs and microsystems are also evolving from low-power to micro-power.

Regarding the physical platform of the medical devices, from the traditional titanium can, many different platforms are making an entrance to the market, and generally speaking, those new platforms tend to require further miniaturization,

11 Trend: Evolution and Diversification

towards micro-scale. Extremely small biomedical ICs and microsystems provide a perfect solution for new emerging physical platforms.

Note that these two trends of power scaling and physical-form-factor scaling are closely interrelated. Considering that a significant portion of the device volume is occupied by the energy source, as the form factor of medical devices shrinks, the size of energy sources also reduces. Therefore, the form-factor scaling makes the reduction of the available amount of energy even more drastic and the need for ultra-low-power biomedical IC and microsystem extremely significant.

12 Towards Bioelectronic Medicine

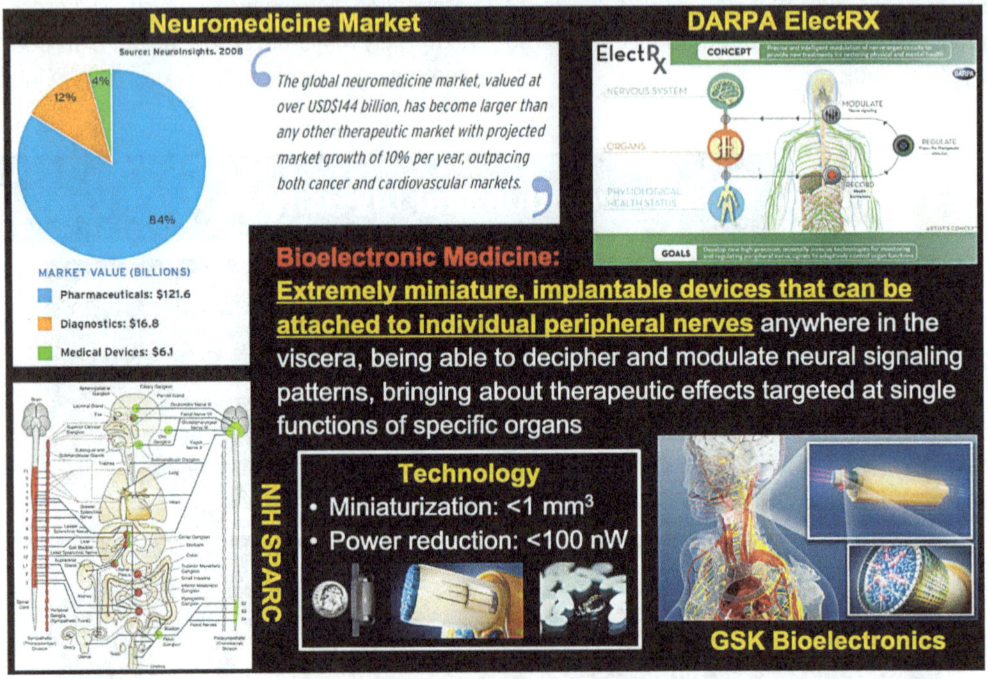

Continuing the development for such power reduction and miniaturization of medical devices, researchers believe that a new form of medicine, called bioelectronic medicine or electroceuticals, will emerge eventually.

Bioelectronic medicine is a new approach to diagnose and treat diseases of the human body in ways that current pharmaceutical interventions cannot. It is based on extremely miniaturized implantable devices that can be attached to individual peripheral nerves anywhere in the viscera, being able to decipher and modulate neural signaling patterns, bringing about therapeutic effects targeted at single functions of specific organs.

To realize this ambitious vision, the medical device technology should advance much more to achieve extreme miniaturization as well as power reduction down to the level of sub-mm^3 and sub-100 nW, respectively. Once developed, the bioelectronic medicine will cure diseases without the complicated side effects of pharmaceuticals and revolutionize the disease treatment and management, especially in cases where pharmaceuticals are unavailable or insufficiently successful.

13 BRAIN Mapping, Decoding, and Interfacing (I)

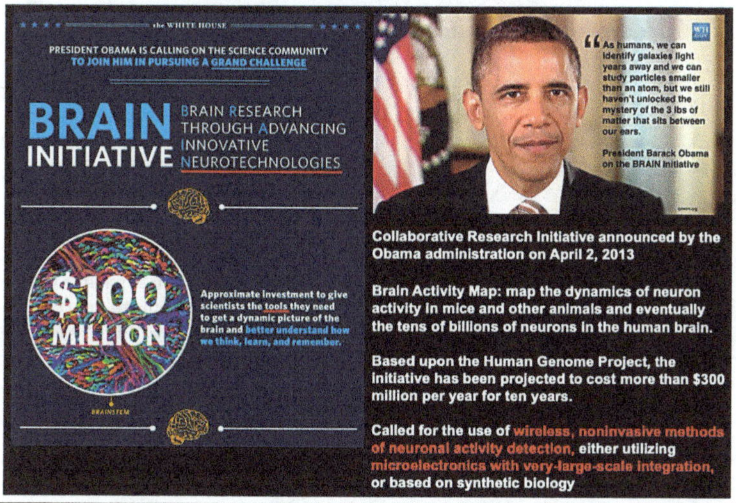

Another area where the biomedical ICs and microsystems can make significant contributions is brain mapping, decoding, and interfacing.

By obtaining a complete and dynamic picture of the human brain in action and comprehensive understanding of its functions, the mysteries of brain disorders such as Alzheimer's and Parkinson's diseases, depression, and traumatic brain injury can be resolved. Not only that, creating the functional activity map of the human brain can be regarded as a reverse-engineering process, which enables the construction of human-made electronics with an ultimate form of intelligence eventually.

However, the unimaginable complexity and dynamic characteristics of human brain impose a great challenge in unlocking its mysteries. Considering that the human brain contains nearly one hundred billion neurons with hundreds of trillions of dynamic connections, a huge leap in neurotechnology is fundamental prerequisite to provide unprecedented tools that can be used for investigating human brain in any useful way.

To promote the brain research through advancing innovative neurotechnologies in the United States, the Obama administration announced the Brain Initiative and called for the use of wireless, noninvasive methods of neuronal activity detection, either utilizing microelectronics with very-large-scale integration or based on synthetic biology.

14 BRAIN Mapping, Decoding, and Interfacing (II)

Neuroengineers are interested in increasing the number of neural recording channels incessantly with the ultimate goal of mapping the entire brain. Considering the number of neurons and synaptic connections in our brain, the current state-of-the-art systems monitor only a tiny fraction of neurons and their connections.

Over the last five decades, progress in neural recording techniques has allowed the number of simultaneously recorded neurons to double approximately every 7 years, mimicking Moore's law. Keeping this pace, it will take 200 more years to cover the full human brain, and hence, disruptive innovations of neurotechnologies are required.

The extremely scaled biomedical ICs and microsystems with ultra-high-density integration and ultra-low power consumption can be a crucial enabler for introducing a totally different pace of increase in the number of simultaneously recorded neurons.

14 BRAIN Mapping, Decoding, and Interfacing (II)

Moore's Law in Neuroscience

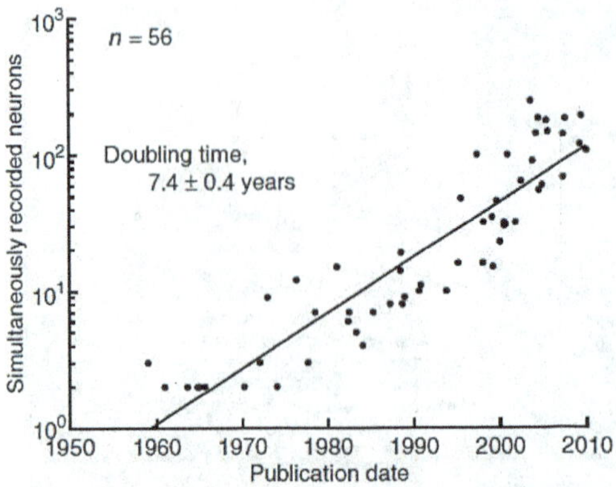

Over the last five decades, progress in neural recording techniques has allowed the **number of simultaneously recorded neurons to double approximately every 7 years**, mimicking Moore's law.

Keeping this pace, it will take **200 more years to cover the full human brain.**

Stevenson & Kording, "How advances in neural recording affect data analysis," Nature Neuroscience

DESIGN CONSIDERATIONS AND INTEGRATION CHALLENGES

MICROSYSTEMS

15 Device Size (Form Factor) (I)

- **Determined by the size of the physical object where the device is embedded**

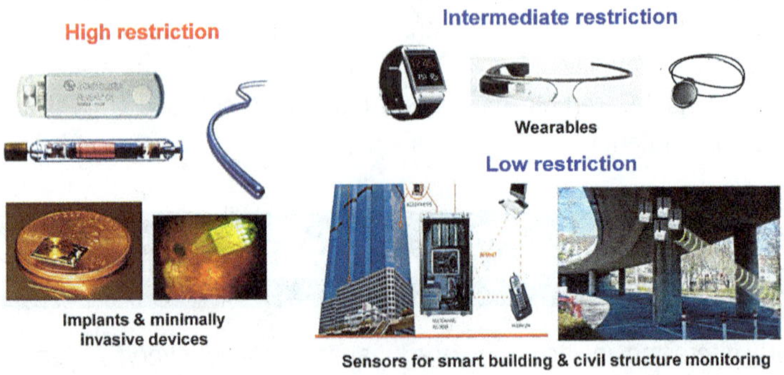

The device size or form factor is usually determined by the size of the physical object where the device is embedded.

For example, the sensor nodes for smart building and civil structure monitoring don't have a strict restriction in their size.

However, medical devices such as implants and minimally invasive devices have an extreme restriction.

For wearables either for healthcare or other lifestyle application, a small form factor is essential to let users feel comfortable when they wear those devices, but the requirement is not as stringent as the implants and minimally invasive devices.

16 Device Size (Form Factor) (II)

- **Determines available power and energy**

Battery

~600mWh/cc
~300mWh/g

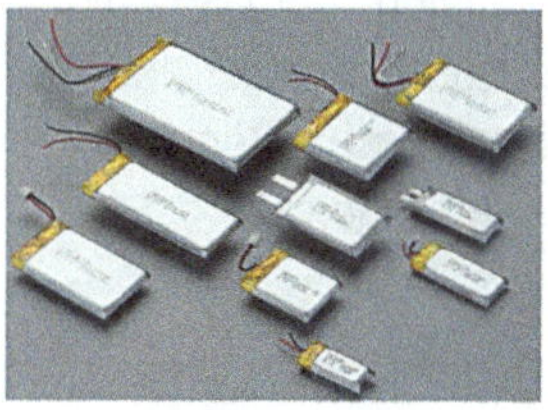

250-730mWh/cc
100-265mWh/g

Ambient Energy Source

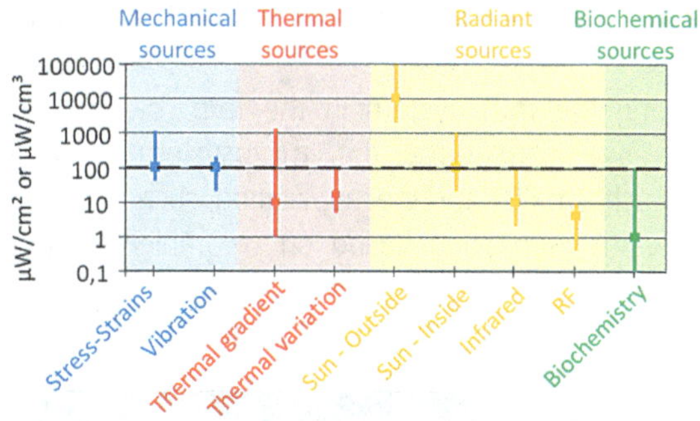

The device size that is constrained by its application, in turn, determines the available capacity of power and energy.

Typical primary lithium button cell battery provides 600 mWh of energy per cm^3, while rechargeable lithium-polymer cell has the capacity in the range of 250 to 730 mWh per cm^3. Therefore, with current battery technology, if a sensor microsystem of about 1-cm^3 size consumes 1-mW power on average, the microsystem can operate for several hundreds of hours before battery replacement or recharge.

If ambient energy sources are tapped to harvest necessary energy for microsystem operation, roughly about 1 to 100 μW per cm^3 of power is available depending on the kind of sources and environmental condition. Consequently, to realize an autonomous microsystem having a form factor less than 1 cm^3, the average power consumed for the overall operation of such a system should be lower than a few μW to be on the safe side.

17 Device Size (Form Factor) (III)

Low restriction

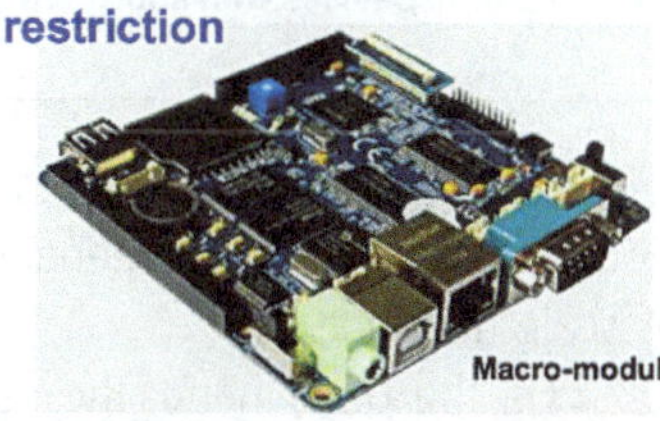

Macro-module

Intermediate restriction

Micro-module

System-on-chip

High restriction

Advanced packaging

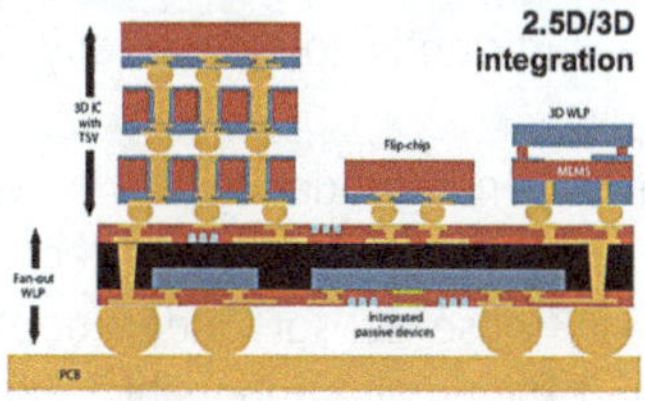

17 Device Size (Form Factor) (III)

The device form factor also determines the microsystem integration scheme.

When there is no or little constraint in the device size as in the case of bench-top medical instrument, the microsystem can be integrated in the form of macro-module.

By integrating the microsystem in the form of micro-module, we can make the microsystem form factor much smaller for the application such as wireless health monitoring based on wearable devices.

When there is a high restriction on the device form factor as in the case of implants, advanced integration technologies have to be used to achieve extreme miniaturization. System-on-chip approach, advanced packaging technology such as wafer-level chip-scale packaging, and 2.5D/3D IC approach based on through-silicon interposer and through-silicon-via technologies are such integration technologies for extreme miniaturization.

18 Device (or Power Source) Lifetime (I)

- **Determined by usage scenario, accessibility, and number of deployed devices**

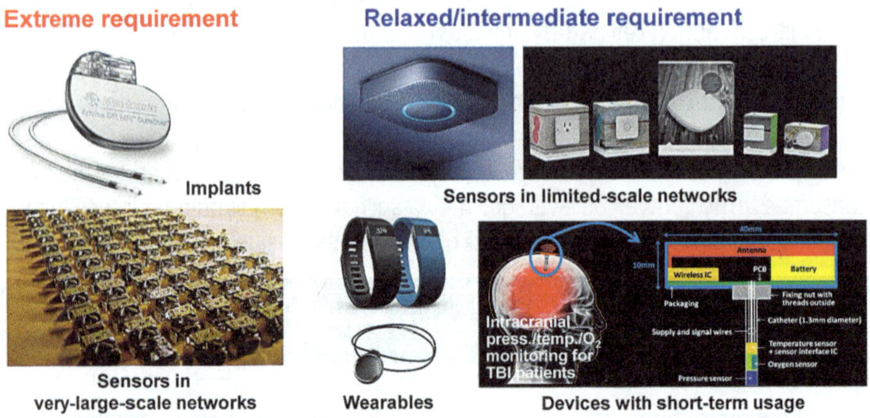

The power source or device lifetime is determined by the device usage scenario, accessibility of the installed device, and the number of deployed devices.

Since the implant is placed inside a patient's body and surgical procedures are involved to access, its lifetime should be sufficiently long, typically well over 10 years.

As the number of sensor nodes in the network grows larger, frequent replacement or recharge of power sources in those sensor nodes become more expensive and eventually impossible. Energy autonomous sensor nodes offer nearly infinite lifetime and zero maintenance cost for such large-scale networks.

On the other hand, the sensor nodes in limited-scale networks that we can find in the present smart home and wireless health monitoring applications for example, wearable devices, and devices with short-term usage have less strict requirement on their power source lifetime. Of course, even in those applications, longer lifetime is preferred, though not forced.

19 Device (or Power Source) Lifetime (II)

- **Determines allowable power consumption and power sourcing/management strategy**

Energy-autonomous sensor node

$P_{avg} < P_{harvest}$

Indoor PV harvesting:

- $1cm^2 \rightarrow \sim 10\mu W > P_{avg}$
- $1mm^2 \rightarrow \sim 100nW > P_{avg}$

Pacemaker

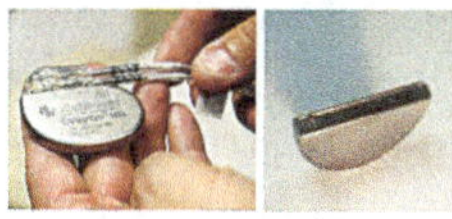

$P_{avg} \times T_{life} = E_{battery}$

Li-I_2 battery:

- $E_{battery} > 1Wh$
- $P_{avg} < 10\mu W$
- $T_{life} > 10$ years

Smart watch

$P_{avg} \times T_{op} = E_{battery}$

Rechargeable battery:

- $E_{battery} = 300mAh$
- $P_{avg} < 15mA$
- $T_{op} > 20$ hours

The required power source or device lifetime determines allowable power consumption and power sourcing/management strategy.

For the energy-autonomous sensor node, the average power consumption should be smaller than the harvested power throughput. When powered by indoor photovoltaic harvesting for example, the sensor node with a size of 1 cm³ has to consume the average power lower than 10 μW to achieve energy autonomy. If the size is 1 mm³, less than 100 nW should be consumed on average.

The implantable cardiac pacemaker with a lithium iodine battery of about 1-Wh capacity should operate with less than 10-μW average power consumption so that the lifetime longer than 10 years can be achieved.

To operate the smart watch for longer than 20 hours without a recharge, the average consumption needs to be lower than 15 mA when it is powered by the rechargeable battery having a capacity of 300 mAh.

20 Physical Interface with Surroundings (I)

- **Provides isolation barrier between the device and the surroundings to protect one against the other**

Hermetic encapsulation

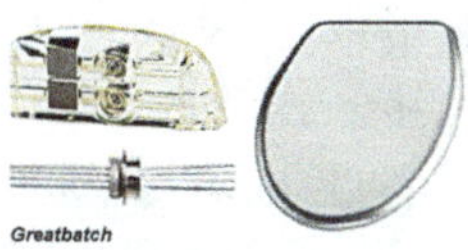

Implants: body fluids

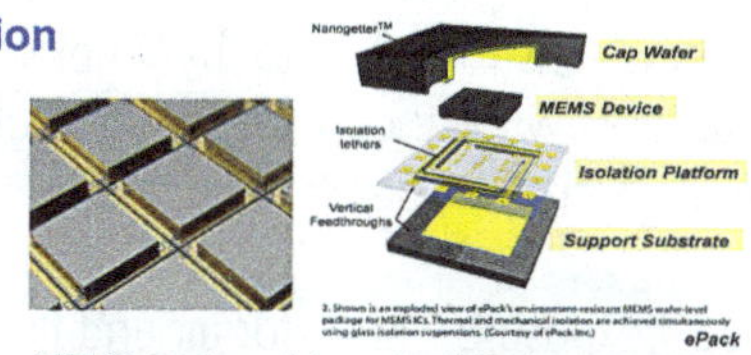

MEMS: dirt, moisture, (ambient pressure)

EMI shielding

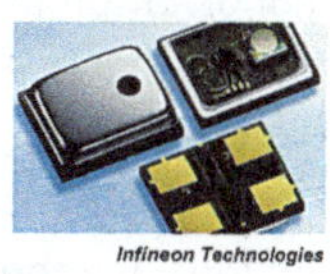

Biocompatible packaging

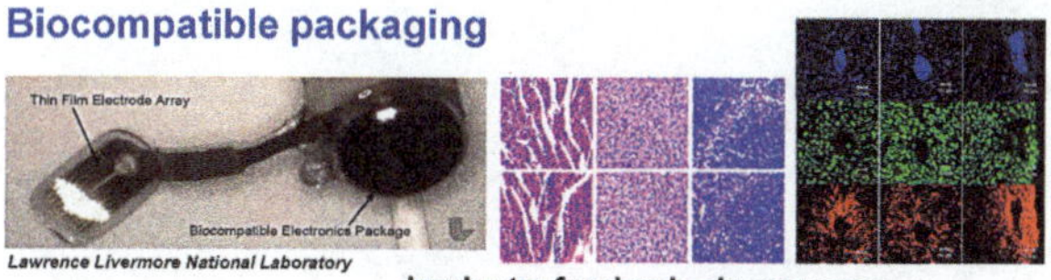

Implants: foreign body response

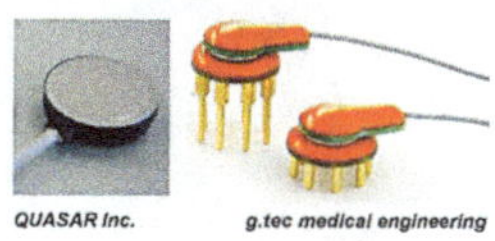

On-body electrodes: skin irritation

20 Physical Interface with Surroundings (I)

The physical interface of the microsystem with its surroundings should provide an effective isolation barrier between the device and surroundings to protect one against the other.

Hermetic encapsulation isolates the implantable device from surrounding fluid or protects the MEMS sensors against dirt, moisture, and ambient pressure when the sensor is designed to operate in a vacuum.

Biocompatible packaging is an essential requirement for the implants and on-body devices to minimize the foreign body response and harmful effects on surrounding tissue and cells.

Sometimes the package provides EMI shielding if the device operation and performance are sensitive to electromagnetic interference. The MEMS microphone is such an example.

21 Physical Interface with Surroundings (II)

- **Provides proper interface for interaction (sensing and actuation) with the surroundings**

Minimal controlled exposure to the surroundings

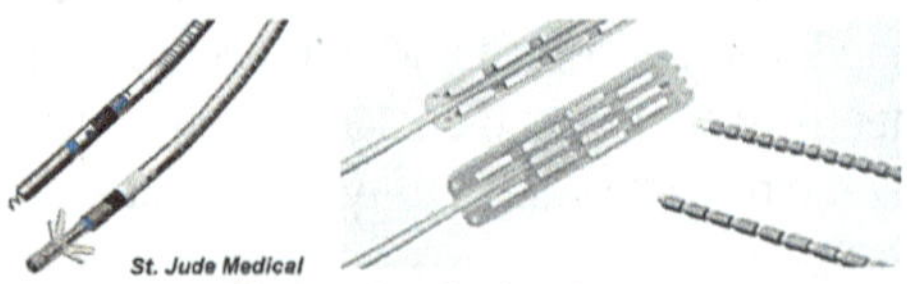

Electrodes for implants

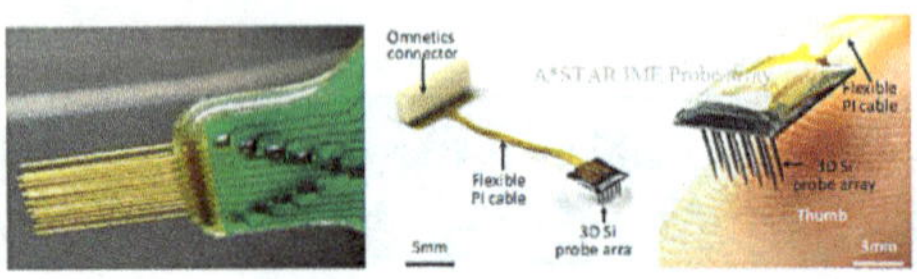

Chronic neural probes

Inlet and guiding structures for air and fluid

Gas sensors

MEMS Microphones

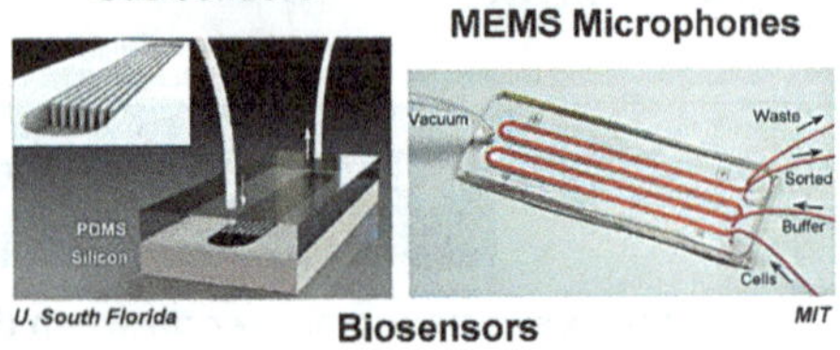

Biosensors

While the effective isolation barrier is constructed on one hand, a proper interface with the surroundings for sensing and actuation has to be provided on the other hand.

To do so, minimal controlled exposure of sensors or actuators to the surroundings needs to be implemented somehow, as being done for implantable electrodes, probes, and sensors.

For gas sensors, biosensors and MEMS microphones, proper inlet and guiding microstructures for air and fluid need to be embedded.

SENSOR AND ACTUATOR INTERFACE

22 Unidirectional vs. Bidirectional (I)

- Determined by
  - Need of driving for sensing
  - Need of closed-loop operation

Cardiac pacemaker

Closed-loop operation required

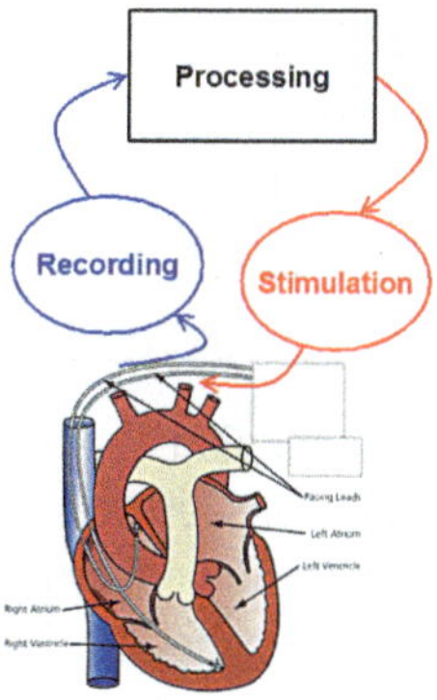

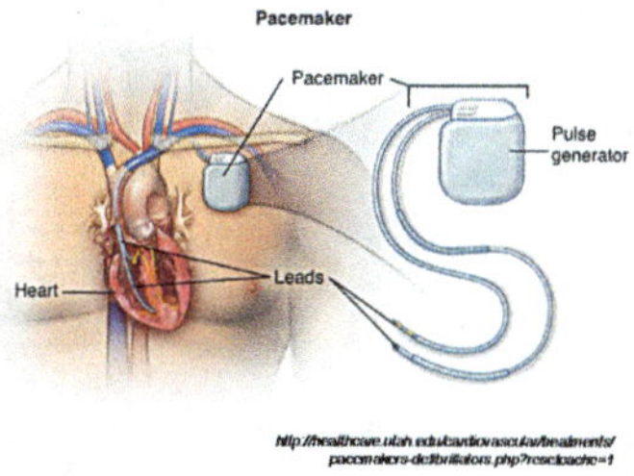

- ECG Signal Sensing
- Signal Processing and Decision of Stimulation Parameters
- Stimulation for Pacing

Impedance spectroscopy

Driving needed for sensing

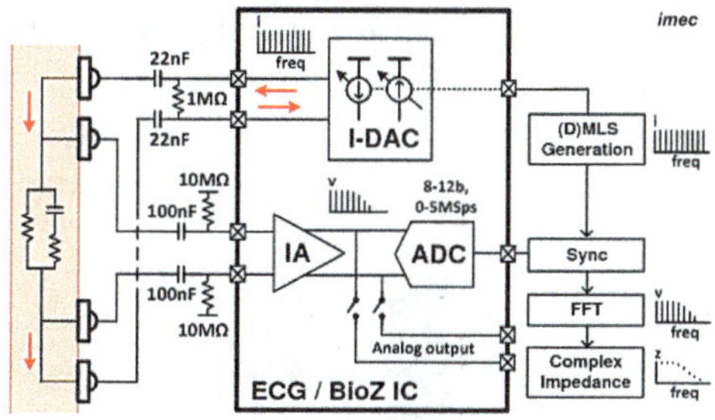

Ultrasound imaging

Driving needed for sensing

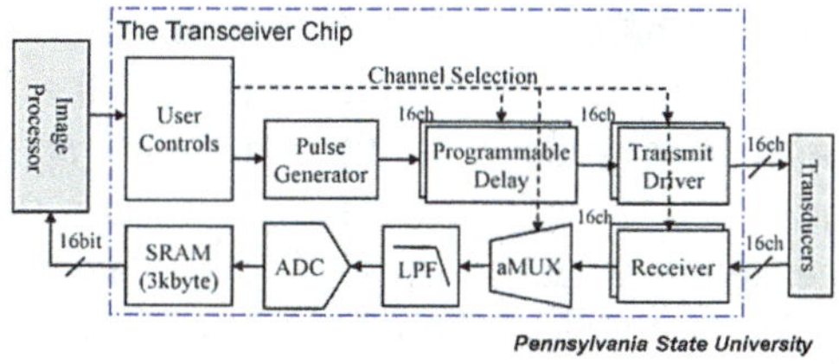

When the sensor and actuator interface is designed for biomedical microsystems, it should be determined whether the unidirectional interface is sufficient or the interface needs to be bidirectional.

This decision can be made considering the need of the driving function for sensing and the need for closed-loop operation.

As in the cases of bioimpedance measurement and ultrasound imaging, if the driving function is required for the execution of targeted sensing function, a bidirectional interface has to be implemented. In bioimpedance measurement, the sensing target should be driven with known current (or voltage), and the resulted voltage across the target (or current flowing through the target) is measured. In ultrasound imaging, the ultrasonic wave is transmitted by driving the ultrasound transducer first, and then the returning wave, which is reflected by the imaging target and incident back to the transducer, is sensed.

For the implementation of closed-loop operation also, the bidirectional interface is required. In the cardiac pacemaker, for example, the ECG signal is sensed and processed to stimulate the heart with appropriate pacing parameters electrically. Here, the bidirectional interface performs the biopotential recording for ECG signal sensing as well as the electrical stimulation for cardiac pacing.

23 Unidirectional vs. Bidirectional (II)

- **Determines**
  - **Need of duplexing and duplexing type (half-duplexing or full-duplexing)**
  - **Need of high-voltage protection for sensing circuits**
  - **Required dynamic range**

Ultrasound imaging

Half-duplexing & high-voltage protection

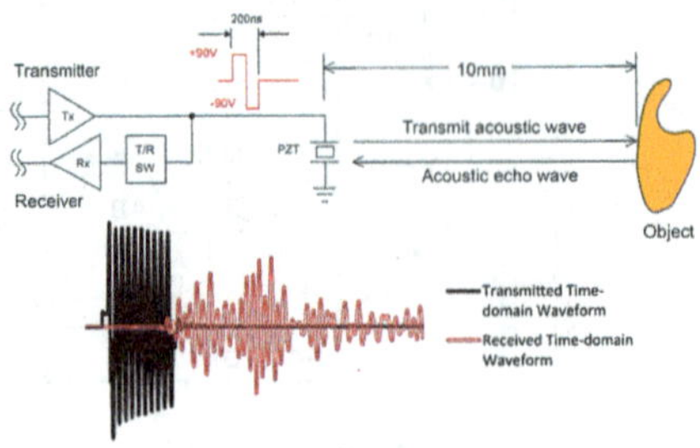

Closed-loop neuromodulation

Half/full-duplexing, HV protection, wide DR

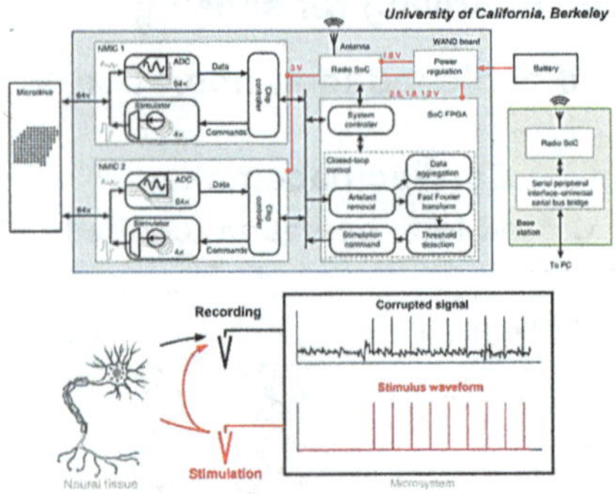

Once it is determined to implement a bidirectional interface, we need to consider whether the sensing and actuating functions can operate simultaneously or should take turns. The signal generated by actuation is usually much larger than the signal sensed by the interface, and hence, the simultaneous operation of the sensing and actuating functions may lead to the saturation of the sensing circuit chain.

To avoid this issue, in ultrasound imaging, the half-duplex operation is employed. The ultrasound transducer is electrically actuated by the transmitter circuit first to transmit the acoustic wave, and then the wave reflected by the target object is sensed by the receiver circuit after being translated into the electrical signal through the transducer. When the transmitter circuit operates and during some time after the transmission, the receiver circuit doesn't operate to avoid saturation.

In the case of closed-loop neuromodulation, the same issue exists because the stimulation artifact is much larger than the neural signal to be recorded. If the recording circuit chain has a limited dynamic range, it is easily saturated by the stimulation artifact, and hence the half-duplex operation is required. However, in the half-duplex operation, a portion of the neural signal information can be lost due to the discontinuously performed recording function. The amount of information loss can be significant, especially when there are a large number of stimulation channels, and hence the recording function should stop for a long time or too often. On the other hand, if the recording circuit chain has a sufficiently broad dynamic range to accommodate the stimulation artifact without saturation, the full-duplex operation becomes possible.

Another critical factor to consider when implementing the bidirectional interface is the high-voltage protection for the sensing circuits. Since the actuation circuit deals with the high-voltage signal, it is implemented using a special type of transistors which can withstand high voltages applied across their terminals. However, the transistors used in the sensing circuit are normal ones. Therefore, if the sensing circuit is directly exposed to the high-voltage signal generated by the actuation circuit without any protective means, the transistors in the sensing circuit may experience an immediate breakdown or a significantly degraded lifetime.

24 Unimodal vs. Multimodal

- **Determined by**
  - **Need of holistic monitoring and analysis**
  - **Need of full-duplexing bidirectional interface operation**

Wearable health monitoring

Multimodal sensing and holistic analysis

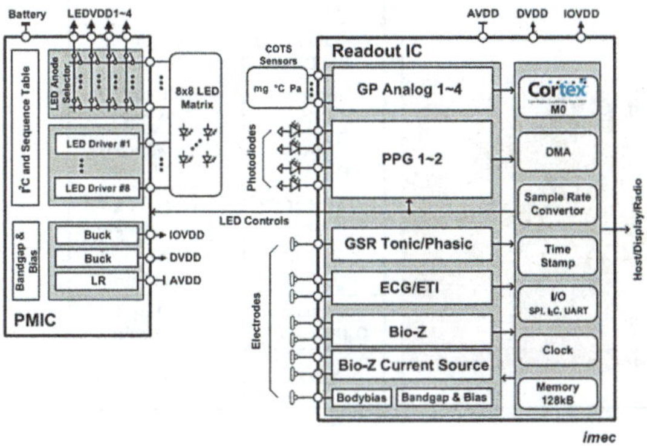

Multichannel bidirectional interface

Full-duplexing multimodal neural interface

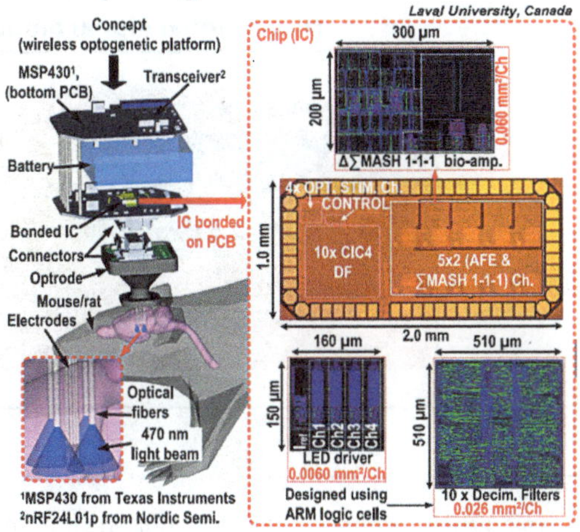

Modalities employed for sensing and actuation functions can be another factor to consider strategically.

In simple sensing applications such as conventional inertial sensing and gas sensing, a single modality of sensing is sufficient to achieve a goal given by the applications. However, the need for multimodal sensing and actuation is growing in various applications.

One such case is the applications where a holistic analysis based on abundant and heterogeneous data is required.

For monitoring patients' health using wearable sensors, for example, data from many sensors in different modalities such as ECG, bioimpedance, PPG (photoplethysmogram) and GSR (galvanic skin response), can be instrumental. Holistic environmental monitoring can be another example.

Multimodalities of sensing and actuation functions can also play an essential role in bidirectional interfaces.

When performing neural recording and stimulation simultaneously for neurological study, or closed-loop neuromodulation, employing multiple modalities can mitigate the challenge posed by the presence of strong stimulation artifacts. For example, the stimulation can be performed in the optical domain utilizing the principle of optogenetics, while the recording is carried out in the electrical domain. In such a way, the modalities of recording and stimulation can be set differently, and thus the stimulation artifacts hardly affect the recording operation and performance.

25 Minimum Detection Limit

- Determined by
  - Intrinsic signal strength
  - Distance from the signal origin
  - Signal level of interest

Biopotential signal amplitude

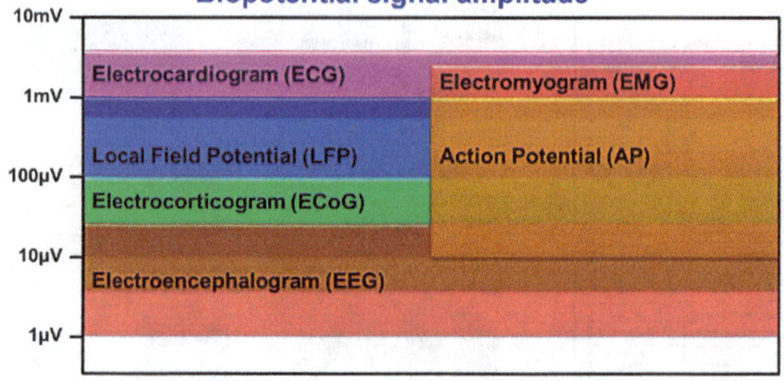

Indoor air quality guidelines

VOCs	Guideline Values
Formaldehyde HCHO	30-120 μg/m^3 (25-100ppb)
Acetaldehyde CH_3CHO	48 μg/m^3 (27 ppb)
Benzene C_6H_6	16-110 μg/m^3 (5-34 ppb)
Toluene $C_6H_5CH_3$	260-1092 μg/m^3 (69-290 ppb)
Ethylbenzene $C_6H_5CH_2CH_3$	1447-3880 μg/m^3 (264-892 ppb)
Xylene $C_6H_4(CH_3)_2$	870-1447 μg/m^3 (200-333 ppb)
Styrene $C_6H_5CH{=}CH_2$	30-300 μg/m^3 (7-70 ppb)
TVOC	200-3000 μg/m^3

The minimum detection limit required for the sensing function is determined by the intrinsic strength of the signal that needs to be sensed, the distance from the signal origin, and the signal level of interest.

Recording of various biopotential signals illustrates these points very well.

Local field potential, ECoG, and EEG have the same signal origin. However, the signal amplitudes are not the same because the distance between the location we record and the origin of the signal is different. The closer to the origin, the larger amplitude we obtain, resulting in more relaxed requirement on the minimum detection limit.

Compared to local field potential and action potential, the ECG signal usually has a larger amplitude because of the difference in their intrinsic signal strengths. The local field potential and action potential are originated from the activities of neurons in the brain, while the ECG signal is from the activities of electrogenic cells in cardiac muscles.

The table on the right-hand side shows the indoor air quality guidelines, especially for volatile organic compounds harming human health significantly. Since the harmful effect is caused even with a very little amount of such gases, the required detection limit reaches as low as several ppb levels.

26 Bandwidth

Required sensing bandwidth is mainly determined by the intrinsic signal characteristics and the distance between the signal origin and the sensing site.

Except some applications sensing images and videos, the bandwidth requirement is not usually so high but tends to be higher when the sensor is placed closer to the signal origin.

For example, although neural signals such as action potential, local field potential, ECoG, and EEG share the same signal origin, the signal bandwidth for invasive neural recording is larger than that for non-invasive EEG recording.

26 Bandwidth

- Determined by
  - Intrinsic signal characteristics
  - Distance from the signal origin

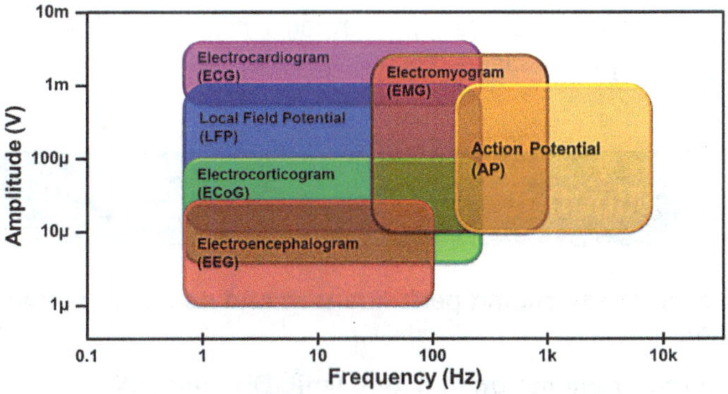

27 Dynamic Range

- Determined by
  - Intrinsic signal (or sensor) characteristics
  - Effect from interferences and disturbances
  - Signal range of interest

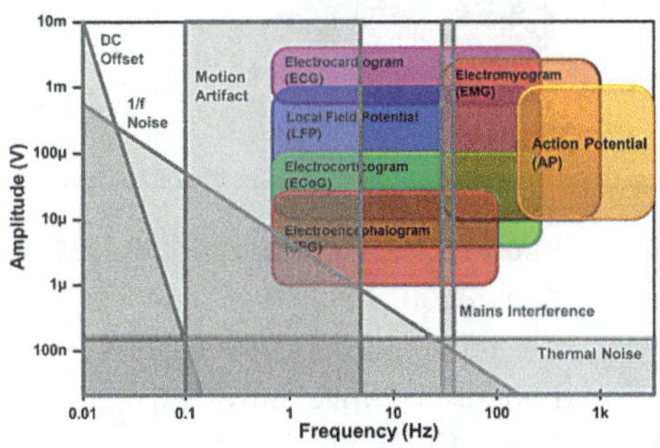

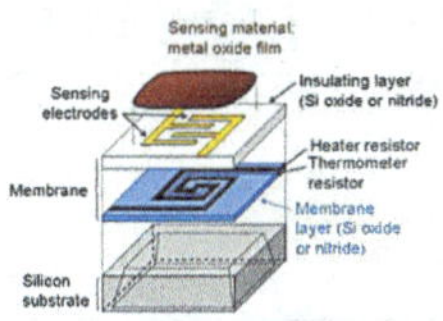

Metal-oxide thin-film gas sensors

DR ~140dB

Inaccuracy <1%

Accelerometers

- Consumer/automotive grade
- Industrial grade
- Tactical grade
- Navigation grade (FS: >15g, DR: >110dB, BW: >300Hz)

The dynamic range requirement is determined by the Intrinsic signal or sensor characteristics, the effect from various interferences and disturbances, and the signal range of interest.

For example, assume that we record the local field potential and spike signals simultaneously.

Since the spikes are with an amplitude of tens of µV added on the local field potential with an amplitudes of about 2 mV, if an input-referred noise of 4 μV_{rms} is needed to meet the signal-to-noise ratio requirement of the spike signal, the dynamic range requirement becomes 54 dB, resulting in a 9-bit resolution.

On top of that, if you consider interferences as large as a few hundreds of mV, the required dynamic range may reach nearly 100 dB. To avoid that, a good

27 Dynamic Range

common-mode rejection performance needs to be guaranteed.

Some sensors such as gas sensors and inertial sensors for navigation require a very large dynamic range.

Metal-oxide thin-film gas sensors require the interface circuit having a large dynamic range over 140 dB to accommodate high sensitivity, process spread, drift over time, and large variation in sensor resistance values for different doping types and levels used to detect different target gases.

The navigation-grade accelerometer requires the dynamic range larger than 100 dB.

28 Min. Det. Limit, DR, and BW

- **Determines required sensing performances and necessary power consumption**
  - **Stringent requirements on min. det. limit, DR, and BW**
    - → **Significant power consumption**
    - → **Stronger needs on low-power circuit techniques**
  - **Calibration strategy: another important factor to consider**

Low-cost sensors

Delft U. Tech.

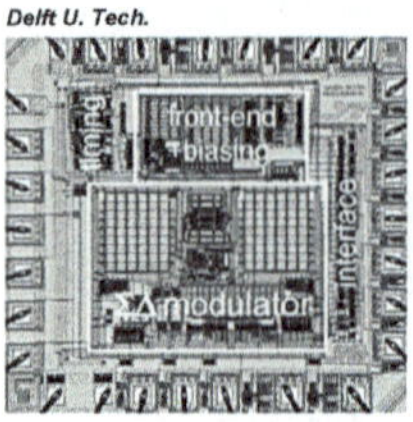

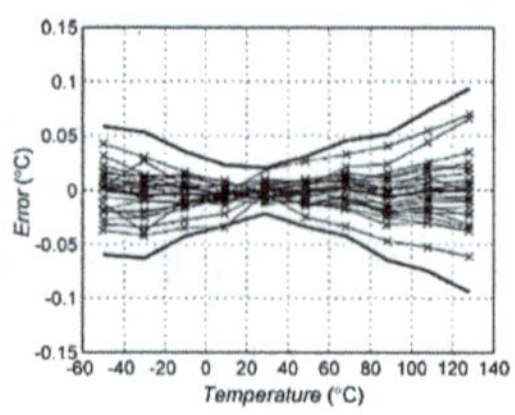

Nano-scale sensors (large variations)

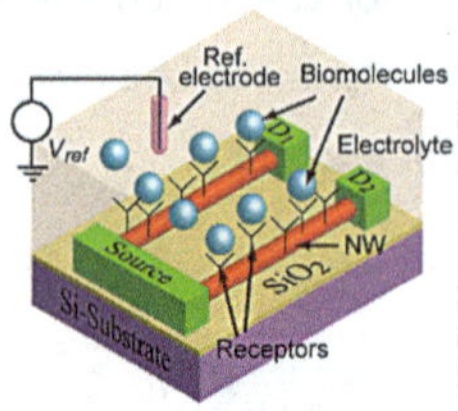

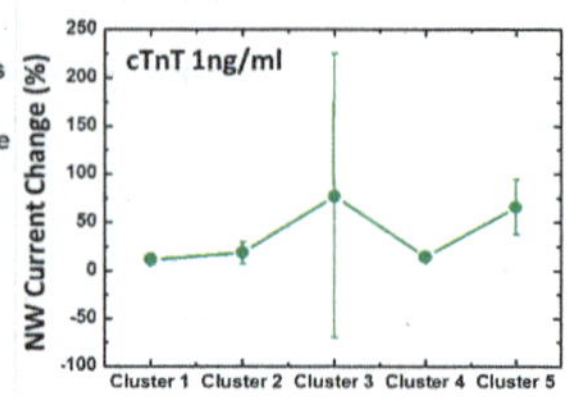

The requirements on minimum detection limit, bandwidth, and dynamic range determine the necessary sensing performances and power consumption.

Stringent requirements on minimum detection limit, bandwidth, and dynamic range usually lead to significant power consumption. So, low-power circuit techniques are needed more strongly.

To provide necessary performances, the sensor and its interface circuit often need to be calibrated.

Therefore, the calibration strategy is another important factor to consider. Especially when the application requires a low-cost sensing solution, the time and effort spent for calibration need to be minimized while providing necessary performance after calibration.

In the case of nano-scale sensors, they tend to have significant variations in their characteristics, although ultra-high sensitivity is offered. So, an effective and efficient calibration method needs to be developed together.

29

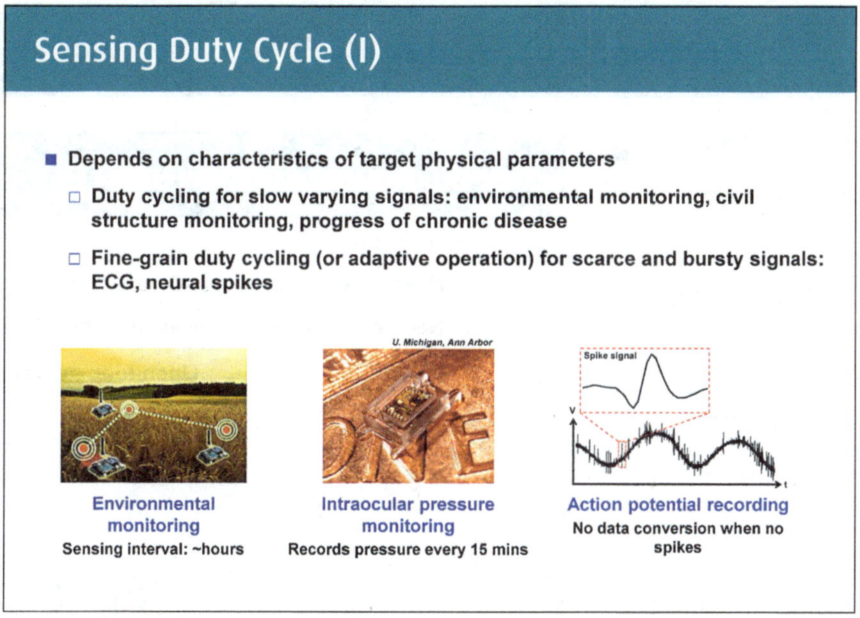

Depending on the characteristics of target physical parameters, the sensing operation can be duty-cycled to minimize the overall energy consumption as well as the amount of data that need to be transmitted.

Low-duty-cycle sensing can be utilized when slow varying signals are monitored, as in the cases of environmental monitoring, civil structure monitoring, and progress monitoring of chronic diseases. In environmental monitoring, the sensing interval can be hours, and for intraocular pressure monitoring, the pressure may be recorded every 15 minutes, for example.

When we sense scarce and bursty signals such as ECG signal and neural spikes, fine-grain duty cycling or adaptive operation can be employed. For example, in neural spike recording, by implementing a spike detection function, we may execute the data conversion process only when spikes are detected.

30

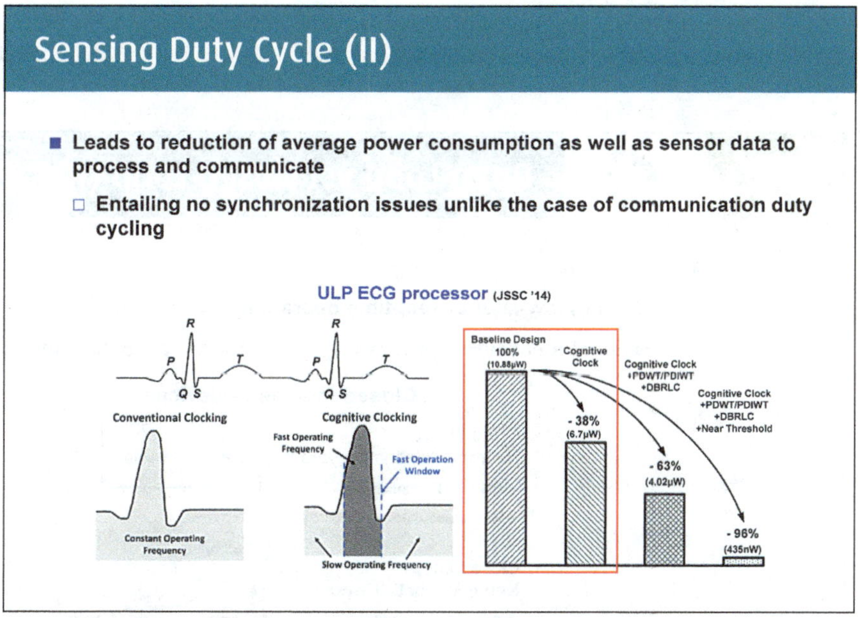

Such duty-cycled sensing operation leads to the reduction of average power consumption as well as sensor data.

Good news is sensing duty cycling does not entail any synchronization issues, unlike communication duty cycling, where the synchronization between sensor nodes or between a sensor node and a host device poses a significant challenge.

This figure shows the example of adaptive or cognitive clocking scheme used for ultra-low-power ECG signal monitoring and processing. By using this technique, the average power consumption is reduced by about 40%, without causing any timing or synchronization issues.

DIGITAL CONTROL AND SIGNAL PROCESSING

31 Open Loop vs. Closed Loop (I)

- **Determines**
  - **Need of low-latency real-time operation**
  - **Need of local signal processing and wireless raw data transmission**

Open-loop neurodevices

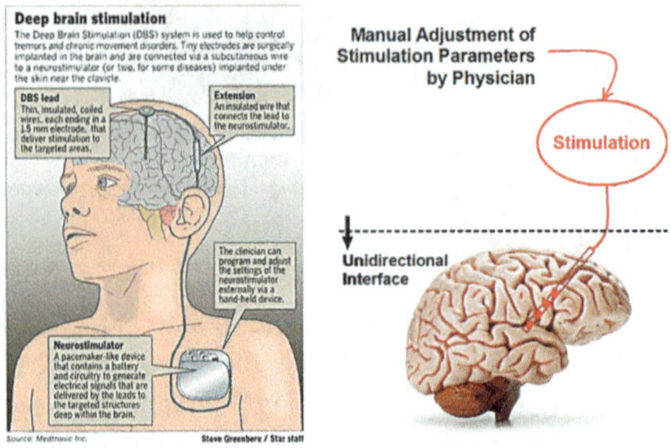

Whether the biomedical microsystem operates in open loop or closed loop can affect the design requirements of its digital control and signal processing circuits significantly.

The deep brain stimulator (DBS) is the most popular type of neurodevice applied to the human central nervous system.

Although it helps a large number of patients suffering from essential tremors and movement disorders, it operates in an open-loop manner using unidirectional neural interface and its stimulation parameters are manually adjusted by physicians. The DBS device in its current form is too simple to implement comprehensive therapeutics with high stimulation efficacy, energy efficiency, and patient safety.

32 Open Loop vs. Closed Loop (II)

- **Determines**
  - **Need of low-latency real-time operation**
  - **Need of local signal processing and wireless raw data transmission**

Closed-loop neurodevices

- Closed-Loop Neuromodulation Devices
  - DBS for Parkinson's
  - Neuromodulator for Epilepsy
- Closed-Loop Neuroprosthetic Devices
  - Cortical Neuromotor Prosthesis
  - EMG-based Limb Prosthesis
  - Peripheral Nerve Prosthesis

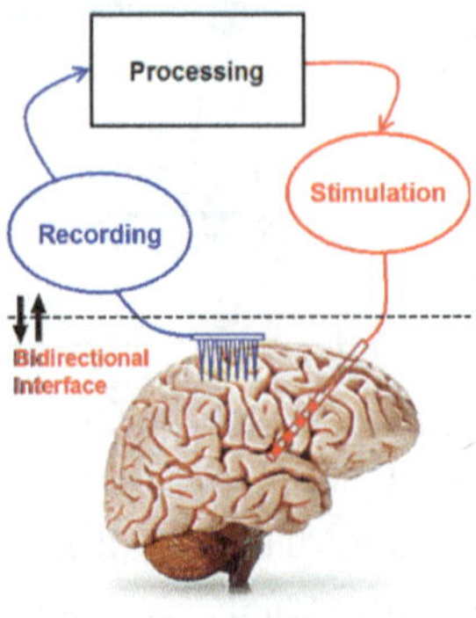

32 Open Loop vs. Closed Loop (II)

Observing the great success of cardiac rhythm management devices which operate in a closed-loop manner, neurodevices are being developed toward the closed-loop operation.

However, compared to the cardiac devices, closing the loop in neurodevices is much more challenging. Neurodevices have to record and process the multi-channel neural signal, unlike cardiac devices. The stimulation can be through multiple channels as well. Not only that, the signal to process is much more ambiguous and harder to analyze than the signal processed in the cardiac devices.

However, overcoming all these challenges certainly rewards the efforts by enabling the next-generation closed-loop neuromodulation and neuroprosthetics.

For the closed-loop operation of biomedical microsystems, a low-latency real-time signal processing is often required because generating therapeutic stimulation for patients immediately after analyzing the sensed signal is critical in some applications such as epileptic seizure suppression and essential tremor suppression.

Depending on the application, the signal processing has to be carried out locally (inside the microsystem) with low latency but limited accuracy, or the raw data of the sensed signal can be transmitted wirelessly to the remote site, where the more exhaustive and accurate signal processing is executed.

33 Processing-Heavy vs. Communication-Heavy

- **Determined by**
  - **Device operation scenario: open loop or closed loop**
  - **Need of low-latency real-time operation**
  - **Required analysis accuracy**
  - **Power budget**

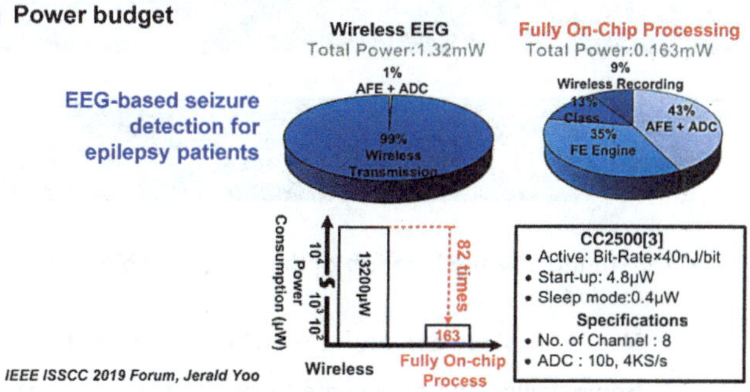

The biomedical microsystem is designed to be either processing-heavy or communication-heavy, and it is determined by several factors such as the device operation scenario, need of low-latency real-time operation, required analysis accuracy, and power budget.

If the device operates in open loop, the latency in obtaining the processed results is not usually critical, and hence the device can be designed as a communication-heavy system to achieve higher processing accuracy by utilizing sufficient computing resources present in the remote site. However, for the closed-loop operation, due to the need for low-latency real-time operation, the microsystem design becomes processing-heavy.

If the given power budget is strictly limited, it is preferred to minimize the amount of wirelessly transmitted data because the wireless transceiver consumes the highest power in typical microsystems. The data should, therefore, be processed locally as much as possible before transmission, leading to a processing-heavy system. Of course, if we try to perform the local signal processing with very high accuracy, the processing circuit can also consume excessively high power. The required accuracy and performance of the local signal processing should be carefully considered so that the overall system power consumption can stay within the budget.

34 Machine Learning vs. Neural Network

- Determined by
  - Need of low-latency real-time operation
  - Required analysis accuracy
  - Availability of sufficient training datasets
  - Power budget

Traditional Machine Learning	Neural Network
Good Accuracy	High Accuracy
Low Latency	High Latency
High Energy Efficiency	Low Energy Efficiency
Limited Training Datasets	Sufficient Training Datasets
Patient-Specific	Patient-Nonspecific
Real-Time Closed-Loop Operation	Non-Real-Time Analysis

If the microsystem requires a local signal processing function capable of learning from the data, either traditional machine learning or recent neural network approach can be employed.

The traditional machine learning approach may provide a reasonably good accuracy but not as high accuracy as the neural network approach. However, if the available amount of training datasets is significantly limited, the machine learning approach can provide even better performance than the neural network approach, which typically requires a massive amount of training datasets to achieve targeted high accuracy. Therefore, the machine-learning-based processing function can be trained well by using a limited amount of patient-specific datasets, while the neural-network-based processing function can be trained with patient-nonspecific datasets to obtain a sufficiently large amount for training. Also, the machine learning approach provides lower latency and higher energy efficiency than the neural network approach, in general.

COMMUNICATION

35 Medium (I)

- Determined by the location of the physical object the microsystem is embedded.
  - Most of devices communicate wirelessly over the air.
  - Some devices communicate through other media such as human body and wires.

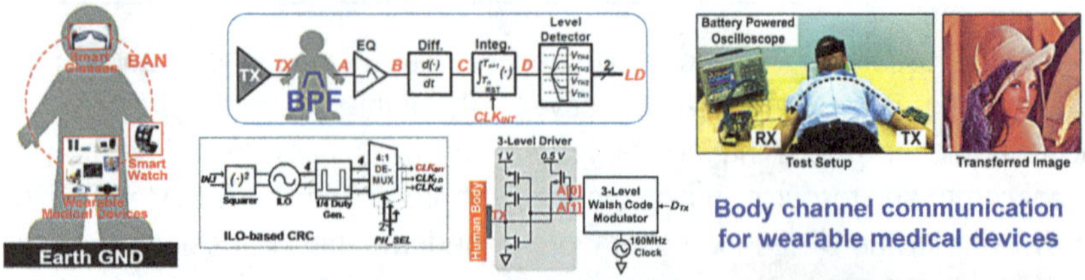

Body channel communication for wearable medical devices

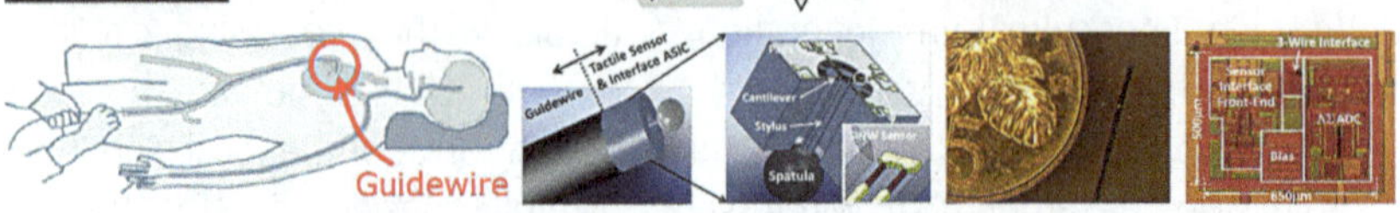

Tactile sensor microsystem for guidewires in minimally invasive surgical operations

35 Medium (I)

The medium used for communication is determined by the location of the physical object where the microsystem is embedded.

In most cases, communication is performed wirelessly through free air.

However, some devices may communicate through other communication media.

Wearables and implants may use the human body as a communication channel to achieve higher energy efficiency compared to the communication over the air.

The communication can also be performed over a limited number of wires in the application, such as the tactile sensor microsystem for guidewires used in minimally invasive surgical operations. Inside the coiled guidewire structure, a couple of thin wires can be embedded, and those wires are used to supply power to the sensor microsystem as well as to communicate the command and sensor data by using powerline communication techniques.

36 Medium (II)

- **Determines communication frequency, bandwidth, and achievable data rate**

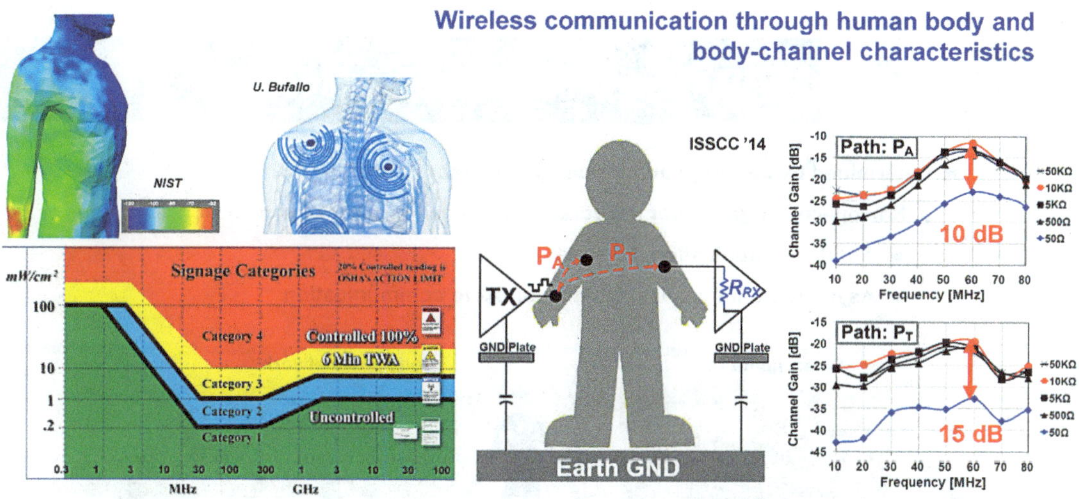

The communication medium, in turn, determines the communication frequency, bandwidth and achievable data rate.

When we communicate through the human body, the frequency band from 40MHz to 70MHz is usually used, as the human body has band-pass characteristics. Also, the safety is another important concern.

37

Distance, Symmetry, and Protocol (I)

- **Determined by coverage and topology of sensor networks**
  - **Wide-coverage sensor networks with a large number of sensor nodes:**
    - **Long communication distance**
    - **Symmetric transceiver for multi-hop/ad-hoc communication**

Smart grid

Oil & natural gas pipeline leak detection

Other factors to consider are the communication distance, symmetry, and network topology.

Wide-coverage sensor networks with a large number of sensor nodes require long communication distance and symmetric transceivers for multi-hop and ad-hoc communication. Such examples are the smart grid and the sensor network for oil pipeline leak detection.

The biomedical applications usually employ narrow-coverage sensor networks with a small number of sensor nodes.

38

Distance, Symmetry, and Protocol (II)

- **Determined by coverage and topology of sensor networks**
  - **Narrow-coverage sensor networks with a small number of sensor nodes:**
    - **Short communication distance**
    - **Asymmetric transceiver for master-slave communication**

Smart home

Body area network

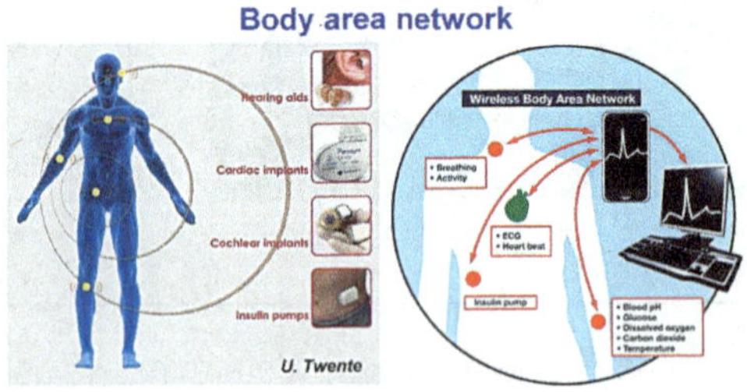

Narrow-coverage sensor networks with a small number of sensor nodes operate with short communication distance and asymmetric transceiver for master-slave communication.

By doing so, the energy consumption of the slave node can be minimized, while burning more power in the master side operating with sufficient energy source typically.

Examples are the smart home network and the body area network for biomedical applications.

39 Data Rate

- **Determined by the amount of sensor data to communicate**
  - **Data rate requirement is not usually so high.**
  - **Some devices dealing with image/video data and information from large-scale sensor array require high-data-rate communication.**

Wireless capsule endoscopy

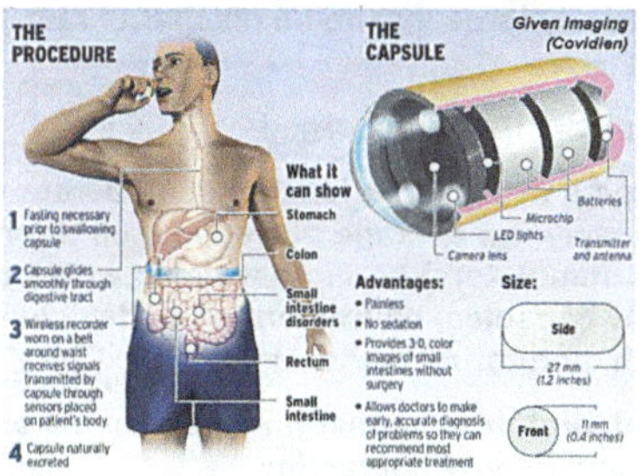

Multichannel neural recording

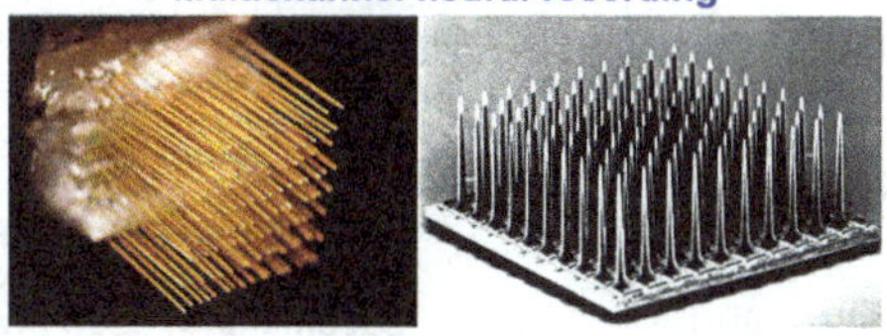

The communication data rate is determined by the amount of sensor data to communicate.

The data rate requirement is not usually so high for most of the biomedical microsystems.

However, some devices dealing with the image/video data such as capsule endoscopes and the information from large-scale sensor arrays such as multichannel neural recording microsystems require high-data-rate communication.

40 Communication Duty Cycle

- **Determined by characteristics of sensed physical parameters and usage scenario**
  - **Wireless transceiver: usually most power-hungry**
  - **Duty cycling: reduce P_{avg} significantly**
  - **Synchronization issue mitigation:**
    - **Low-power high-accuracy real-time clock**
    - **Low-power wake-up receiver**
    - **Combination of both**

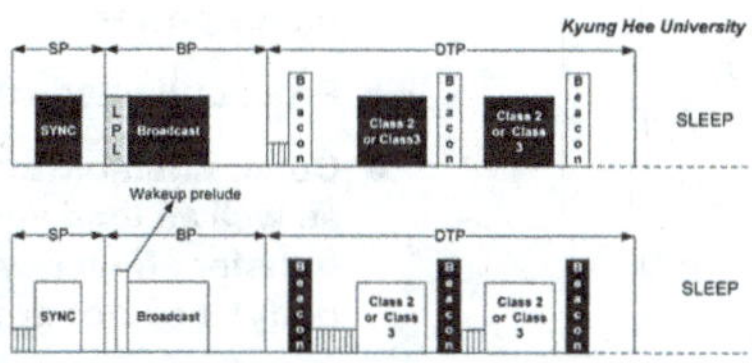

The communication duty cycle is determined by the characteristics of sensed physical parameters and usage scenario.

Since the communication transceiver is usually the most power hungry block in sensor microsystems, a significant amount of energy can be saved by duty cycling the communication function.

However, it entails synchronization issues, which can be mitigated by using a low-power high-accuracy clock generator, a low-power wake-up receiver, or a combination of both.

ENERGY SUPPLY AND MANAGEMENT

41

Battery, WPT, and Harvesting (I)

- **Choice of energy source**
  - **Battery**
    - **Pros: stable supply, relatively large maximum discharge rate, highly efficient use of energy**
    - **Cons: limited energy capacity if not recharged**
    - **Applications: microsystems with simple functions, moderate performances, and extremely low average power consumption (e.g., pacemakers, deep brain stimulators) or microsystems with limited lifetime (e.g., wireless endoscopy capsules, neuromonitoring devices for TBI patients, wearable health monitoring devices)**
    - **Mitigation of cons: co-integration of recharging mechanism such as wireless power transfer and energy harvesting**

The choice of the energy source for biomedical microsystems can be various, including the battery, wireless power transfer (WPT), and energy harvesting (EH).

The battery is the type of energy source which can provide stable supply and relatively large maximum discharging rate as well as allows highly efficient use of energy. However, without a means for recharging it, the capacity of the energy it can provide is severely limited, especially when the available room for the installation of the battery inside the microsystem is significantly small in volume.

The battery can be used as an energy source for the microsystems with simple functions, moderate performances, and very low average power consumption, such as cardiac pacemakers and deep brain stimulators, or the microsystems with limited lifetime such as wireless endoscopy capsules, neuromonitoring devices of traumatic-brain-injury (TBI) patients, and wearable health monitoring devices.

To overcome the disadvantage of limited energy capacity, a recharging mechanism such as the WPT and EH can be co-integrated in the microsystem.

42

Battery, WPT, and Harvesting (II)

- **Choice of energy source**
  - **Wireless power transfer**
    - **Pros: unlimited energy capacity as long as power transmission continued**
    - **Cons: unstable supply (sensitive to coil-to-coil distance and orientation as well as load variation), moderate maximum discharge rate, low energy-transfer efficiency (especially for implants placed deep inside human body), need of power-transmitting device**
    - **Applications: advanced microsystems with complex functions, high performances, high operation duty cycle, long lifetime, and relatively large average power consumption (e.g. multichannel neural-recording microsystems, closed-loop neuromodulation microsystems)**
    - **Mitigation of cons: use of back-up battery to stabilize energy supply, use of capacitor (or battery) to increase maximum discharge rate, use of acoustic wave rather than electromagnetic wave to improve energy-transfer efficiency for deep-in-body devices**

The WPT can provide unlimited energy capacity as long as power transmission from the external devices is continued. However, the power supply based on WPT can be unstable because the WPT efficiency is sensitive to the coil-to-coil distance, orientation, and alignment, as well as the load variation. It

42 Battery, WPT, and Harvesting (II)

provides a moderate discharge rate depending on the wireless charging rate and the device used for storing the charged energy. The efficiency of WPT can be significantly low, especially for the implants placed deep inside the human body. To operate the microsystem with WPT, the power-transmitting device is required externally, and it can be regarded as a disadvantage.

The WPT can be applied to advanced microsystems with complex functions, high performances, high operation duty cycle, long lifetime, and relatively large average power consumption, such as multichannel neural-recording microsystems and closed-loop neuromodulation microsystems, retinal prosthesis microsystems, and cochlear implants.

The disadvantages of WPT can be mitigated by employing several techniques. A back-up battery can be used to stabilize its energy supply, and the use of the capacitor or battery can help to increase its maximum discharging rate. To improve its energy-transfer efficiency for the implantable devices deep inside the body, the acoustic wave can be used as a power carrier, rather than the conventional electromagnetic wave.

43 Battery, WPT, and Harvesting (III)

- **Choice of energy source**
  - **Energy harvesting**
    - **Pros: unlimited energy capacity as long as ambient energy source available**
    - **Cons: unstable supply (sensitive to ambient conditions), usually very low maximum discharge rate, low energy-harvesting efficiency (especially for small harvesting devices)**
    - **Applications: microsystems with low operation duty cycle, long lifetime, and relatively low average power consumption (e.g. monitoring devices for chronic diseases, bioelectronic medicine, wearable health monitoring devices)**
    - **Mitigation of cons: use of back-up battery to stabilize energy supply, use of multiple ambient sources to stabilize energy supply, use of capacitor (or battery) to increase maximum discharge rate**

The EH can provide unlimited energy capacity as long as the ambient energy source is available. However, the power it supplies can be unstable due to its sensitivity to ambient conditions, the maximum discharge rate is usually very low, and the energy-harvesting efficiency is low, especially for the harvesting device having a small volume.

The EH can be used for microsystems with low operation duty cycle, long lifetime, and relatively low average power consumption such as monitoring devices for chronic disease, bioelectronic medicine, and wearable health monitoring devices.

To stabilize the energy supply, a back-up battery can be used, and the energy can be harvested from multiple kinds of ambient sources. A capacitor can be used to increase the maximum instantaneous discharge rate.

44 Battery, WPT, and Harvesting (IV)

- **Choice of energy source**
  - **Summary**

		Battery	WPT	Harvesting
Characteristics	Energy Capacity	Limited	Unlimited	Unlimited
	Energy Supply Stability	High	Medium	Low
	Max. Discharge Rate	High	Medium	Low
	Energy Efficiency	High	Medium – Low	Low
	Need of Additional Device	No	Yes	No
Applications	Avg. Power Consumption	Low – Medium	Medium – High	Low
	Device Lifetime	Short – Long	Long	Long
	Operation Duty Cycle	Low – High	High	Low

What we have discussed for the battery, WPT, and EH are summarized in this table.

45 Operation Scenario and Management Strategy

- **Duty cycling is key.**

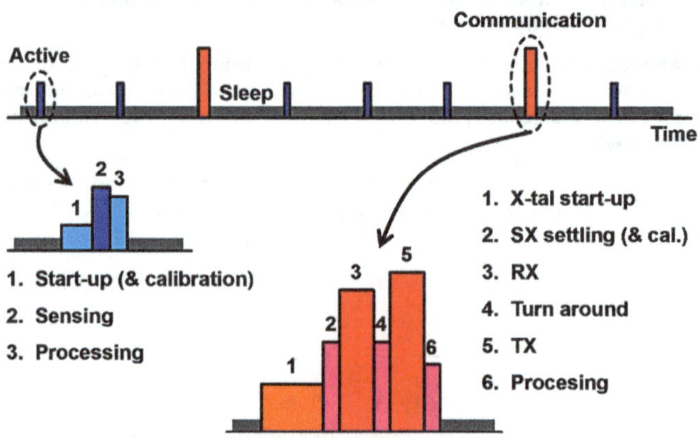

IEEE ISSCC 2015 Tutorial, Allison Burdett

In operating biomedical microsystems, while providing a stable energy source is important, managing the energy usage to be as efficient as possible is also essential.

The most effective and widely used technique for cutting down the average power consumption of the microsystem is duty cycling. In a duty-cycled operation, the microsystem stays in sleep mode by default and wakes up only when it needs to perform certain operations such as periodic sensing and wireless communication.

In sleep mode, most of the circuit blocks in the microsystem is turned off, except a few blocks such as the timer and state-retention memory. In the microsystem operating with a low duty cycle and hence staying in sleep mode for most of the time, the power consumption of the always-on circuit blocks determines the overall power consumption

Operation Scenario and Management Strategy

of the microsystem. Therefore, it is important to develop an ultra-low-power timer circuit which can provide a high timing accuracy as well as an ultra-low-power state-retention memory circuit with robust performances.

In active mode, the microsystem starts up necessary circuits, and performs the operation of sensing and processing, before it enters the sleep mode back. In communication mode, the crystal oscillator, which plays a role of the reference clock, starts up first, and the frequency synthesizer (SX) is locked to the reference clock after finite settling time. Then, the receiver (RX) operates for receiving the commands and data from the external device. After changing the communication function from reception to transmission, the transmitter (TX) sends out the sensed and processed data. Finally, it performs necessary processing operations and enters the sleep mode again.

Since the power consumption required for wireless communication is most significant, the average power consumption can be reduced significantly by minimizing the frequency and duration of operating in the communication mode.

CONCLUSION

- **Advancement of biomedical IC & microsystem technology**
  - **Enable the healthcare in patients' daily life at low cost,**
  - **Reduce the burden of the healthcare system,**
  - **Enable preventive care, early detection and effective disease management,**
  - **Improve the quality of patients' life dramatically by enabling the disabled,**
  - **Often save the patients' life,**
  - **Make the treatment procedure less invasive and more effective.**
- **Biomedical IC & microsystem capabilities enable the emerging biomedical applications including:**
  - **Implantable medical devices**
  - **Minimally invasive surgical operation**
  - **Point-of-care diagnosis**
  - **Wireless health monitoring**
- **Unlike personal computers and mobile phones, biomedical microsystems have many different applications, which lead to different approaches for design, implementation, and integration.**
  - **For different applications and usage scenarios, different strategies for sensing, processing, communication, powering, and microsystem integration need to be employed to provide optimum solutions.**
  - **It's an important prerequisite to have a good understanding of the applications where the designed circuits and systems are deployed.**
  - **While improving circuit-level techniques on one hand, optimally crafting system designs on the other hand can draw the maximum out of currently available technologies.**
  - **IC designers in this field play a role of solution providers who develop best solutions for the problems posed by specific applications, leveraging advanced IC design techniques.**

CHAPTER 02

Biomedical Sensor Interface Circuits

Nick van Helleputte

imec, Belgium

While there are many diverse biomedical systems and application, ultra-low-power electronic sensor interface circuits are a key part of all of them. They are responsible for amplifying and digitizing the various physiological signals that need to be recorded. Area-efficient, high-performance and low-power sensor interface circuit design is thus an important prerequisite for a successful biomedical system. This chapter will address exactly these topics. A brief overview of various sensor types and how to interface them will be discussed. The unique design challenges for sensor interfaces will be addressed. The chapter will discuss in-depth state-of-the-art instrumentation amplifier designs as well as trans-impedance amplifier designs both of which are at the heart of a lot of sensor interfaces. Advanced circuit design techniques like chopping, bootstrapping, offset cancellation will be explained. Finally, a look into novel and promising research areas like time-domain sensor interface design will conclude the chapter.

INTRODUCTION

1 What is a sensor

Device that senses a characteristic or physical property and generally provides a quantitative indication.

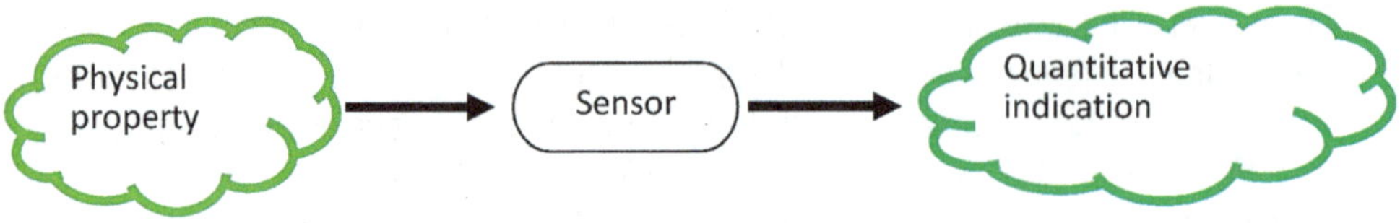

When we talk about a sensor, what are we actually talking about? In its most general definition, a sensor is something that translates a physical property into a quantitative indication. The physical properties can be very diverse, including light, sound, pressure, force, motion, and so on. They can also include physiological signals like heart rate, muscle movement, blood oxygen saturation, brain activity etc. But even chemical compounds, specific molecules and even DNA strands are all physical properties that can be sensed.

2 Sensor evolution

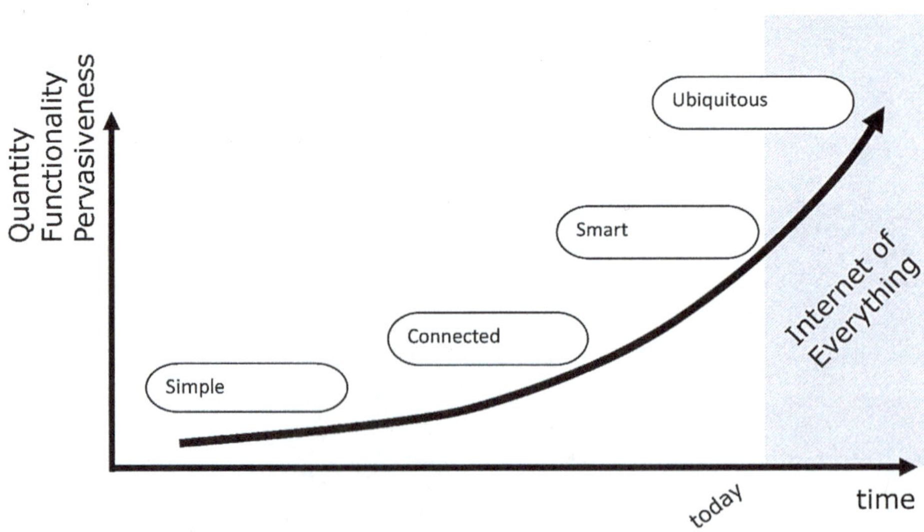

2 Sensor evolution

The deployment of sensors has seen a very drastic evolution in the last decades. Early sensor devices were fairly basic, providing a quantitative output for basic physical properties. As technical advancement progressed, sensors devices were used more and more in very diverse applications. They became connected allowing remote monitoring of complex processes in so called sensor networks. As CMOS technology advanced and allowed for ever lower power data computation, the sensor nodes became smart. This means local signal processing allowed for interpretation of the data and in some cases even reaction. An example is for example a pacemaker. This is a "sensor" device that can measure the heart rhythm. But it does so much more. It is able to interpret the heart rhythm and detect abnormalities. It can then also stimulate and pace the heart to bring the rhythm back to a normal pace. Now in the ear of IoT, sensors are becoming ubiquitious. We see smart, connected sensors being embedded everywhere in our surroundings.

3 Smart sensor

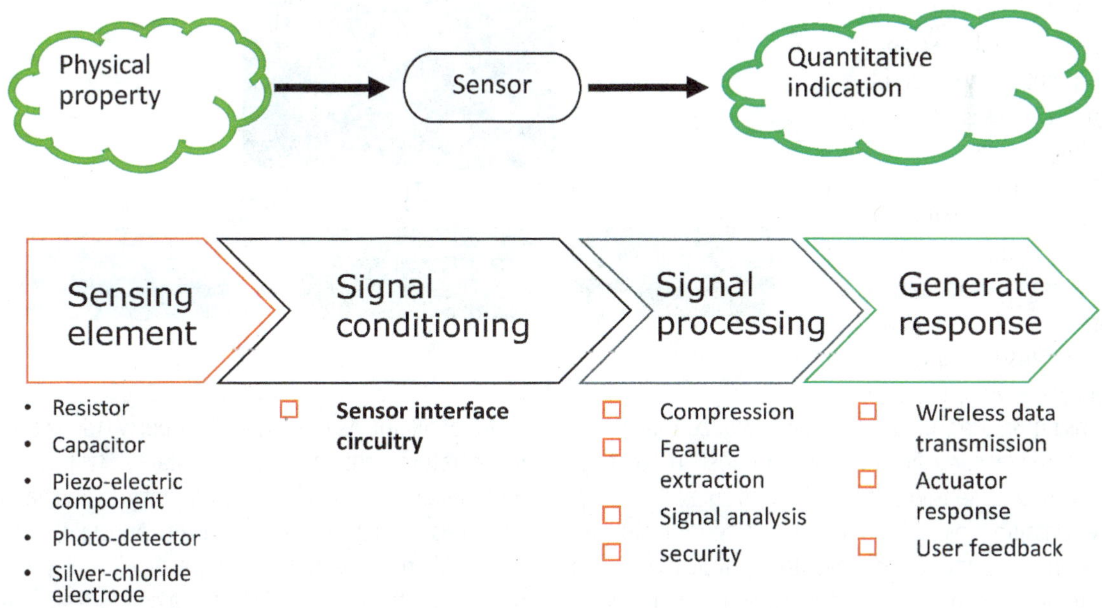

3 Smart sensor

A smart sensor typically has 4 main parts. The sensing element is responsible for converting the physical property to an electrical signal. There are many sensor elements and they are specifically designed for the property they need to measure. They can be basic electrical components like a resistor or a capacitor possible embedded in a MEMS structure where external force will change the resistance or capacitance. Notice that the sensing element can also be just an electrode. For example to record electrophysiological signals like electro-cardiogram (ECG), a silver-chloride electrode is attached to the human body which can directly pick up the weak voltage signals generated by the human body whenever the heart beats. The second part consists of signal conditioning circuitry including impedance buffering, signal amplification and filtering and analog-to-digital conversion. The 3rd part is (digital) signal processing where the sensed signal is processed and interpreted. Finally a suitable response is generated. This can be again quite diverse depending on the sensor application. The rest of this chapter will focus primarily on the 2nd part, namely the analog sensor interface circuitry.

4 Electrical sensing mechanisms

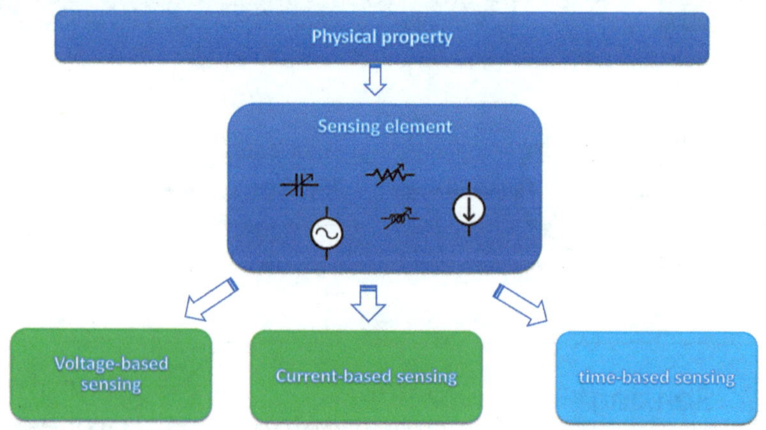

While there are a huge amount of physical properties that can be sensed each via special sensing elements, the outputs are always either a voltage or current that shows a well known relationship with the physical property. Hence there are 2 main categories of sensor interface circuits, voltage-based sensing or current-based sensing. Notice that recent years have seen an increased interest in so-called time-based sensing. While in principle, this could be classified as either voltage or current sensing, the main principle is that the physical property will be turned into a time-varying signal where the information that is to be sensed is in the time domain. An example is a resistive sensor that is being used together with a well known capacitor in an RC-oscillator. As the physical property (for example temperature) changes, the resistance will change and hence the oscillation frequency will change. Time-based circuits operate on this principle, where they will measure not a voltage or current per se, but rather the time-variant behavior. Because the circuits are very different from traditional voltage/current based sensing, they will be briefly discussed separately.

5

Voltage-based sensing (I)

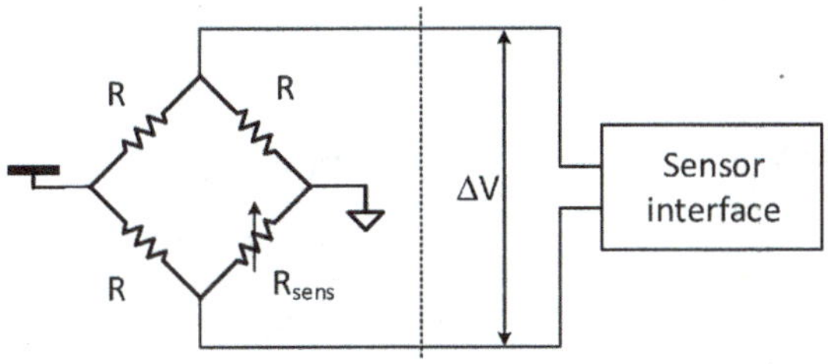

$$\Delta V = \left(\frac{1}{2} - \frac{R_{sens}}{R + R_{sens}}\right) VDD$$

? Need to measure ΔV without impacting the outcome

The main concept behind voltage sensing is that the sensing element will generate a usually very small voltage that is often proportional to the physical property to be sensed. A well known example is a bridge readout, where a variable resistor (Rsens) is placed in a bridge configuration together with known (and well matched) resistors R. If the bridge is biased with VDD, it is easy to derive the deltaV signal. Notice that this is just one example, voltage based sensing certainly doesn't always require a bridge readout. In case of electrophysiological sensing, the voltage signal itself is the information that needs to be sensed. This is the case for electrocardiogram (ECG), electroencephalogram (EEG) or electromyogram (EMG) for example, which can all be recorded directly using electrodes placed on the human body. The main challenge for the sensor interface circuit designer is how to sense this very small voltage reliably, without impacting the outcome.

6

Voltage-based sensing (II)

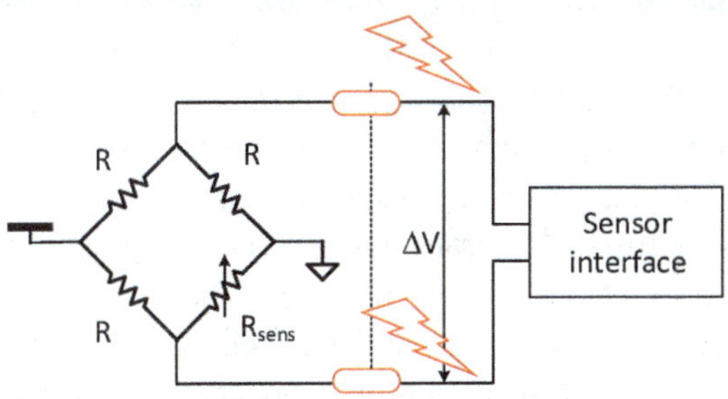

- Limited drive capabilities
- Interconnection elements are not always perfect
  - Series impedance
  - Offset voltages
- Interconnection can pick up noise/interference

While this would seem trivial, the main challenges are that the sensing element almost always has very limited drive capability. This means that the sensor interface circuit cannot pose a low impedance. Furthermore the interconnection elements are not always perfect. There can be a noticeable series impedance or offset voltage which might be mismatched to make matters worse. This is especially the case in biomedical recording as will be shown further. Finally, the interconnection can pick up noise and interference.

7 Voltage-based sensing (III)

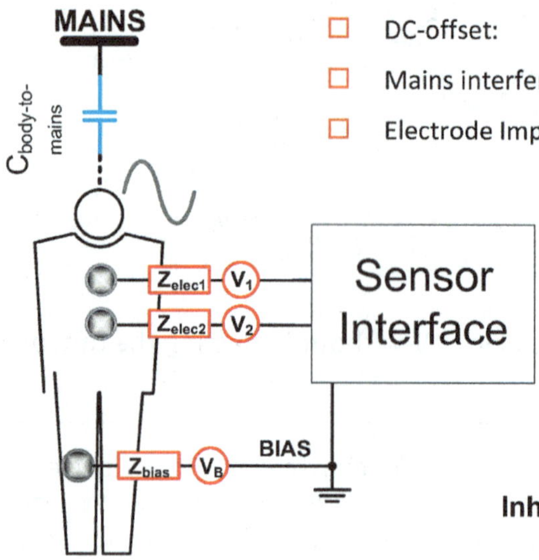

- DC-offset: $V_{DC} = |V_1 - V_2|$
- Mains interference (V_{CM}): $V_{CM} = I_C \times Z_{BIAS}$
- Electrode Impedance Mismatch converts V_{CM} into V_{DM}

$$V_+ = V_{CM} \times \frac{Z_{IN}}{Z_{IN} + Z_{E1}}$$

$$V_- = V_{CM} \times \frac{Z_{IN}}{Z_{IN} + Z_{E2}}$$

Inherent CMRR $\frac{\Delta V}{V_{CM}} = \frac{\Delta Z}{Z_{IN}}$

This slide shows some of the challenges specifically for electrophysiological sensing. The example shows ECG, but it is equally applicable to EEG or EMG. ECG is typically recorded with 2 sensing electrodes. To properly bias the body with respect to the sensor interface, a 3 bias electrode is usually provided. The electrodes used are typically silver-chloride electrodes. When these come into contact with the skin and the human body, a chemical half-cell is generated which results in a certain offset voltage. This offset voltage can be different for each electrode and can change slowly over time. The means that the signal of interest (ECG) is superimposed on a very slowly varying baseline drift. Due to the very slow nature of this drift, it is usually referred to as a DC offset. This DC offset can be several 100s of mV whereas typical ECG amplitudes are a few mV maximum. Another problem of the electrodes is that they form a complex series impedance. In the case of the gel electrodes, this impedance can be as small as a few kOhm, but in case of dry electrodes, this impedance can go up to several Mohm! This impedance will be in series with the input impedance of the readout circuitry. This has 2 main problems. First of it can generate a signal attenuation due to the resistive divider effect. Secondly, the mismatch in the electrode impedance will turn common-mode interference (i.e. due to main coupling into the body) into a differential mode signal. To achieve a good CMRR, the circuit designer must design a readout circuit with very high input impedance.

8 Voltage-based sensing (IV)

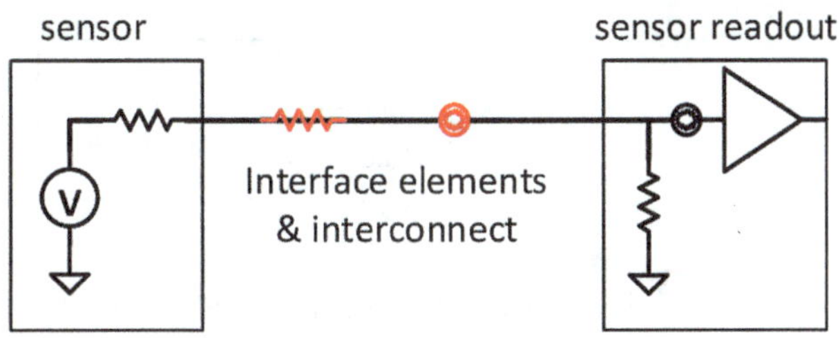

- Good sensor minimizes effects of
  - interfacing elements
  - source impedance
  - noise sources

In summary, most sensing elements can be considered as a voltage source. However they typically are quite "bad" voltage sources. This can be simply modelled by an output impedance. The circuit designer is responsible for designer a sensor readout that minimizes the effects of the interfacing elements, source impedance (and mismatch thereof in case of a differential readout) and noise sources.

9 Voltage-based sensing (V)

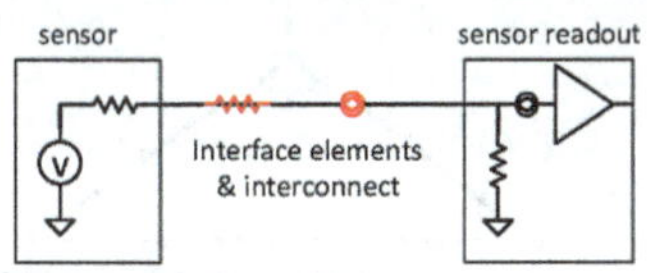

• High input impedance	> 100MΩ to GΩ range
• Low thermal noise	nV/√Hz – μV/√Hz
• Low 1/f noise	<1Hz
• Large dynamic range	70 -100dB
• Low power	μW range
• High CMRR	>100dB
• Low offset	μV range

⟹ **Precision Instrumentation Amplifiers**

This slide shows typical electrical specifications for electrophysiological sensing. The requirements can vary significantly depending on the application, but almost always there will be a need for very high input impedance, very low noise, high dynamic range and high CMRR. The most suitable analog circuit for voltage based sensing is hence usually a high-precision instrumentation amplifier. We will investigate a few circuit implementation further in this chapter.

10

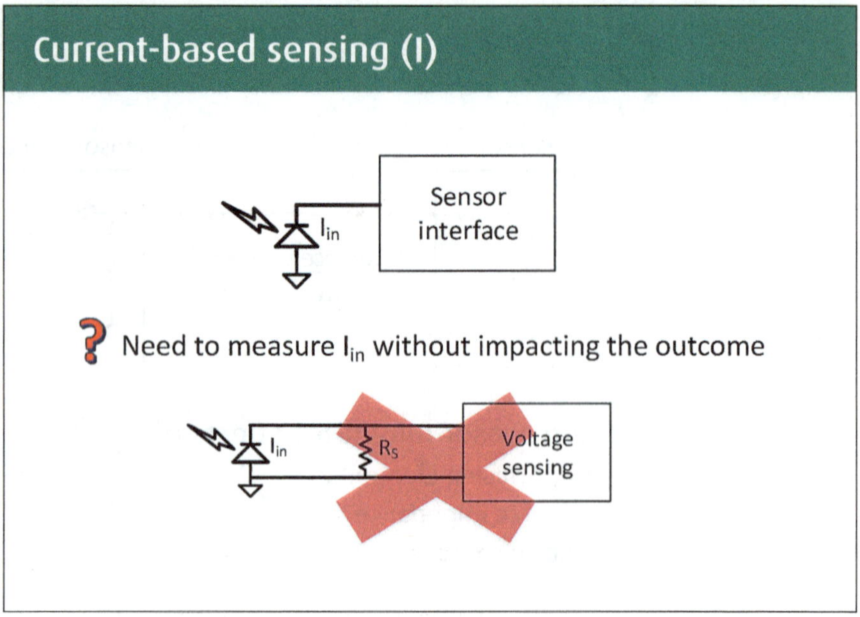

The dual case of voltage sensing, is obviously current-based sensing. An example is a photodector which generates a current proportional to light. This circuit is for example used in optical physiological recording like photophlethysmogram (PPG). The circuit designer must now design a circuit that can sense minute currents reliably without impacting the outcome. One could be tempted to turn this problem into a voltage based sensing problem. We all know that for example by a current signal can be converted into a voltage signal with a simple resistor. While this might be applicable in some cases, there are a number of important drawback to that method.

11

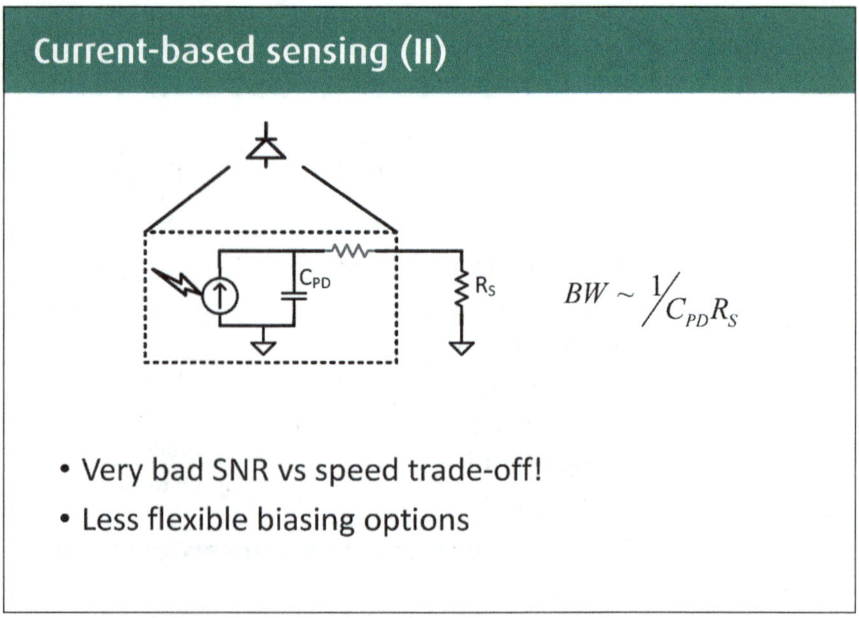

One of the problems, is that , similarly as with voltage based sensing, the sensing element usually does not act as an ideal current source. There usually will be non-ideal components like an noticeable output impedance present. If we simply connect a resistor Rs to the sensing element, it is clear that the signal amplitude will be proportional to Rs, but the bandwidth will be inversely proportional to Rs. So this method usually presents a bad SNR vs speed trade-off. Furthermore, it offer less flexible biasing options. In most cases the sensing element must be properly biased in order to work. Think for example back to the photodetector case, where the performance of the diode depends on the voltage bias across it.

12 Current-based sensing (III)

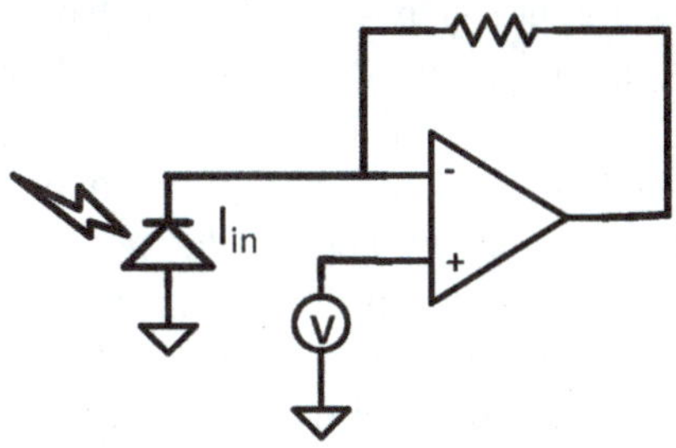

- Trans-Impedance amplifier architecture (TIA)
- High-BW
- Flexible gain
- Flexible biasing

Hence a much better solution is to use special current-input circuitry. One of the obvious choices is a trans-impedance amplifier. Because the input of the amplifiers are at virtual ground, this structure can achieve much higher BW, provides flexible gain options and allows for flexible biasing options of the photodetector.

13 Phased-based sensing

- Pros
  - Low supply voltage operation
  - Easy on-chip integration (low area)
  - scalable architecture (digital components)
- Cons
  - Phase noise of VCO is usually limiting factor to performance
  - Linear tuning range of VCO might be limited
  - Absolute accuracy of VCO → often calibration/tuning required.

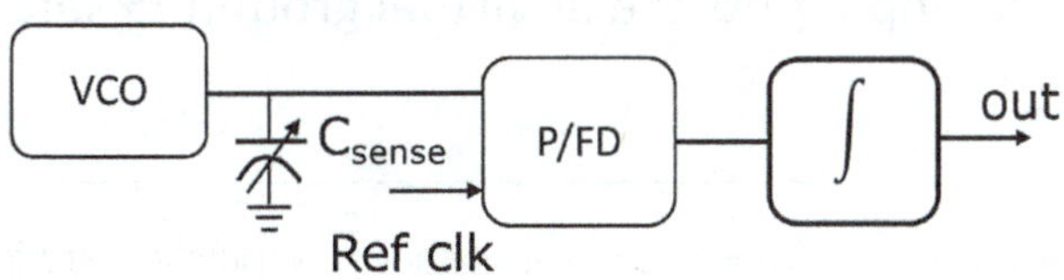

13 Phased-based sensing

Finally a brief note on phase (or time-) based sensing. As mentioned earlier, the main principle here is to put the sensing element in an oscillator circuit where it will modify the phase. This has a number of benefits. One of the more interesting benefits is that this can operate on very low supply voltages, while still in principle allowing large dynamic ranges. With technology scaling comes also supply voltage scaling. In traditional voltage/current-based sensing circuits, this poses a challenge for the dynamic range. However, in time-based circuits, the information is encoded in the phase domain. Hence in principle, the dynamic range is not dependent on the voltage supply. Furthermore, these type of circuits tend to be rather small and are more digital-like, which makes them interesting for deep sub-micron technologies. Of course this type of readouts have their own challenges. Since the information is in the phase domain, the phase noise of the reference clock (VCO) is usually the limiting factor in the performance. Furthermore, the linear tuning range of the oscillating circuit might be limited, often requiring complex calibration methods for large dynamic range readouts. Finally the absolute accuracy of a VCO (free-running) is challenging to achieve. Hence this type of readouts often require calibration or tuning. While these circuits are worth exploring more, they fall outside the scope o this introductory book-chapter.

PRECISION IA ARCHITECTURES

14 Single amplifier - pros

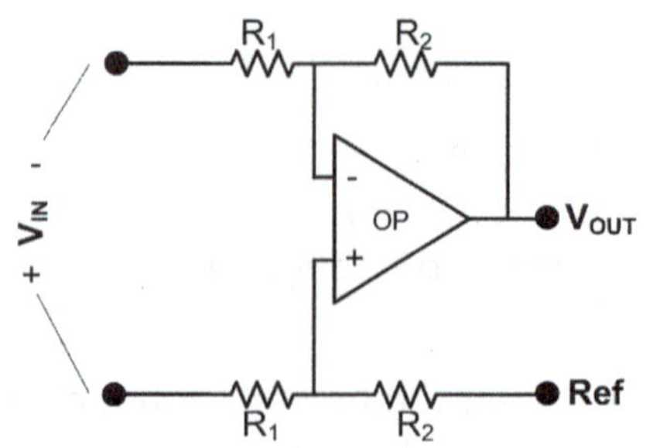

$$\text{Differential Gain} = \left(\frac{R_2}{R_1}\right)$$

- Simple structure
- Opamp inputs are at virtual ground → large input range

There are many ways to amplify a small voltage signal. The most straightforward one is a classical resistive feedback amplifier. It is certainly a well-known and simple structure to design. The opamp inputs are at virtual ground which means that it not difficult to design a readout circuit with large input range. However it hardly makes for a good sensor readout.

15 Single amplifier - cons

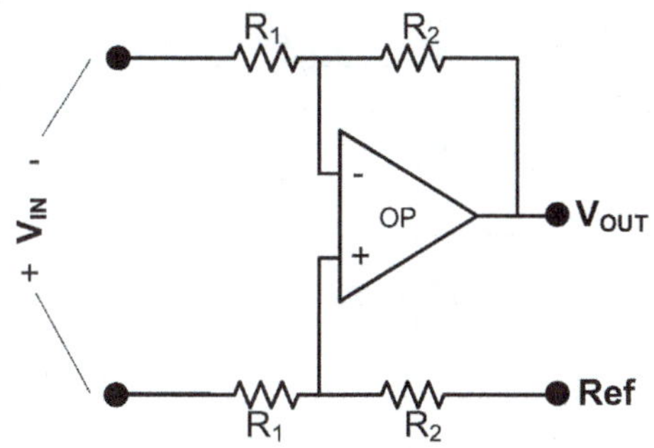

- Very limited input impedance (R1)
- Resistor matching required for good CMRR

The obvious problem is that is present a very limited input resistance (R1), which can't be made to large because of area and noise concerns. Furthermore it relies an good matching to achieve a good CMRR. Truth be told, in modern technologies, quite decent resistor matching can be achieved with careful layout.

16 Capacitive gain elements

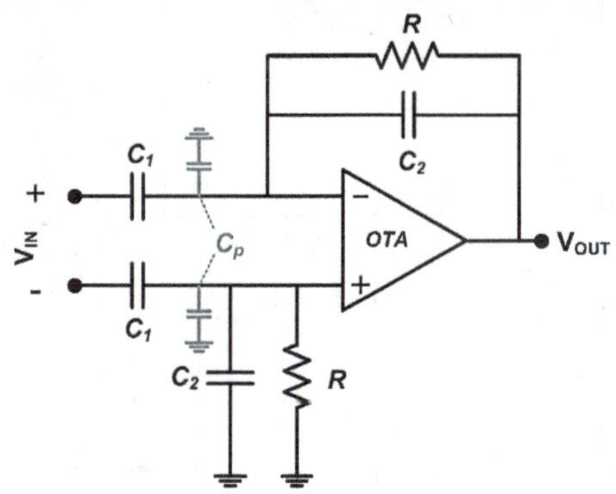

[Harrison2003]
[Olsson2003]

- Low power (only OTA as active component)
- AC-coupled

Of course, an alternative is a capacitive feedback circuit. This can be implemented quite low power (not resistive load). Furthermore is implements inherent AC-coupling which is very interesting if the sensing element exhibit undesired DC-offsets (as is often the case in electrophysiological sensing). As long as the capacitance is kept low, the input impedance can be quite high, into the tens of Mohms range. This circuit is indeed used quite often, especially where power and area are important.

17

Capacitive gain elements - cons

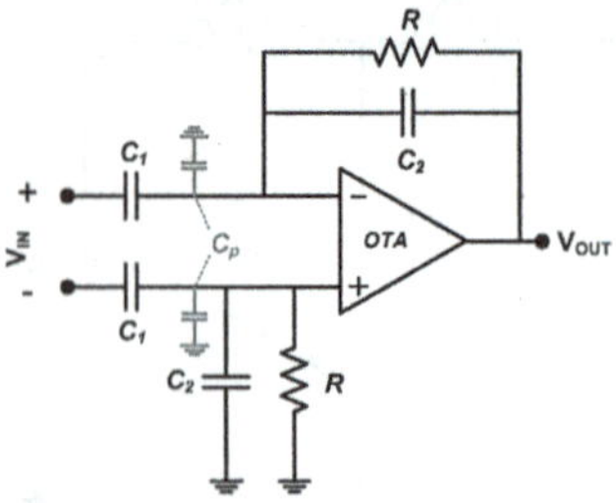

Input Impedance $\approx \dfrac{1}{jwC_1}$

Noise PSD (RTI):
Thermal + 1/f Noise

$$\left(\frac{C_1 + C_2 + C_P}{C_1}\right)^2 \times \overline{v_{OTA}^2}$$

Limited CMRR $\dfrac{1}{\sqrt{\sigma_{C_1}^2 + \sigma_{C_2}^2 + \sigma_{Cp}^2}}$

- Conflicting design requirements!

However it too is not without its flaws. As is clear from the formula, the input impedance is inversely proportional to C1. Hence as long as C1 is kept small, quite large input impedances can be achieved. However, this comes with a few drawbacks. It can be readily shown that in the presence of parasitic input capacitance C1), the input referred noise of the OTA (V_{OTA}) actually increases. This is known as the noise amplification effect and can be intuitively understood by the capacitive divider effect formed by C1 and Cp which will reduce the input signal. Finally, the CMRR depends on the matching of the passives. For small C1, this matching will lead to poor CMRR values. Hence this circuit topology comes with important design tradeoffs.

18

Capacitive gain elements

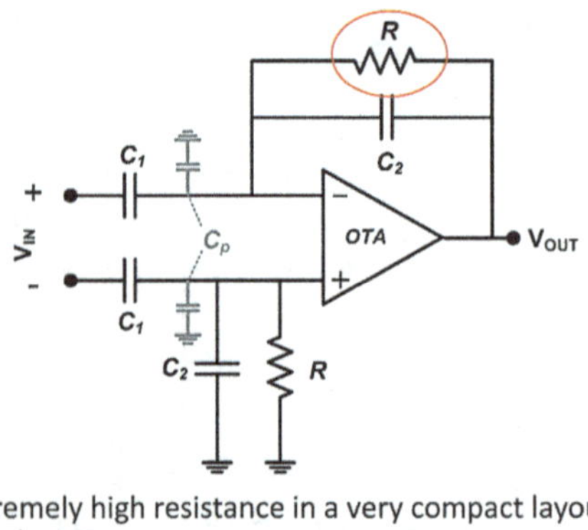

Pseudo-resistor

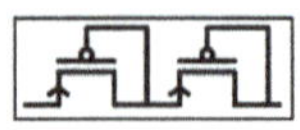

- Extremely high resistance in a very compact layout, but:
- Highly PVT-sensitive
- Light sensitive

It is worth noting that one must take care of proper DC biasing. The most straightforward approach is to put a resistor in parallel with C2 which will bias the opamp inputs. Of course this will result in a high-pass filter behavior. Often sensor signals have very low bandwidth of interest, which means that the cut-off filter must be sufficiently low. In biomedical applications, this can often be <1Hz. Of course to achieve such low cut-off frequencies with reasonably sized (i.e. pF range) capacitors, one must implement extremely high resistive elements. One way to achieve this in an area-efficient manner is with so-called pseudo-resistors.

19 Pseudo-resistors

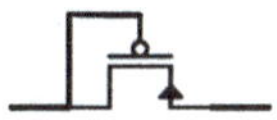

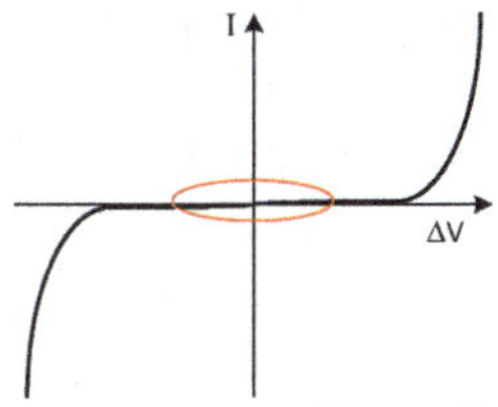

Diode-connected MOS biased in cut-off

- dV/dI is extremely small for small ΔV
- Equivalent resistance values: > 100GΩ is possible
- Very small area

- Very non-linear behavior
- Very PVT sensitive
- Light-sensitive
- Leakage sensitive

→ Equivalent resistance value can easily change orders of magnitude!

There are many ways to implement them, but essentially they are a combination of transistors biased in deep cut-off region. This can provide extremely high equivalent resistance, even into the terra-Ohm region at very compact areas. However these structures are highly light sensitive (since they are in essence diodes) and highly PVT sensitive. The HPF corner can change over 2 orders of magnitude over PVT corners! Furthermore, they behave non-linear and due to the very small current involved, one must take great care about leakage. Even small leakage currents can create a huge offset in over these pseudo-resistors. There are several advanced pseudo-resistor structures out there that try to compensate the behavior over corners and prevent offsets due to leakage. A detailed discussion is outside the scope of this book chapter, but the reader is encouraged to further research this topic.

I.e. Current Biased Pseudo-Resistor for Implantable Neural Signal Recording Applications

Sitong Yuan, Louis G. Johnson, Cheng C. Liu, Chris Hutchens

20 Classic IA

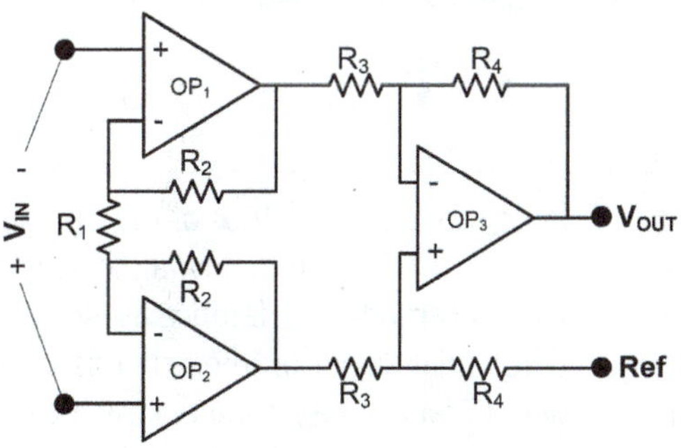

[Smither1989]
[VanRijn1990]
[Burke2000]

$$\text{Differential Gain} = \left(1 + \frac{2 \times R_2}{R_1}\right)\left(\frac{R_4}{R_3}\right)$$

- High input impedance
- Easy gain control via R1

20 Classic IA

Another well known and often used architecture is the classical instrumentation amplifier based on three opamps. The main principle is that the differential signal of interest is copied across R1 and amplified by the high-gain opamps OP1 & OP2. If a single-ended output is desired, this can be achieved as shown in the slide using OP3. This structure is very interesting because it provides very high input impedance. In fact OP1 & OP2 primarily act as impedance buffers. Gain can be easily changed vis resistor R1, which often can be external to allow an easy gain modification.

21 Classic IA – cons (I)

- CMRR requirements

$A_{CM} \approx 1$

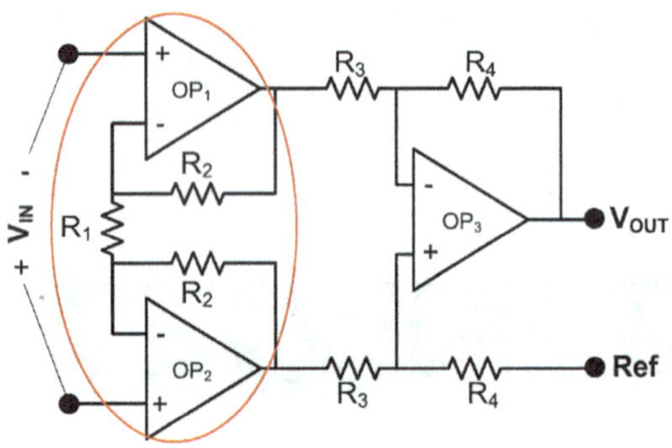

→ Output DR requirement must be designed to handle CM signals

→ 2nd stage must provide also decent CMRR

→ Relies on matching of passives

However going into more detail, this structure also has a number of drawbacks. While the CMRR can be high and is primarily dependent on matching of resistors, it is important to note that the first stage formed by OP1 & OP2, doesn't actually suppress common-mode signal, but rather has a common-mode gain of 1. Since, especially in biomedical applications, the common-mode interference can large (up to several 100s of mV), the first stage must have a sufficient output dynamic range to handle this common-mode interference as well as the amplified differential signal of interest. This usually limits the gain to fairly small numbers. As a consequence, the subsequent stages (in the slide above OP3) usually have strict noise and power requirements also.

22 Classic IA – cons (II)

- Noise considerations
- Main noise contributors:

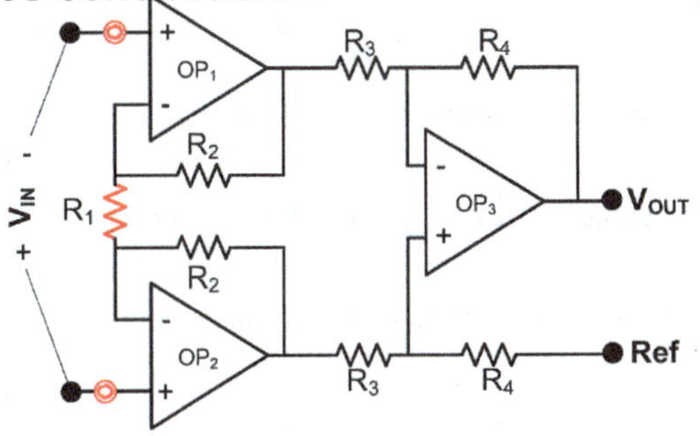

→ Low-noise opamp
- ❑ Power!

→ Small resistor
- ❑ Power!

In fact, we can have a more detailed look at power and noise. If we focus now solely on the first stage, it is obvious that the primary noise contributors will be the input-referred noise of OP1/OP2 and the noise of resistor R1.

23 Classic IA – cons (III)

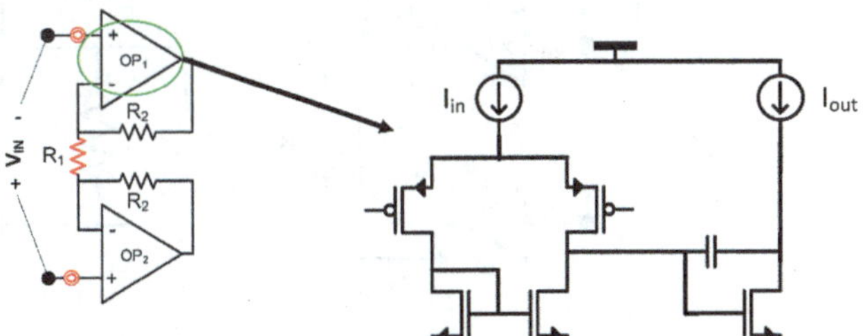

- ❑ Low noise opamp:
  - $VN_{in}^2 \sim 1/I_{in}$
    → requires opamp with large power **input** stage
- ❑ Small resistor
  - $V_{in,MAX} = R_1 \times I_{out,MAX}$
    → requires opamp with large power **output** stage

In order to achieve low noise performance, the designer must minimize both noise contributions. For an opamp, the main noise contribution usually comes from the input differential pair. This noise (at least the thermal part) is inversely proportional to the current. Hence to lower the noise, as a designer, you will need to burn power in the input stage of the opamps. However, to reduce the noise contributions of R1, a small resistor must be chosen. But in order to still have a sufficient large input range, this means that the output current must be sufficiently large. Hence this structure requires opamps with both a high power input stage and a high power output stage. While the classical IA is a very interesting and often sued structure, it does not provide the optimal power-noise tradeoff.

24 Current-mode IA (I)

[Toumazou1989]

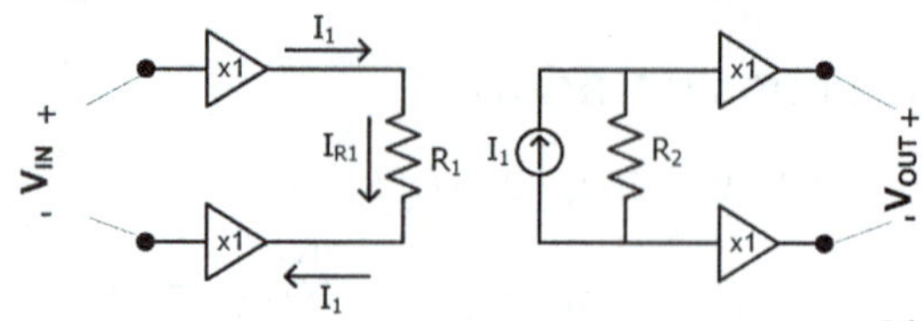

- Transcondutance stage + transimpedance stage
- Input stage acts as impedance buffer
- CMRR primarily limited by input buffers matching

Gain $\approx \frac{R_2}{R_1}$

Input Impedance $\approx \frac{1}{jwC_p}$

CMRR $\approx \frac{1}{\sigma_{buf}}$

Another class of instrumentation amplifiers are the so-called current mode or current feedback IAs. They typically consist of a transconductance (TC) stage which converts the input signal into a current. This current is then converted to an output voltage with a trans-impedance stage. Notice that the classical IA structure does something quite similar conceptually. Because instrumentation amplifiers must have a high input impedance, some kind of buffer circuit is needed at the input. This buffer circuit can present a very large input impedance, (typically a parasitic routing and gate capacitance). However, mismatch in this buffer circuit will typically be the dominant factor limiting the CMRR.

25 Current-mode IA (II)

[Toumazou1989]

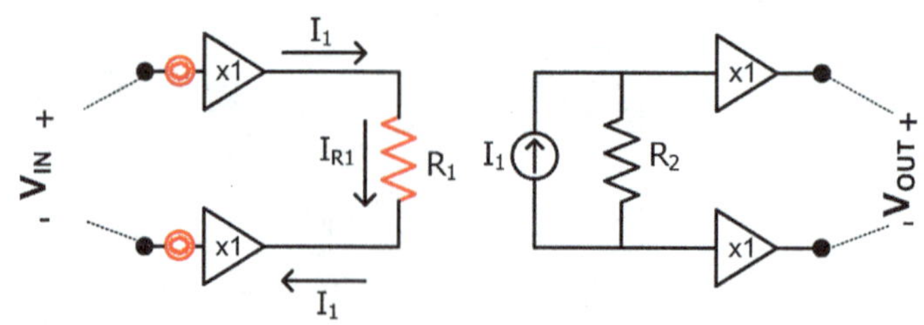

- Main noise contributors only in input stage
  - Clever design techniques can be used to require only 1 high current stage to optimize noise

Noise PSD (RTI): $2\times\overline{v_{BUF}^2}+\overline{v_{R_1}^2}+\frac{v_{R_2}^2}{Av^2}$

Looking at the noise though, the dominant noise sources are again, the input-referred noise of the buffers and the resistor noise of R1. The resistor noise of R2 is typically negligible assuming reasonable gain levels. At first sight, this doesn't seem to be very different from the classical instrumentation amplifier architecture. However, as will be shown further, there are clever design techniques and circuit topologies that will result in a much more attractive power-noise tradeoff.

26

Current-mode IA (III)

[NvH2012]

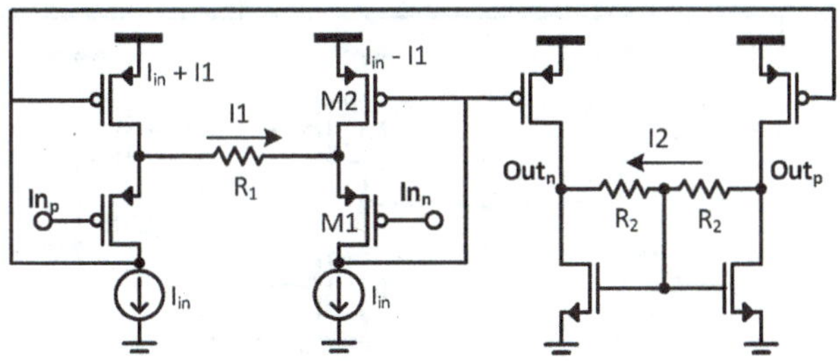

$$I1 = \frac{V_{in}}{R_1}$$

$$V_{out} = I2 \times 2R_2$$

$$\frac{V_{out}}{V_{in}} = \frac{2R_2}{R_1}$$

This slide shows an example of how such an instrumentation amplifier can be implemented. The input stage is based on the flipped-voltage follower structure. The input transistors M1 act as source followers since they have a fixed current Iin flowing through them. Hence the input signal is copied across the resistor R1 and a signal dependent current I1 flows into the load devices M2. Indeed devices M1/M2 and current sources Iin form a transconductance stage. The input buffers as are created by a single transistor M1! The current I1 is mirrored to the output stage, where it is again converted into a voltage by output resistances R2. The overall transfer function is simply given by the resistor ratio (assuming a current mirror ratio of 1:1 between TC and TI stage).

27

Current-mode IA (IV)

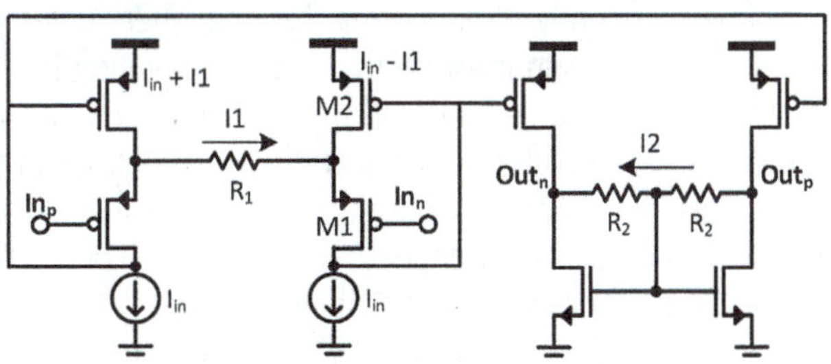

- Input noise: $\sim\sqrt{2dV_{ieqM1}^2 + dR_1^2}$
- Input range: $\sim R_1 x\, I_{in}$

I_{in} minimizes noise & maximizes input range!

If we analyze the noise in more detail, it will be obvious that the dominant noise sources will be the noise of the input devices M1 and of the input resistor R1. However it is obvious now that to minimize noise and maximize the input range, the designer only has to increase the current in the input stage Iin. Contrary to the classical instrumentation amplifier architecture, the output stage doesn't need high output current drive capabilities for noise or input range considerations.

28 Current-mode IA (V)

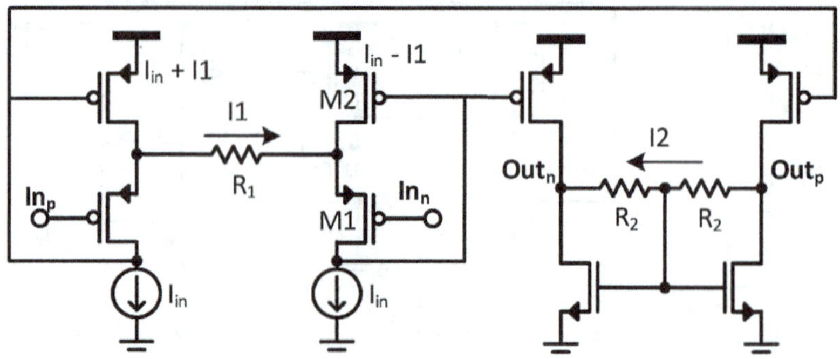

- Input transconductance stage rejects common-mode signals
  → Less strict DR requirements on output stage

A second major benefit is that this architecture already reject and suppresses common-mode signals even at the TC stage. Remember that the classical instrumentation architecture has a common-mode gain of 1 which limits the output dynamic range and consequently often means that the instrumentation amplifier gain must be to avoid saturation in the presence of large common-mode variation. This doesn't happen with this architecture. Common-mode signals are effectively attenuated and the output common-mode signal depends on a common-mode feedback circuit. In the slide above, this is formed by a simple tail current sources who's gate is connected to the output common-mode signal obtained at the midpoint of output resistors R2. Of course more elaborate common-mode feedback circuits can be implemented for more precise common-mode control. The important realization is that this architecture actively suppresses the input common-mode signal and hence can achieve much higher differential gain factors without causing saturation in the presence of strong common-mode interference. Hence the subsequent programmable gain blocks in the signal chain can have significantly more relaxed noise (and hence power) requirements.

Of course there is a drawback to this circuit as well. Since the input buffer is formed by a simple source follower, these architectures tend to have worse linearity than the classical instrumentation amplifier architectures.

TRANSIMPEDANCE AMPLIFIERS

29 Single resistive feedback

[Yang2015]

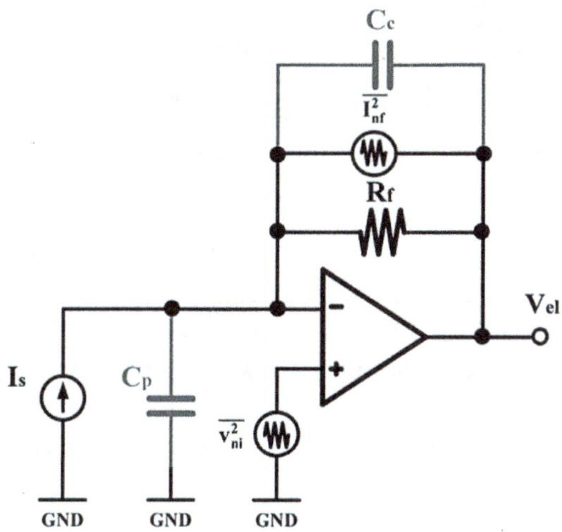

- Parameters
  - Gain $R_{eq}=R_f$
  - Bandwidth $BW = \frac{1}{R_{eq} C_f}$
  - Input noise $\overline{I_n^2} = 4k_B T / R_{eq}$
- Advantages
  - Compact
  - Low noise/power
- Disadvantages
  - Compensation capacitor C_c may limit the bandwidth

Let's make a switch to current based sensing circuits. A typical example would be an optical readout for example for photoplethysmogram (PPG). In PPG recording, a light source (often from an LED) is used to shine light into the human body. Various components in the blood will absorb light at specific wavelengths differently. More specifically oxygen-rich blood will show a different absorption at specific wavelengths than oxygen-poor blood. Hence the reflected (or transmitted) light will be modulated by the heart rate. By measuring (often with a photodector) this modulated light, one can measure in a non-obtrusive manner the heart rate at regions of the body with superficial blood flow (i.e. finger tips, ear lobes, wrist, ...).

As already mention in the introductory chapters, one of the most used architectures to read out a photodector is a trans-impedance architecture as shown above. The formulas for the gain, bandwidth and noise are shown in the slide and are easy to derive. In most cases this architecture is perfectly suitable. The most notable drawback is that the capacitance Cc is often needed for stability purposes. Indeed the dominant pole is found at Rf, but the parasitic capacitance Cp of the photodetector can be significant enough to warrant a large Cc and hence a low BW.

Yang Zhao; Jian Zhao; Xi Wang; Guo Ming Xia; An Ping Qiu; Yan Su; Yong Ping Xu, "A Sub-μg Bias-Instability MEMS Oscillating Accelerometer With an Ultra-Low-Noise Read-Out Circuit in CMOS," in *Solid-State Circuits, IEEE Journal of* , vol.50, no.9, pp.2113-2126, Sept. 2015

30 TIA: T-resistive feedback

[Yang2015]

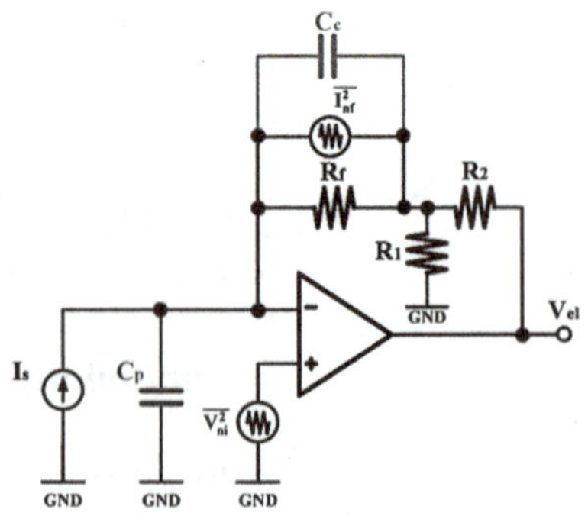

- Parameters
  - Gain $R_{eq} = R_f R_2 / R_1$
  - Bandwidth $BW = \frac{1}{R_{eq} C_f} \frac{R_2}{R_1}$
  - Input noise $\overline{I_n^2} = \left(\frac{R_2}{R_1}\right)^2 4k_B T / R_{eq}$
- Advantages
  - High gain/bandwidth
  - Less gain-bandwidth trade-off due to C_f
- Disadvantages
  - High noise

There are multiple ways to provide different design tradeoff by using more advanced feedback networks. Shown above is a T-resistive feedback network. Again the formulas for gain, bandwith and noise are shown above. As is clear this architecture can provide high gain and high bandwidth if ones choses a large R2/R1 ratio. However it tends to result in larger input referred noise levels. There is no free lunch in circuit design.

Notice that to build a proper PPG circuit, more advanced topics must be addressed, most notably ambient light cancellation. However those circuit topologies fall outside the scope of this book chapter.

Yang Zhao; Jian Zhao; Xi Wang; Guo Ming Xia; An Ping Qiu; Yan Su; Yong Ping Xu, "A Sub-μg Bias-Instability MEMS Oscillating Accelerometer With an Ultra-Low-Noise Read-Out Circuit in CMOS," in Solid-State Circuits, IEEE Journal of , vol.50, no.9, pp.2113-2126, Sept. 2015

31 Advanced circuit techniques

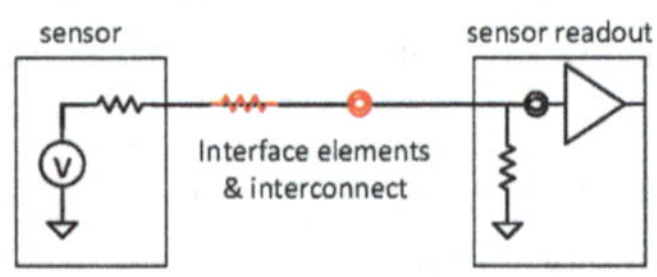

Application requirements

- High input impedance
- Low thermal noise
- Low 1/f noise
- Large dynamic range
- Low power
- High CMRR
- Low offset

Design techniques

- Chopper modulation
- Correlated double-sampling
- DC-servo
- Bootstrapping
- Design in weak inversion

In the remainder of this book chapter, we will discuss a few advanced circuit techniques that are often used in (biomedical) sensor interface circuits. The slide above shows again some of the major application requirements a designer will be faced with. Fortunately there are a number of advanced circuit techniques that the designer can use.

32

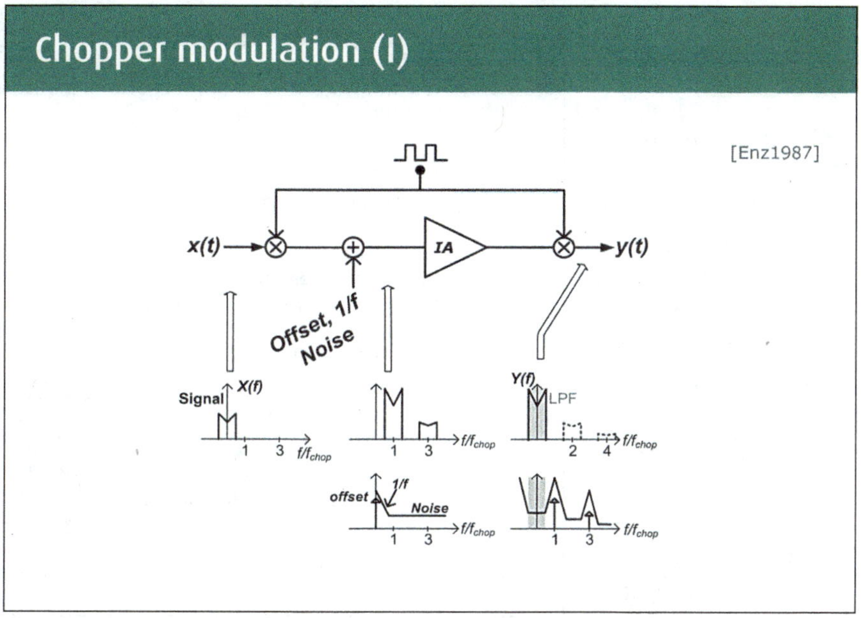

Chopper modulation is a very powerful circuit technique to handle low-frequency or static non-idealities like 1/f noise, DC-offset and mismatch. When dealing with low-frequency signals of interest, as is often the case in biomedical and sensor applications, mismatch, DC-offset and 1/f noise are all going to corrupt the measurement since they manifest themselves in the frequency band of interest. The main principle of chopper modulation is to modulate the signal of interest to a high frequency band before any active amplification. Hence those non-idealities don't overlap with the signal of interest. After amplification, a chopper downmodulates the amplified signal of interest back to baseband, while the non-idealities are modulated to a higher frequency. They can then be removed by a low-pass filter.

33

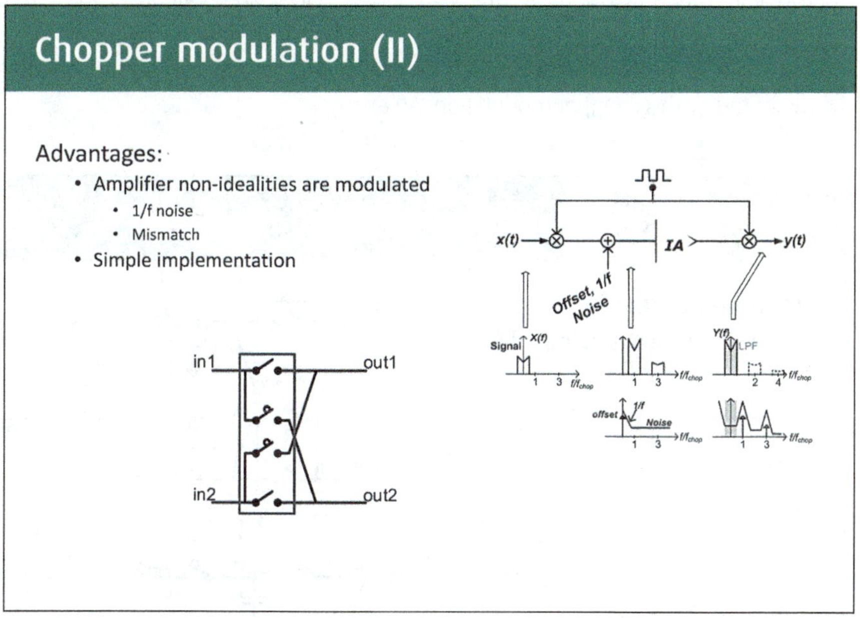

The benefits are clear. In principle, chopper modulation can completely mitigate 1/f noise, DC-offset and effects of mismatch. In practice there are a number of reasons why complete mitigation is difficult to achieve – as will be discussed further. Furthermore, the modulation can be achieved very simple. In fact passive mixers are most often used. They are relatively noise-less, small area and don't consume any power other than the controlling clocks. The passive mixers or choppers shown in the slide above basically switch the differential signal between the positive and negative input nodes effectively generating sign inversals. As such they mathematically corresponds to a multiplication with +1 and -1 alternatively.

34 Chopper modulation (III)

Advantages:

- Amplifier non-idealities are modulated
  - 1/f noise
  - Mismatch
- Simple implementation
- Splits CM and DM into separate bands
  - Improved CMRR! (assuming linear system)

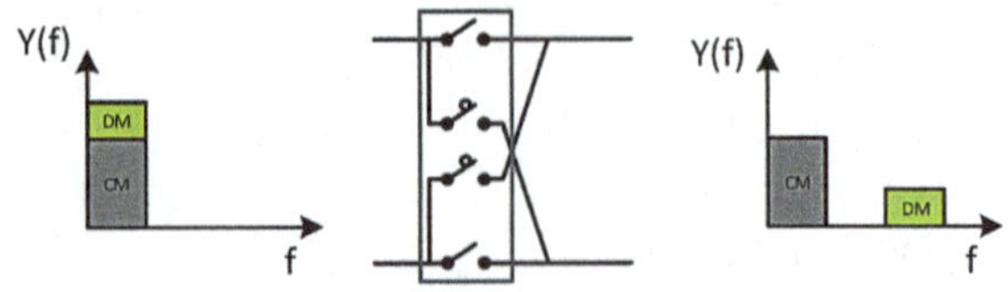

Another interesting aspect of choppers is that they only have an effect on differential mode signals. It is obvious that for common-mode signals, choppers have no effect. For high CMRR applications, typical in biomedical recordings for example, this is very beneficial. This means that after the choppers, the CM signals are out-of-band with respect to the differential signal of interest. If the IA has a limited CMRR, the common-mode signals will convert to differential signals, but at baseband. Hence chopper modulation can also positively impact the CMRR and mitigate mismatches that would otherwise lead to reduced CMRR. Of course, this assumes a linear system. Non-linear behavior due to large common-mode signals will not be mitigated by this chopper modulation. An example is a gain dependency on the input-common-mode level. Basically every amplifier will have a limited input range, and when driving the amplifier outside the input range, the gain will rapidly drop. Chopper modulation will not be able to mitigate these effects and will lead to a in-band CMRR reduction. The reader is encouraged to perform a simple simulation to further understand this mechanism.

35 Chopper modulation (IV)

Disadvantages:

- Reduced input impedance

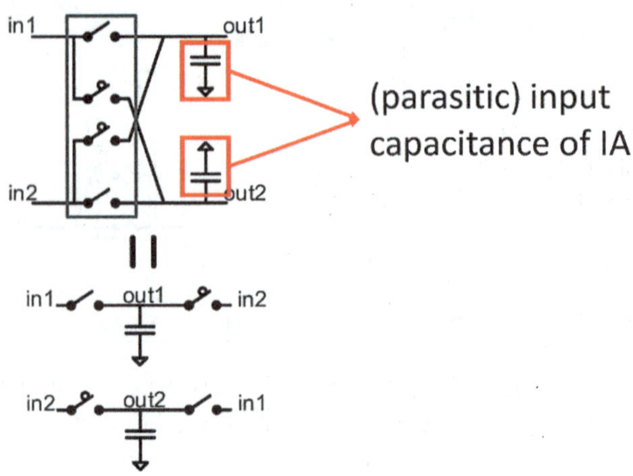

35 Chopper modulation (IV)

Unfortunately, chopper modulation has also a few drawbacks. The most important one is probably that it reduces the input impedance. Once can redraw the chopper circuit as shown in the slide. While it is exactly the same circuit, if drawn like that, it is easy to recognize that choppers actually form a switched-capacitor resistor where the capacitor represents the input capacitance of the amplifier (parasitic routing and gate capacitance). Where without chopping the input impedance is only the parasitic capacitance Cp, this reduces by a factor equal to the chopper frequency when the circuit is chopped. Hence one must chop at the lowest possible frequency. In practice one often tries to chop at or very near to the 1/f noise corner.

36 Chopper modulation (V)

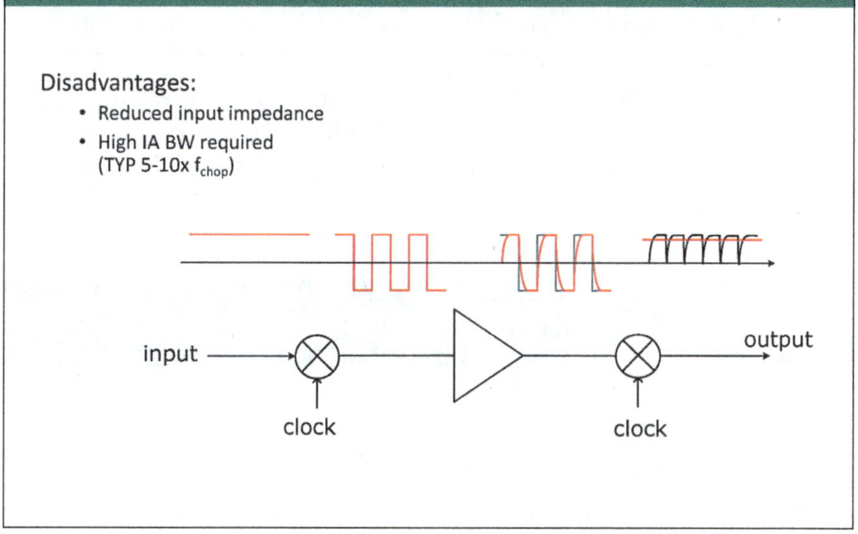

Another drawback is of course that your amplifier must amplify high-frequency signals. In practice however, this is not too challenging, since typical chopper frequencies are in the range of kHz – to maintain a high enough input impedance - which is seldom a problem for modern CMOS technology. Nevertheless, it is important to realize that the finite BW will result in an error. This is shown in the slide above. Assuming a DC input signal. After chopping, this signal will look like a square wave signal as shown in the slide. Due to the limited BW and slew rate of the amplifier, the amplified signal will no longer represent a ideal square wave, but rather will exhibit some finite slewing/settling behavior. When this signal is chopped again, high frequency chopper glitches appear in the output signal. While these can be filtered out with a low-pass filter, the resulting signal will actually be slightly smaller than what one would expect.

Another way to understand this effect is that chopping implements in the frequency domain a convolution with a pulse train. Hence the chopper modulated input signal will have a lot of higher order harmonics. Due the limited BW of the amplifier, those higher order harmonics are filtered out. After downconverting (again a convolution with a pulse train in the frequency domain), the energy in the higher order harmonics is lost which results in a smaller output signal.

To mitigate this effect, a high-BW IA is needed. A good rule of thumb is to make sure that the IA BW is at least 5x to 10x higher than the chopper frequency. Since most of the signal energy will be present in these harmonics, the gain reduction is usually acceptable. However, this obviously depends a lot on your application!

37 Chopper modulation (VI)

Disadvantages:

- Reduced input impedance
- High IA BW required → chop at low-impedance nodes (TYP 5-10x f_{chop})

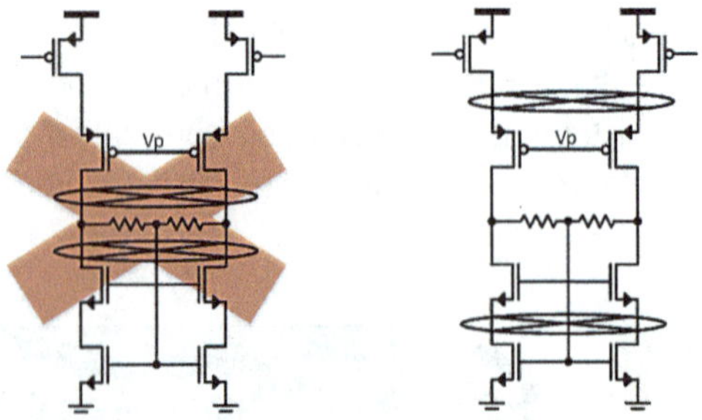

Due this effect, it is very critical to chop at nodes where you have high BW available. In the picture above, you can see a typical cascaded output stage of an amplifier. If you want to implement the output chopper in such an amplifier, as a designer you have 2 choices. You can either chop really at the high-impedance output nodes, or you can chop at the low impedance nodes before the cascodes. It is obvious that the circuit has a much higher BW at the low-impedance nodes which is where you should implement the choppers to reduce the chopper glitches explained in the previous slide.

38 Chopper modulation (VII)

Disadvantages: [Enz1996]

- Reduced input impedance
- High IA BW required (TYP 5-10x f_{chop})
- Input spikes and chopper clock feedthrough lead to residual offset

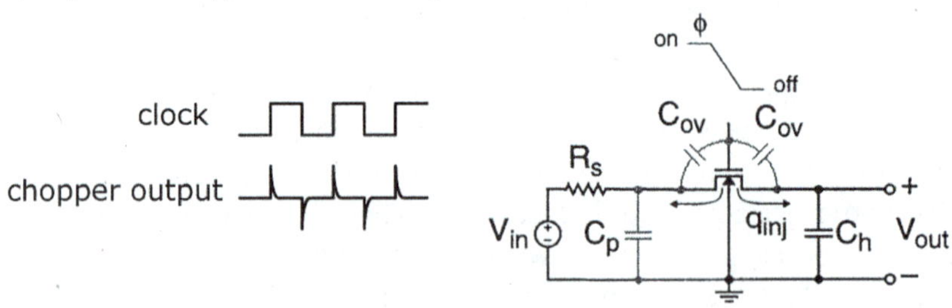

There are other drawbacks to choppers also. These are clock feed through resulting in another form of chopper glitches. This occurs via the parasitic gate-source and gate-drain capacitances. To mitigate these effects, one can employ dummy switches, charge injection insensitive switch architectures and complementary drive circuits. An in-depth discussion is out of the scope of this book chapter, but the reader is encouraged to research these topics further.

39

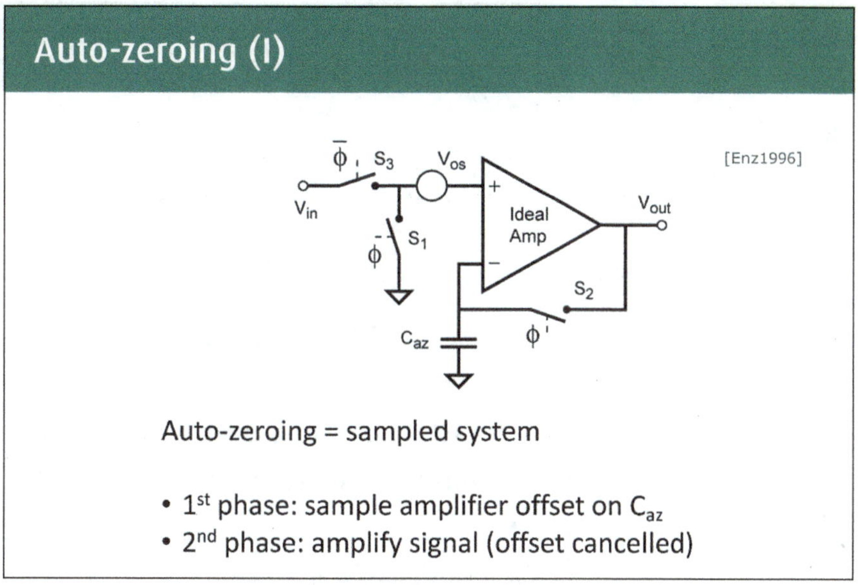

Auto-zeroing is another technique used to mitigate 1/f noise and DC-offset. It is a sampled technique though. The main principle is that the 1/f noise and DC-offset are first sampled and stored on a capacitor Caz. IN this phase S1 and S2 are closed, while the input is disconnected (S3 open). Because the amplifier is in buffer configuration, the non-idealities (represented by the source Vos) will be sampled on Caz. In the amplification phase, S1/S2 are opened and S3 is closed. The input signal and Vos are amplified, but since Vos is now also present on the negative terminal, Vos is a common-mode signal and will be rejected by the amplifier. This obviously only work if Vos can be considered stable over the period of both phases. This is why is only work for low-frequency non-idealities.

40

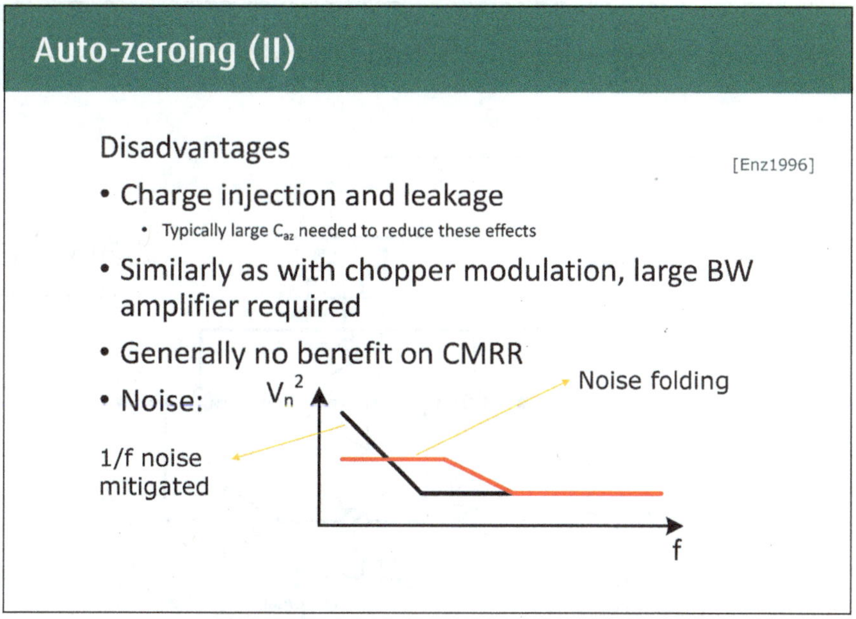

Compared to chopping, auto-zeroing suffers from similar drawbacks. There are a number of specific drawbacks to auto-zeroing though. Charge injection and leakage will typically dictate a large Caz (area!). Furthermore it generally has no effect on CMRR. Finally since it is a sampled system, noise folding will occur. So while in principle the 1/f noise can be mitigated with auto-zeroing, and hence the very low-frequency noise can be improved, due to noise folding, the final noise at intermediate frequencies can actually INCREASE with auto-zeroing as is shown in the plot above.

Auto-zeroing is usually not more beneficial than chopping for most biomedical sensor interface circuits. However it lends itself very well to naturally sampled systems. Examples are for example image sensors where auto-zeroing is very often applied.

41 DC-offset compensation (I)

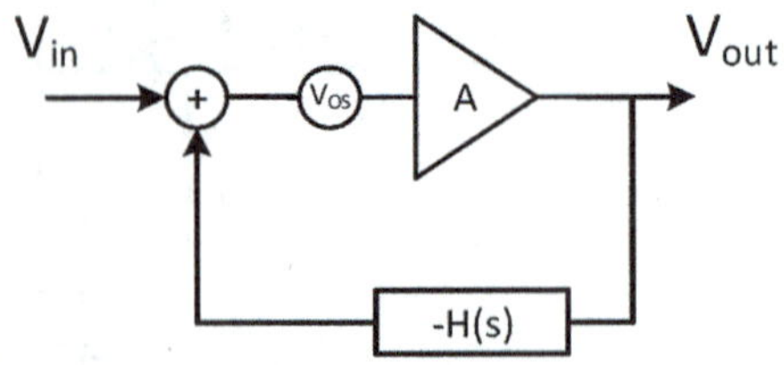

$$\frac{V_{out}}{V_{in}} = \frac{A}{1 - AH(s)}$$

- If H(s) has a low-pass behavior → implements a HPF
- Active DC-offset filter

Another main challenge often found in biomedical interface circuits is dealing with large DC-offsets. As mentioned in the introductory sections, this DC-offset can originate from the actual sensor side. In case of biomedical applications, it is often the consequence of the polarization voltages of electrodes for example. So while chopping can mitigate effects of amplifier DC-offset, it will not be beneficial for DC-offset that find their origin outside the ASIC. To handle this, DC-compensation circuitry is needed. While of course a passive high-pass filter can be used in the input, it tends to significantly lower the input impedance to tens of Mohms range typically. Active DC-offset compensation is usually achieved in a feedback architecture as shown above. The DC output signal is extracted with a low pass filter and fed back negatively.

42 DC-offset compensation (II)

[Yazicioglu2006]
[Denison2007]

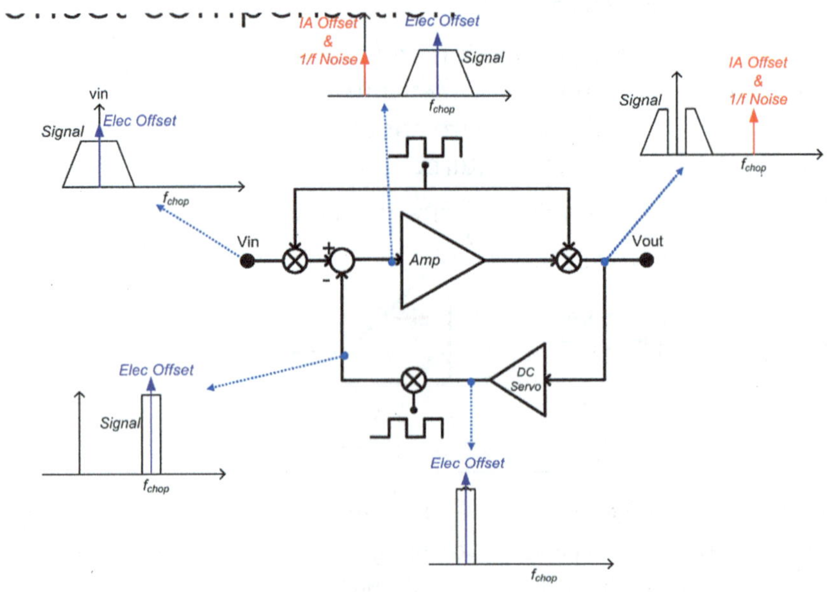

42 DC-offset compensation (II)

Of course, this becomes a bit more complicated when combined with chopping. The designer must be sure also properly put choppers in the feedback path. The correct way to do this is shown in the figure above. The actual DC servo (the active low pass filter) must be placed outside the choppers. The feedback signal must then be upmodulated before being fed back negatively. As is clear, this properly implements a HPF behavior. It is recommended to always consider the frequency domain signals and realize where the signal of interest is (in baseband or at the chopper frequency) in the various parts of the circuit. The reader is encouraged to try to understand how this circuit would behave if one would omit the chopper at the output of the DC servo.

Looking at the circuit above, it is clear that we somehow must add circuitry in the input lines to realize the subtraction. Typically you wouldn't want to put any circuitry in the sensitive input lines since that tends to negatively impact input impedance, CMRR and/or noise.

43 DC-offset compensation (III)

[Yazicioglu2006]
[NvH2012]

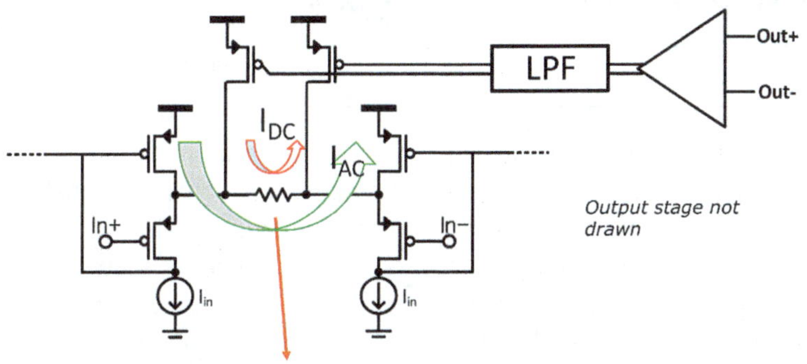

Subtract DC-error on input resistor

Bad power-range trade-off! $V_{off,max} = I_{in}R$

This slide shows how one can practically implement that feedback circuit without negatively impacting input impedance, noise or CMRR. For this example, we use again the CFBIA transconductance stage explained earlier. The circuit above will copy the input voltage signal across the input resistor, generating a signal dependent current ($I_{AC} + I_{DC}$). which is normally copied to the output stage. The slide shows how we can subtract the DC current by differential current source controlled by the DC servo. As a result only the AC current is copied to the output stage.

Since this circuit doesn't act directly on the input lines, it doesn't negatively impact input impedance, noise or CMRR. However, it isn't perfect. The input devices M1 are still subjected to the large DC offset and the full single appear across the resistor R1. As we already discussed earlier, the input range of this circuit scales with Iin and R. To handle large DC offsets, one would need either a large R (large noise) or large Iin (large power).

44 Active HPF (I)

[NvH2014]

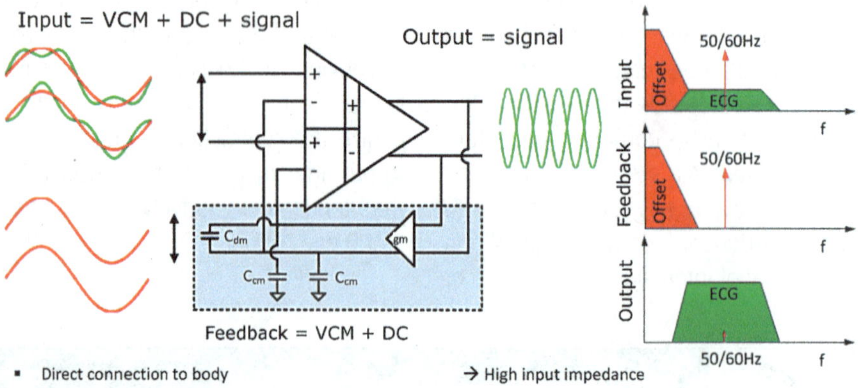

- Direct connection to body → High input impedance
- Pseudo differential IA (DDA alike input structure) → Can handle large offsets
- DC-offset subtracted at input → High gain possible
- DC-servo transparent for CM signals → High CMRR → Only 1 ext. cap required

An alternative structure is shown in the drawing above. Once again, there is no circuitry placed directly in the input leads. This circuit can be connected directly to the body and hence provides large input impedance. It is based on a dual difference amplifier to subtract the output of the DC-servo. The dual difference amplifier structure has 2 input differential pairs. To ensure good performance, the feedback amplifier has a filter that behaves differently for differential signal than is does for common-mode signals. This is very important is will be explained next.

The input signal has 3 main parts.

- large differential DC (unwanted)
- Large in-band common-mode signals (unwanted – i.e. mains interference)
- Small differential signal of interest

The feedback amplifier output will contain both the large differential DC signal, but also the large common-mode signal. This can be achieved by the gm-filter structure as shown above. For differential signals, the BW is determined by Cdm + Ccm. However for common-mode signal, Cdm is shorted and hence the BW only depends on Ccm. This means that the designer can design the feedback network to have independent control over the CM and DM signals. Notice that it is important to also feed back the large common-mode signals, but that way each differential pair has the same operating point. If we would only feed back the large DC signal, there will be large operating point mismatch in the input diff pairs which would seriously limit the input range of this circuit.

45

Active HPF (II)

[NvH2014]

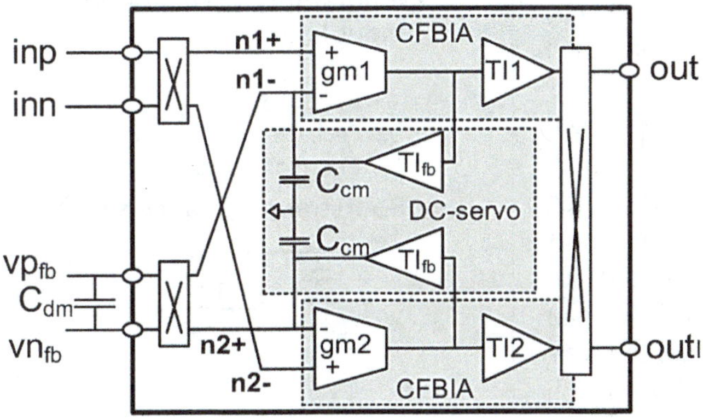

This slide shows how such an amplifier can be implemented. At it's core it contains two CFBIA, where gm1/2 can be implemented in the same manner as the transconductance stages already explained earlier in this book chapter. This slide also shows how this amplifier can be chopped. This structure is very attractive because it combines a number of very interesting benefits. It rejects very large DC offsets while having very high input impedance and low noise. It rejects common-mode and can hence implement large gain factors. The drawback is that it requires 2 transconductance stages which will have its impact on the power consumption.

46

DC cancellation loop in TIA

[Winokur2015]

- Digital feedback loop enables to reconstruct full DR signal

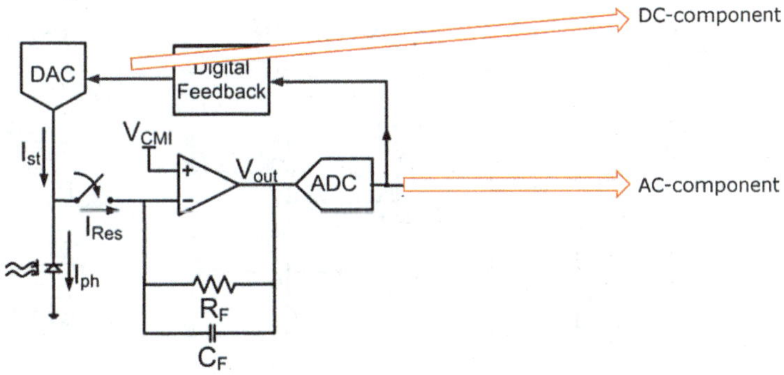

Winokur, E.S.; O'Dwyer, T.; Sodini, C.G., "A Low-Power, Dual-Wavelength Photoplethysmogram (PPG) SoC With Static and Time-Varying Interferer Removal," in *Biomedical Circuits and Systems, IEEE Transactions on* , vol.9, no.4, pp.581-589, Aug. 2015

Notice that DC-cancellation is often also very critical in optical recordings, where the DC level is often significantly larger than the AC signal of interest. This slide shows how such an active DC servo in a TIA structure can be implemented. A current DAC can be used to subtract the DC current from the input. An added benefit is that this current DAC can be digitally controlled. This means that the full range signal can be recreated in the digital domain without needing very large dynamic range TIA and ADCs in the analog domain.

47 Bootstrapping

- Input impedance limited by various parasitic caps in the input line
  - ESD
  - Bondpad
  - Package
  - Board parasitics
- Bootstrapping cancels input capacitance

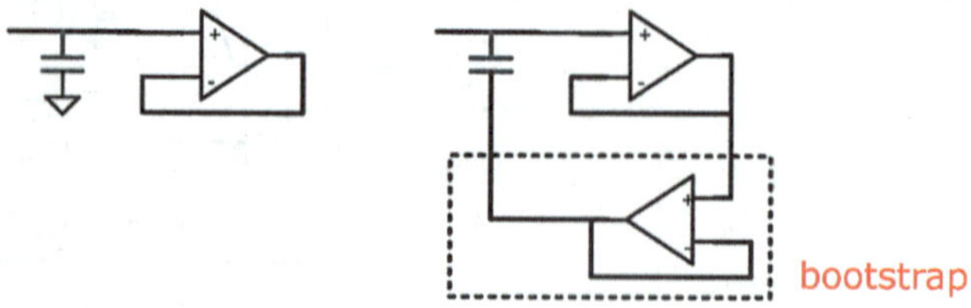

Another very interesting technique often used is bootstrapping. As already discussed earlier, a high input impedance is often needed. In many cases this input impedance is limited by (external) parasitic capacitances. These can originate from ESD structures, bondpad structures, package parasitics, board parasitics, etc. A total parasitic capacitance of a few pF for example will already limit the input impedance to a few 100MOhms at mains frequencies. If very high impedances (i.e. > 1GOhm) are needed at these frequencies, bootstrapping can be used.

The main idea behind bootstrapping is to copy the input signal to the parallel plate of the parasitic capacitance. In this case the current through the capacitor will be zero resulting in a very high input impedance. This can be implemented by providing a driven plane underneath the routing and between the ground plane.

48 Local bootstrap example

[Yazicioglu2006]

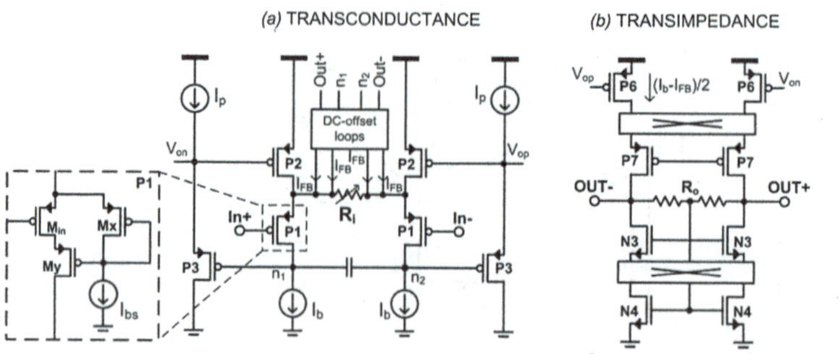

Bootstrap cancels C_{gd} and C_{gs}

→ Large input impedance

$$Gm = \frac{1}{\frac{1}{g_{m,eff}} + Ri} = \frac{1}{\frac{g_{01}}{g_{m1}g_{m2}} + Ri} \approx \frac{1}{Ri}$$

$$Gain \approx \frac{Ro}{Ri}$$

48 Local bootstrap example

Notice that this concept can also be applied within a circuit. The example above shows how one can bootstrap out the parasitic capacitances of an input differential pair. The circuit shows the CFBIA input stage which we discussed already earlier. Device P1 is in fact a combination of 3 transistors and a current source. Remember that P1 is a source follower. This means that the same (AC) signal is present on the source as on the gate. Hence the C_{GS} is already cancelled. The remaining parasitic capacitance – as seen from the input - is C_{GD} and C_{GB}. By adding Mx, My and Ibs, we can also cancel C_{GD}. Indeed both Mx and My also act as source followers (they have a constant current flowing through them) and hence the AC input signal is also copied to the drain – effectively cancelling the C_{GD}. If the well is finally also connected to the source (as can be done with PMOS devices), also the C_{GB} is cancelled. This results in a pretty high input impedance dominated by parasitic routing capacitances and the well-to-bulk capacitances.

49 Bootstrapped IA with trimming

[NvH2014]

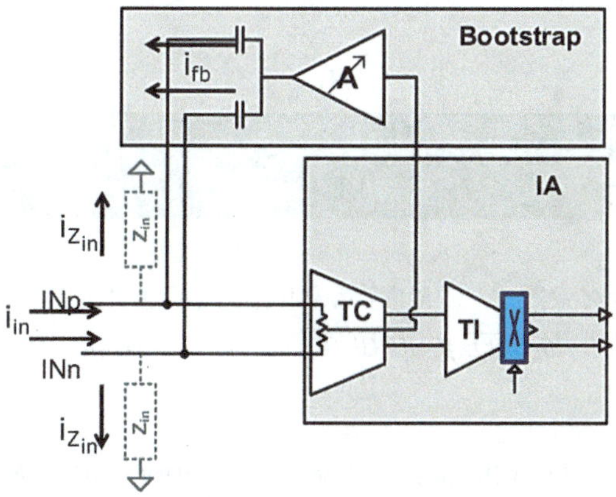

- Adjust A so that $i_{Zin} = i_{fb}$
- $i_{in} = 0$
- → high $Z_{in,cm}$

Bootstrapped input to improve CM input impedance:

Bootstrapping can also be achieved with positive feedback. This slides shows a circuit that improves the common-mode input impedance via positive feedback. The circuit works by sensing the common-mode input and coupling this back to the input lines. The operating principle is that if there would be a finite input impedance as represented by Zin, this will normally limit the input impedance. However, one can tweak the feedback gain in such a way that all the current needed to bias the input impedance is actually provided by feedback network. As soon as $i_{fb} = i_{Zin}$, this means that $i_{in} = 0$. The input impedance as seen from the source will be very high indeed. By making the feedback gain tunable, one could tweak the feedback gain to compensate even for external parasitic like PCB and bondpad parasitics which might not be trivial to bootstrap with the other means described earlier.

50 Conclusion

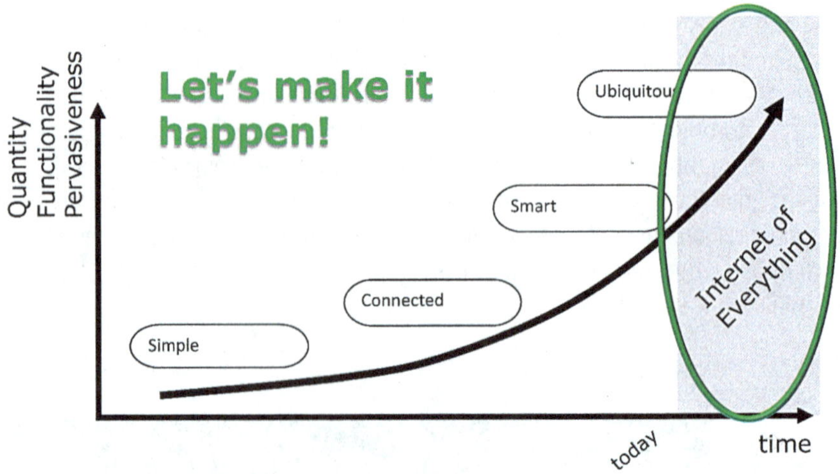

51 References

1. [Yen1982] Yen, R.C.; Gray, P.R., "A MOS switched-capacitor instrumentation amplifier," in *Solid State Circuits, IEEE Journal of*, vol.17, no.6, pp.1008-1013, Dec. 1982

2. [Harrison2003] R. Harrison and C. Charles, "A low-power low-noise CMOS amplifier for neural recording applications," *Solid-State Circuits, IEEE Journal of*, vol. 38, no. 6, pp. 958-965, June 2003.

3. [Olsson2003] R. Olsson, A. Gulari, and K. Wise, "A fully-integrated band pass amplifier for extracellular neural recording," in *Neural Engineering, 2003. Conference Proceedings. First International IEEE EMBS Conference on*, 20-22 March 2003, pp. 165-168.

4. [Smither1989] M.A. Smither, D.R. Pugh, and L.M. Woolard, "CMRR analysis of the 3-op-amp instrumentation amplifier", *IET Electronics Letters*, February 1989.

5. [VanRijn1990] A. C. Metting van Rijn, A. Peper, and Grimbergen, "High-quality recording of bioelectric events; part 1, interference reduction, theory, and practice," *Medical and Biological Engineering and Computing*, vol. 28, pp. 389-397, 1990.]

6. [Burke2000] M. Burke and D. Gleeson, "A micropower dry-electrode ECG preamplifier," *Biomedical Engineering, IEEE Transactions on*, vol. 47, no. 2, pp. 155-162, Feb. 2000.

7. [Toumazou1989] C. Toumazou et. al. "Novel current-mode instrumentation amplifier", *IEE Electronics Letters*, vol. 25, no. 3, 1989.

8. [NvH2012] N. Van Helleputte, S. Kim, Sunyoung, H. Kim, J. P. Kim, J. P. Kim, C. Van Hoof, and F. Yazicioglu, "A 160uA Biopotential Acquisition IC with Fully integrated IA and motion artifact suppression", *Biomedical Circuits and Systems, IEEE Transactions on*, vol.:6 , Iss.: 6, Dec 2012.

9. [Yang2015] Yang Zhao; Jian Zhao; Xi Wang; Guo Ming Xia; An Ping Qiu; Yan Su; Yong Ping Xu, "A Sub-μg Bias-Instability MEMS Oscillating Accelerometer With an Ultra-Low-Noise Read-Out Circuit in CMOS," in *Solid-State Circuits, IEEE Journal of*, vol.50, no.9, pp.2113-2126, Sept. 2015

10. [Wong2008] Wong, A.K.Y.; Kong-Pang Pun; Yuan-Ting Zhang; Ka Nang Leung, "A Low-Power CMOS Front-End for Photoplethysmographic Signal Acquisition With Robust DC Photocurrent Rejection," in *Biomedical Circuits and Systems, IEEE Transactions on*, vol.2, no.4, pp.280-288, Dec. 2008

11. [Winokur2015] Winokur, E.S.; O'Dwyer, T.; Sodini, C.G., "A Low-Power, Dual-Wavelength Photoplethysmogram (PPG) SoC With Static and Time-Varying Interferer Removal," in *Biomedical Circuits and Systems, IEEE Transactions on*, vol.9, no.4, pp.581-589, Aug. 2015

12. [Enz1987] C. Enz, E. Vittoz, and F. Krummenacher, "A CMOS chopper amplifier," *Solid-State Circuits, IEEE Journal of*, vol. 22, no. 3, pp. 335-342, June 1987.

13. [Enz1996] Enz, C.C.; Temes, G.C., "Circuit techniques for reducing the effects of op-amp imperfections: autozeroing, correlated double sampling, and chopper stabilization," in *Proceedings of the IEEE*, vol.84, no.11, pp.1584-1614, Nov 1996

14. [Denison2007] T. Denison, K. Consoer, A. Kelly, A. Hachenburg, and W. Santa, "2.2 μ W 97nV/ √Hz, chopper stabilized instrumentation amplifier for EEG detection in chronic implants," in *Solid-State*

15. [Yazicioglu2006] R. F. Yazicioglu, P. Merken, R. Puers, and C. Van Hoof, "A 60μW 60nV/√Hz readout front-end for portable biopotential acquisition systems," in *IEEE International Solid-State Circuits Conference Dig. Of Tech. Papers*, ISSCC, Feb. 6-9, 2006, pp. 109-118.

16. [Verma2009] N. Verma, A. Shoeb, J. Guttag, and A. P. Chandrakasan, "A Micro-power EEG acquisition SoC with integrated seizure detection processor for continuous patient monitoring," *Symposium on VLSI Circuits*, pp. 62- 63, June 2009.

17. [NvH2014] N. Van Helleputte et. Al., "A Multi-Parameter Signal-Acquisition SoC for Connected Personal Health Applications," *ISSCC*, Feb. 9-13, 2014.

18. [Qinwen2011] Qinwen Fan; Sebastiano, F.; Huijsing, J.H.; Makinwa, K.A.A., "A 1.8 uW 60 nV/rtHz Capacitively-Coupled Chopper Instrumentation Amplifier in 65 nm CMOS for Wireless Sensor Nodes," in *Solid-State Circuits, IEEE Journal of*, vol.46, no.7, pp.1534-1543, July 2011

19. [Anand2015] Anand, T.; Makinwa, K.A.A.; Hanumolu, P.K., "A self-referenced VCO-based temperature sensor with 0.034°C/mV supply sensitivity in 65nm CMOS," in *VLSI Circuits (VLSI Circuits), 2015 Symposium on*, vol., no., pp.C200-C201, 17-19 June 2015 55

20. [Wanyeong2015] Wanyeong Jung; Seokhyeon Jeong; Sechang Oh; Sylvester, D.; Blaauw, D., "27.6 A 0.7pF-to-10nF fully digital capacitanceto-digital converter using iterative delay-chain discharge," in *Solid-State Circuits Conference - (ISSCC), 2015 IEEE International*, vol., no., pp.1-3, 22-26 Feb. 2015

21. [VanRethy2013] Van Rethy, J.; Danneels, H.; De Smedt, V.; Dehaene, W.; Gielen, G.E., "Supply-Noise-Resilient Design of a BBPLL-BasedForce-Balanced Wheatstone Bridge Interface in 130-nm CMOS," in *Solid-State Circuits, IEEE Journal of*, vol.48, no.11, pp.2618-2627, Nov.2013

22. [Kim 2013] Hyejung Kim, et al., "A Configurable and Low-Power Mixed Signal SoC for Portable ECG Monitoring Applications," *IEEE Transactions on Biomedical Circuits and Systems*, 2013

23. [Kim 2010] Hyejung Kim, et al., "ECG Signal Compression and Classification Algorithm with Quad Level Vector for ECG Holter System," *IEEE Trans. on Information Technology in Biomedicine*, vol.14, no.1, Jan. 2010

24. [Gosselin 2009] B. Gosselin and M. Sawan, "Circuits techniques and microsystems assembly for intracortical multichannel ENG recording," in *Proc. 2009 IEEE Custom Integrated Circuits Conf.*, 2009, pp. 97–104.

25. [Kim 2011] Hyejung Kim, et al., "A mixed signal ECG processing platform with an adaptive sampling ADC for portable monitoring applications," *EMBC 2001*, PP.2196 – 2199

26. [Zhang 2013] Jie Zhang, et al., "An Efficient and Compact Compressed Sensing Microsystem for Implantable Neural Recordings", *IEEE Transactions on Biomedical Circuits and systems*, 2013, 56

CHAPTER 03

Neural Stimulation Circuits

Michael Haas
Maurits Ortmanns

University of Ulm
Ulm, Germany

Neural Stimulators have been in use since ancient cultures. Relief of pain or treating epilepsy have been mentioned centuries ago. Today, neural stimulators still follow the same principle, which is the transfer of charge onto electrically excitable tissue, while the methods have obviously changed a lot and benefited majorly from today's integrated circuits technology. Implantable medical devices are now widely employed to restore functions to the impaired individuals suffering from diseases like deafness, blindness, cardiac insufficiency, incontinence, neural disorders, and many more. While such implantable neurostimulators all share similar challenges, they still become increasingly challenging, if a large number of electrodes needs to be realized. In this chapter we will shortly review the history of electrical stimulation and will review the electronic-tissue interface as far as it is needed to understand the following techniques. We will also review the state of the art of electronic implementations of implantable neural stimulators, some recent implementations and their application in system examples such as the retinal implant and neural modulators. The chapter will conclude with showing a necessary safety measure by means of charge balancing to avoid polarization.

1 Implantable medical devices

Implantable medical devices are nowadays widely employed to restore functions to the impaired individuals suffering from diseases like deafness, blindness, cardiac insufficiency, incontinence, neural disorders, and many more. Such implantable systems become increasingly challenging, if a large number of electrodes needs to be realized.

This chapter will review the history of electrical stimulation, give a short look into the electronic-tissue interface as well as state of the art electronic implementations of implantable neural stimulators. It covers applications and system examples such as the retinal implant and neural modulators with high efficiency frontends. It will be concluded with an overview of state of the art charge balancing techniques.

BACKGROUND
History of electrical stimulation

2 History of Electrical Stimulation (I)

- Ancient Greece
- Electrotherapy using torpedo fish
- 30-200V electric disharge

[France, Corse, 09.08.2010, cm 50, by Roberto Pillon, fishbase.org/photos/, CC BY]

[CC BY 4.0, File:Portrait of Hippocrates from Linden, Magni Hippocratis...1665 Wellcome L0014825.jpg]

- Arthritis
- Cephalalgia
- Chronic pain
- Epilepsy

Already ancient Greek and Roman philosophers like Hippocrates, Skivronios Largos, Pliny and Dioscorides referred to the properties of the torpedo fish for medical application. The torpedo fish is capable of producing an electric discharge ranging from 30 to 200 volts.

In the 1st century AD, the Roman physician Scribonius Largus (1–50 AD) describes the curative value of torpedo's electric discharge for alleviating headache and gout [Tsoucalas 2014].

3 History of Electrical Stimulation (II)

Galvani's experiments on bioelectricity come with a popular legend: while skinning a frog leg, his assistant touched an exposed sciatic nerve of the frog with a metal scalpel that had picked up charge. At that moment, the dead frog's leg kicked as if in life. Galvani himself assumed the source of electricity in the body itself.

This observation made Galvani the first investigator on the relationship between electricity and life. Thus, Galvani is properly credited with the discovery of bioelectricity, while his study is nowadays referred to electrophysiology.

3 History of Electrical Stimulation (II)

- 1780 - Galvani's frog leg experiment
  - Accidental discharge of electric charge into a dead animals body
  - Muscle contraction
- Bioelectricity
- A very nice summary of the history of electrical stimulation can be found on:
  - www.ncbi.nlm.nih.gov/pmc/articles/PMC3232561/

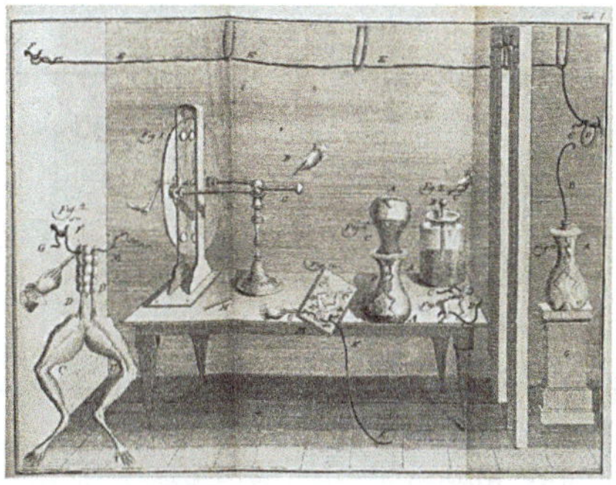

[Figure public domain: commons.wikimedia.org/wiki/File:Luigi_Galvani_Experiment.jpeg]

4 History: 1st Transistor Cardiac Pacemaker (I)

- Earl Bakken: american engineer and co-founder of Medtronic
  - 1957: delivered a battery powered metronome to Dr. Lillehei to replace large, wire powered devices
- Those transistor pacemakers were (externally worn) metronomes

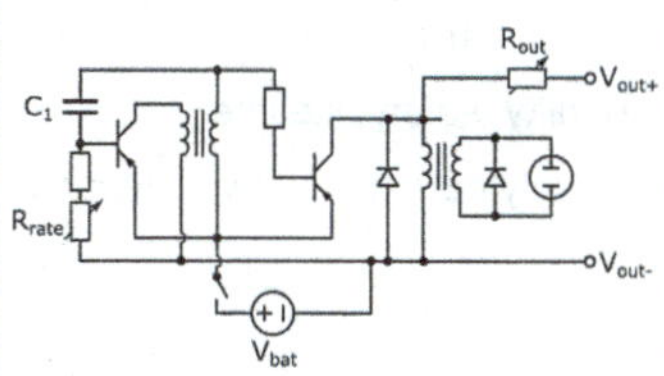

c.f. L.A. Geddes, The "missing link" in the history of cardiac pacing [Retrospectroscope], IEEE Engineering in Medicine and Biology Magazine (Volume: 26, Issue: 4, July-Aug. 2007)

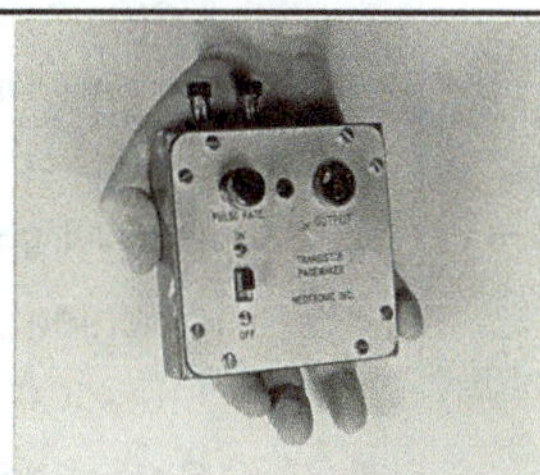

Source: "Transistorized Pacemaker", Box 111, C. Walton Lillehei papers, uarc2000-7, University of Minnesota Archives Image is available on the Open Heart online gallery http://gallery.lib.umn edu/exhibits/show/openheart/item/1879

In the 1950s, Dr. C. Walton Lillehei was performing life-saving surgery on children which often left the children temporarily attached to a pacemaker, which were at that time large – side table sized - devices. Earl Bakken, American engineer and co-founder of Medtronic, was being asked to solve the problem. After drawing a transistor circuit diagram and 4 weeks of manufacturing, Bakken delivered a battery-powered transistorized pacemaker about the size of a hand [Haddad 2006].

5

History: 1st Transistor Cardiac Pacemaker (II)

- Dr. Rune Elmquist, Swedish engineer
  - Delivered an implantable prototype of a cardiac pacemaker
  - Arne Larsson *May 26, 1915, +Dec. 28, 2001
- 1st pacemaker: Oct. 8, 1958 - working for 8h
  - 2 transistors, 1 NiCd battery, 1 coil, etc..

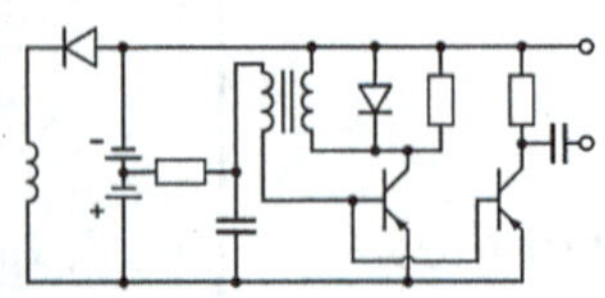

[c.f. Aquilina O., "A brief history of cardiac pacing", Images Paediatr Cardiol 2006;27:17-81]

[Courtesy Siemens Historical Institute]

Arne Larsson was the first human to receive an implantable pacemaker. He had been hospitalized with complete heart block and frequent attacks for the last 6 months, with some 20 to 30 attacks daily. The unit was completely hand-made and it consisted of the NiCd battery, the electronic circuit and the coil recharging antenna. These were encapsulated in a new epoxy resin, which had excellent biocompatibility.

The approximate diameter and thickness became 55 mm and 16 mm respectively, according to the dimensions of the ever so popular shoe polish can from Kiwi: Elmquist had in fact produced two such units using these cans as molds!

The first version, though, unfortunately failed after 8h. The second one was implanted the next day, and it survived a week.

Arne Larsson finally survived both the engineer as well as the surgeon who had saved his life in 1958. He received five lead systems and 22 pulse generators of 11 different models until his death on December 28th 2001 aged 86.

In 1994 Siemens sold its entire pacemaker business to the American company St. Jude Medical.

6

Electrical Neurostimulation Today

- Cardiac pacemaker
  - ½ Mio implants p.a. with >10y lifetime
  - Power, size, functionality, safety, lifetime
  - Intelligent implants vastly determined by IC technology
- Cochlea implants
  - >450,000 worldwide
- Spinal cord stimulators
  - >130,000 worldwide
- Deep brain stimulators
  - >70,000 worldwide
- Vagal nerve stimulators
  - >70,000 worldwide

Today, around ½ Mio. Cardiac pacemakers are implanted per year. They have a lifetime of at least 10 years, their size and safety are manifold better as the ones 50y ago. Their functionality has become way more than just a metronome. Today's intelligent implants adjust to the needs, and a huge part of this improvement towards intelligent, life-long implants was enabled by the progress in IC technology!

INTERFACING ELECTRONICSWITH THE HUMAN BODY

Excitable Tissue, Electrodes, and Safety Limits

7

Characteristics & Challenges for Electrical Stimulation

- Need excitable tissue
  - Charge transfer elicits response
- Need (conducting) electrical interface to the tissue
- Need charge control
  - Avoid electrolysis
- Needs high area and power efficiency
  - Especially for longterm (battery powered) or highly parallel stimulation

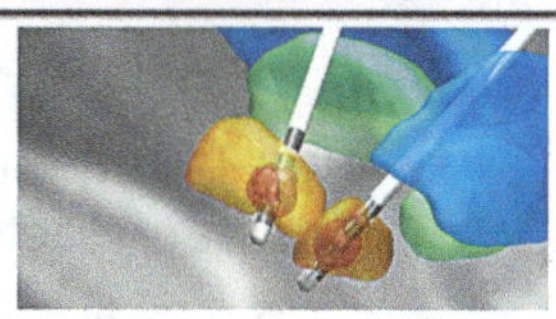

Andreashorn, (commons.wikimedia.org/wiki/File:Deep_brain_stimulation_in_a_Parkinson's_Disease_patient.png), https://creativecommons.org/licenses/by-sa/4.0/legalcode

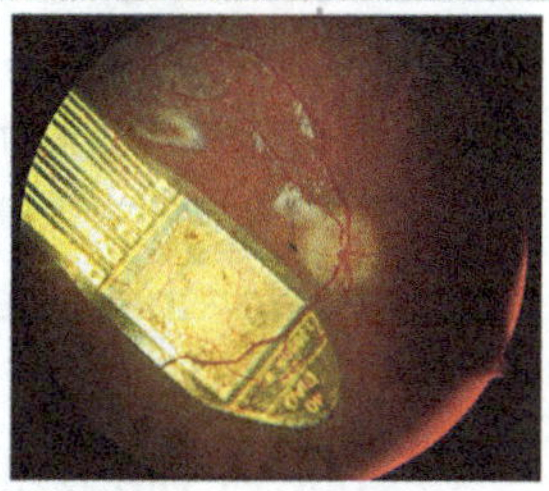

RETINA IMPLANT Alpha AMS. Source: Universitäts-Augenklinik Tübingen / Retina Implant AG

The principle of electrotherapy and stimulation of excitable tissue is always the same: a charge transfer elicits a response. For this charge transfer we typically need a galvanic contact. Thus, a way to conduct charge into the tissue is needed, which raises the concern of electrode safety, electrolysis prevention, charge limits and so on. These requirements are needed irrespective of the application and thus e.g. the number of channels, e.g. a two-lead deep brain stimulator or a 1600 pixel subretinal stimulator.

8

Various Interfaces to the Tissue

- Planar, needle, cuff electrodes
- Single wire electrodes
- Multi electrode arrays
- Silicon
- Polyimide

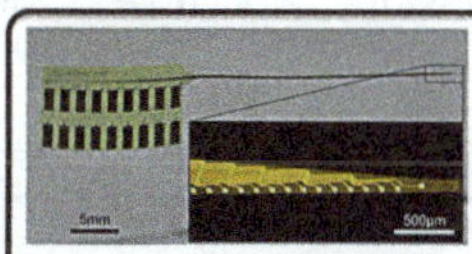

Courtesy: Dr. M. Eickenscheidt, IMTEK, Freiburg University

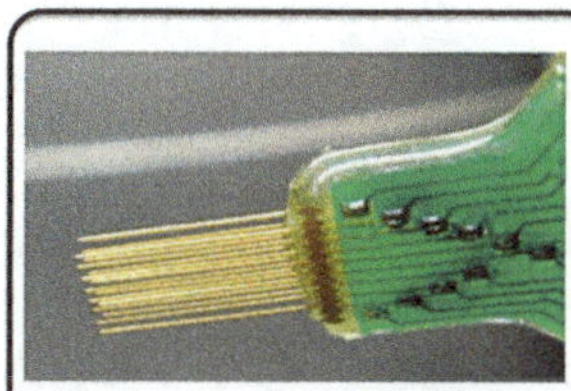

[commons.wikimedia.org/wiki/File:16wire_electrode_array.jpg by ProfessorPotter, Public Domain]

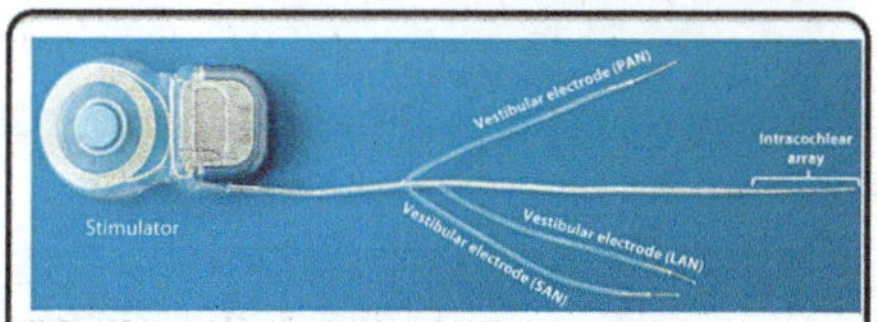

[A. Perez-Fornos, (commons.wikimedia.org/wiki/File:Vestibular_implant_prototype.jpg), „Vestibular implant prototype", https://creativecommons.org/licenses/by-sa/3.0/legalcode]

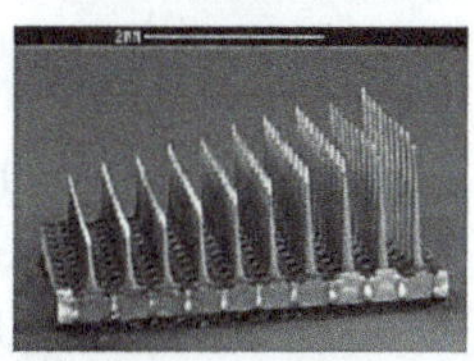

[Oregon State University (commons.wikimedia.org/wiki/File:Utah_(22825989368).jpg), „Utah (22825989368)", https://creativecommons.org/licenses/by-sa/2.0/legalcode]

There exists a huge variety for the electrodes. Silicon based needle type such as the Utah array, polyimide based cuff electrodes, shaft electrodes, single lead electrodes and many more. Together with the stiffness, biocompatibility, impedance, long-term stability, etc. huge scientific and manufacturing expertise is needed for this field. As electronic engineer, we need a basic understanding of this.

9 Background – Electrochemistry

- Two layers of opposite polarity with (molecular) dielectric
  - Helmholtz capacitor C_{DL}
- Equilibrium charge transfer between electrode and solution
  - Reversible potential
- No strong faradaic currents
  - AC linear portion of reversible current defines the charge transfer resistor R_{CT}
- R_S for the electrolyte, tissue, electrode
- (De)-Polarization & reversible reactions

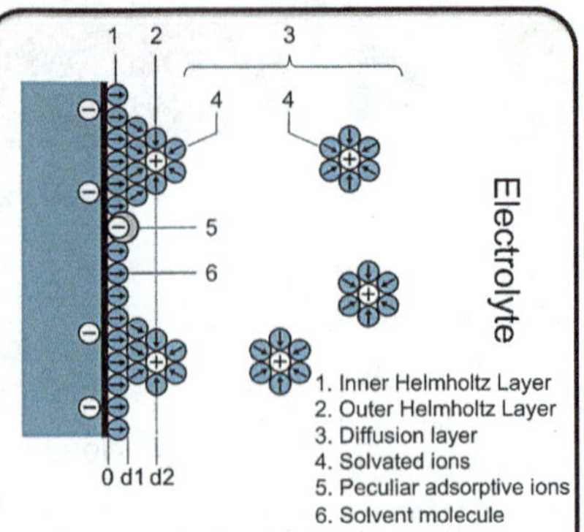

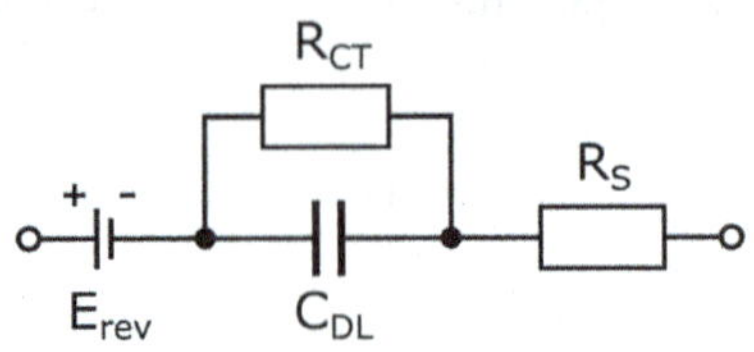

When we bring an electronic conducting material in contact with an ionic conducting solution, a common boundary appears. Helmholtz was the first to discover that charged electrodes immersed in electrolytic solutions repel the co-ions of the charge while they attract the counter-ions to their surface. Two layers of opposite polarity form at this interface. This electrical double layer (DL) is basically a molecular dielectric storing charge electrostatically.

This early model predicted a constant capacitive behavior.

Later, the "Gouy-Chapman" electrode model observed that the capacitance is depended on the applied potential and the ionic concentration in the electrolyte. Many other models followed those early observations.

What is also important to the interface electronics, is that under equilibrium a charge transfer between electrode and solution takes place, which builds up a so called reversible potential.

Reduction and oxidation of electrode ions into the solution or from the solution happens continuously. Basically, a small current is flowing in both directions. At equilibrium, both are equal, and there is no net current! This is called the exchange current.

If we apply a tiny AC voltage to such an electrode, then the AC linear portion of the reversible current defines the so called charge transfer resistor R_{CT}.

Additionally, we usually add a series resistor for the electrolyte, tissue and electrode material impedance R_S, which then builds the Randles cell electrode model. There, often a Warburg element is additionally included.

The Randles equivalent circuit is one of the simplest possible models describing processes at the electrochemical interface [Randles 1947].

Also note, that the reversible potential, which can be calculated using the Nernst equation, varies slightly from electrode to electrode and also over time, e.g. due to concentration change and mismatch in the electrolyte. This is especially important when the electrodes are used for recording applications, where the slowly varying large signals from the electrode typically have to be blocked with significant circuit overhead!

10 Electrochemistry: Safety vs. Electrolysis

- Electrode "polarization"
  - The change of potential of an electrode from its equilibrium potential upon application of DC current / voltage
  - Voltage drop over R_{CT} due to DC current!
  - DC current can not flow through the double layer cap C_{DL}
  - We generally don't want to polarize our electrode

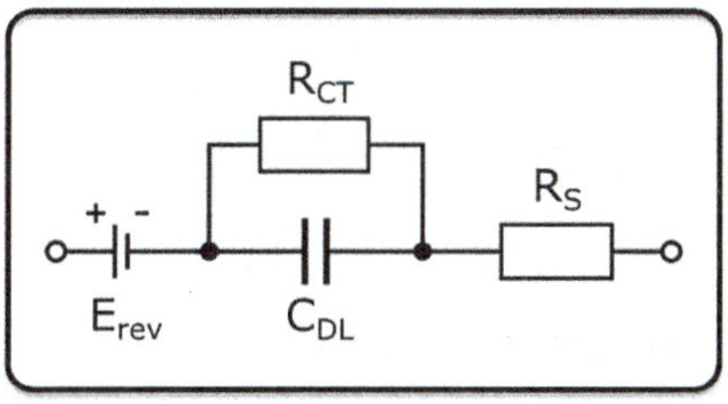

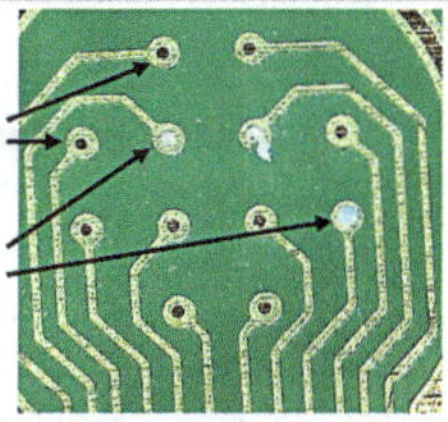

Polarization is the change of potential of an electrode from its equilibrium potential upon the application of a DC current of voltage.

The term polarization originates from the early discoveries that electrolysis causes the elements in an electrolyte to be attracted towards one or the other pole. Thus, polarization was essentially a description of electrolysis itself.

11 Electrochemistry: Electrode & Stimulation

- Stimulation
  - Voltage leveling
    ➔ resistively dominated behavior.
    ➔ Electrochemical reactions
  - Signals are not allowed to level, but we want capacitive coupling!
  - **Very large R_{CT}** helps!
- Use (mostly) noble metals for the electrode
  - "No" reaction, biocompatible, very large R_{CT}
  - Use surface roughening to increase surface area
    - Decrease overall impedance

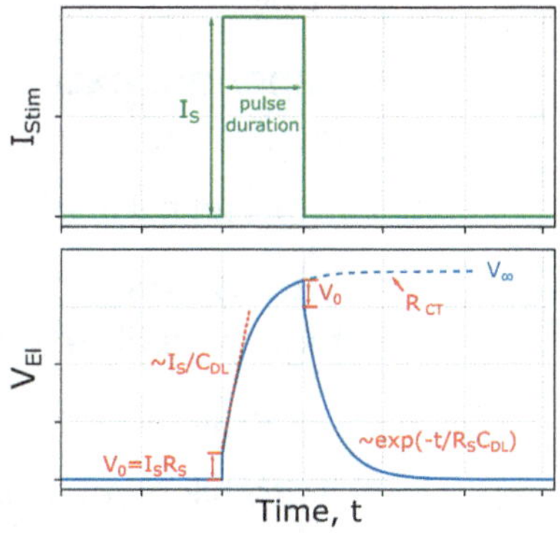

[c.f. E. McAdams, ISSCC 2014, SC1, The Biomedical Electrode-Tissue Interface]

When we apply a current pulse to an electrode, we don't want to reach the voltage leveling region, since then we might enter the strong faradaic reactions. We would like the electrode to behave as capacitive coupling. For this, usually noble metals are used as electrode material, which show little reaction, are mostly biocompatible and have large R_{CT}.

12 Background – Electrophysiology

- Cell communication: combination of **electrical** & **chemical** activity

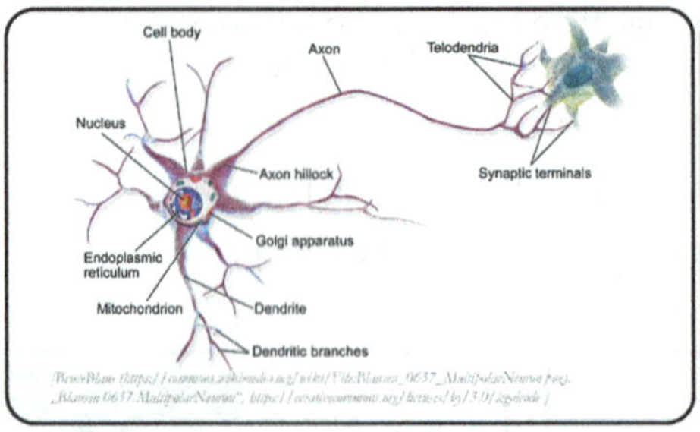

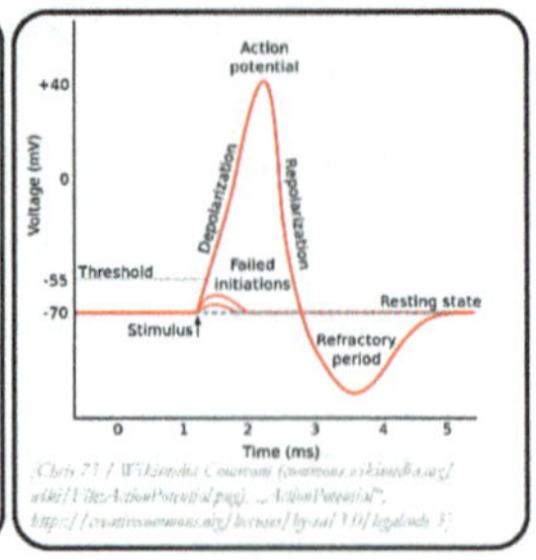

- Electrical stimulation (and also sensing) of neural signals can be done on both the electrical and the chemical activity

Also a minor understanding of the electrophysiology of the excitable cells is important. Dendrites receive input signals through the synapses of other neurons. The soma processes these signals and turns them into an output, which is sent out to other neurons through the axon and synapses. Communication between neurons takes place at synapses. The electrical signal traveling down the axon is known as action potential. When it reaches the end of the axon, the information is transmitted across the synaptic gap using neurotransmitters.

13 Excitation Properties

- Important excitation properties that affect the transmembrane action potential
  - essential to effectively stimulate the neuron cells
- Various requirements on excitation leads to – optimally – highly flexible stimulators
  - to allow various stimulation scenarios
  - to allow multiple safety features
  - to adapt to environmental and patient specific situations

Excitation properties are essential to effectively stimulate the neuron cells: Strength-duration relationship (threshold current for excitation as a function of the pulse duration), charge-duration relationship (amount of charge for stimulation decreases with shorter pulses), current-distance relationship (more current for farer distance), Stimulus polarity (anodic stimulation threshold about 5-8 times higher than for cathodic stimulation).

14

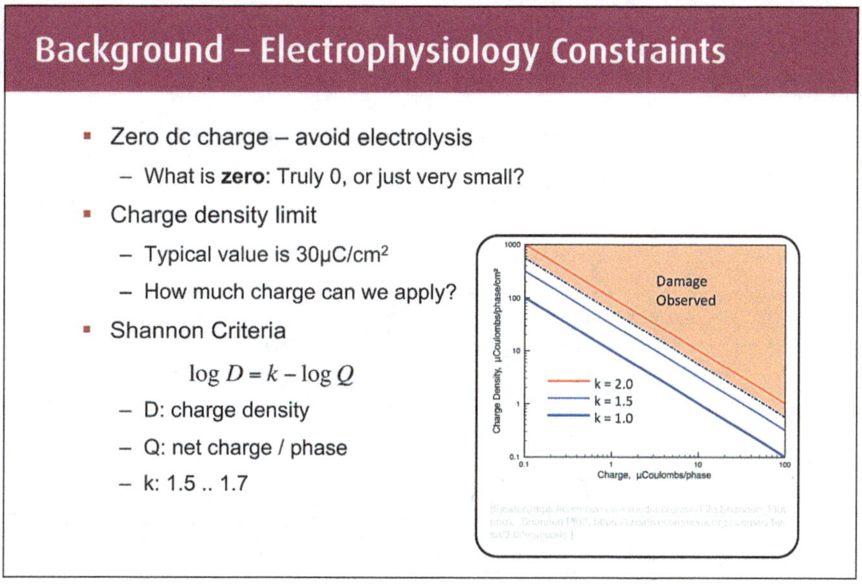

Polarization is prohibited, thus no DC current is allowed to avoid electrolysis. But how close to a real zero does that need to be and how can we achieve it.

Then, how much charge can be applied over an electrode? There is the Shannon criteria [Shannon 1992], which constitutes an empirical rule in neural engineering that is used for the evaluation of possible damage from electrical stimulation to nervous tissue [Grill 2004] [Merrill 2005] .

STIMULATION FRONTENDS

When we look at the basic idea of neural stimulation, the underlying principle is always to transfer charge onto the excitable tissue in order to elicit a physiological response.

15

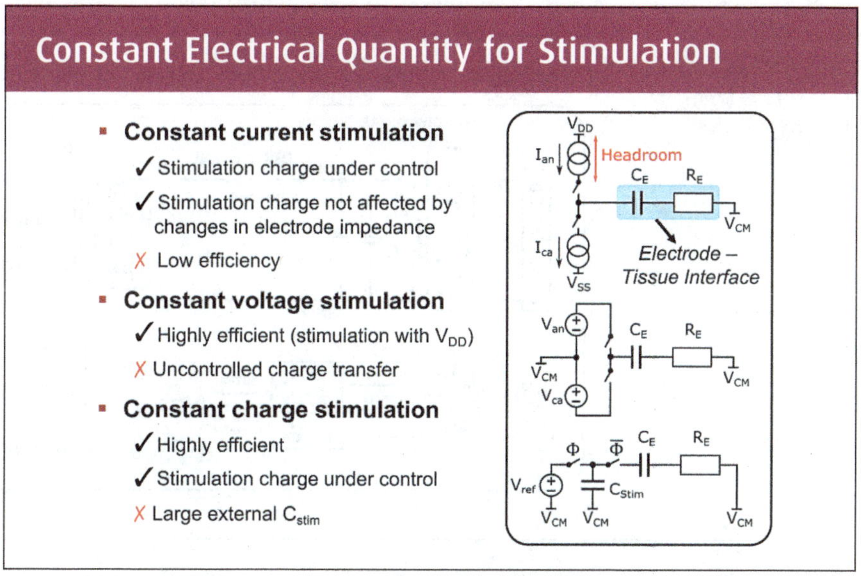

In early implementations, constant voltage stimulation (CVS) was used, because it can be implemented very simple and it is very efficient. Unfortunately, the charge transfer is uncontrolled.

In constant current stimulation (CCS), the stimulation charge is under perfect control and since using integrated circuits, its implementation has become similarly easy. Unfortunately, the efficiency is limited, since supply voltage and electrode voltage can hugely differ. The last technique, which was proposed in the literature, is constant charge stimulation [Simpson 2007].

16 Constant Charge Stimulation I – Efficiency

- Higher efficiency with charge based stimulation
- Avoid voltage compliance problem of Constant Current Stimulation

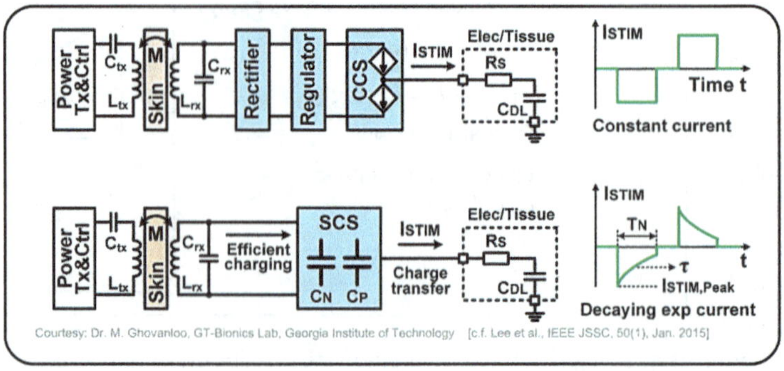

In constant charge stimulation, the charge transfer is principally done as in a charge pump. A capacitor (array) is charged to the available supply, and then discharged into the electrode.

In one of its advantageous implementations [Lee 2015], the charging of the capacitors was being done directly from the inductive link, thus even extending its high efficiency into the power management.

17 Constant Charge Stimulation II – Area

- 2 external caps per stimulation site
- Potentially advantageous implementation for low channel count

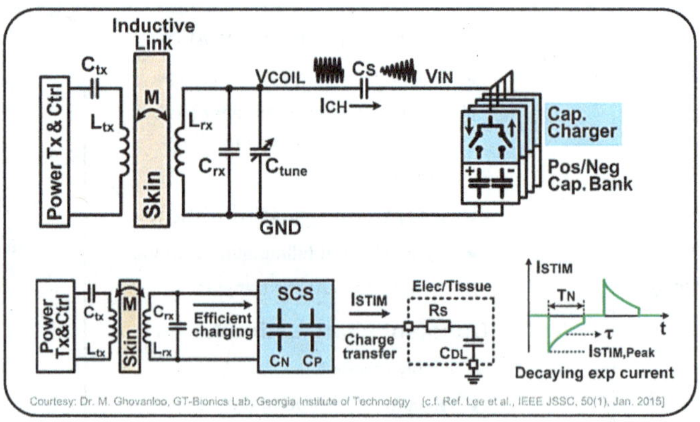

Despite its high efficiency, the technique has not been widely used yet. One obvious problem is that the electrode capacitor is in the µF range, whereas integrated capacitors would be well below 1nF. Thus, in [Lee 2015], two external capacitors were used for a low channel count stimulator.

18 Constant Current Stimulation – Compliance (I)

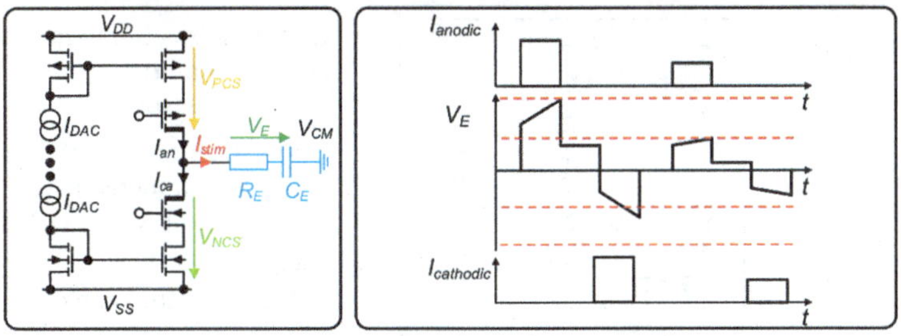

- Constant V_{DD} but variable stimulation current and "load"
  ✗ Large fractions of power wasted
- Voltage compliance ~ $f(I_{max}, Z_E, t_{max})$
  Z_E=10kΩ+100nF, I_{max}=**500µA**, t_{max}=1ms → **V_{DD}-V_{SS}> 20V** ✓ VC sufficient
  Z_E=10kΩ+100nF, I_{max}=**125µA**, t_{max}=1ms → **V_{DD}-V_{SS}> 5V** 75% wasted

Looking at the root causes of the low efficiency of CCS, we have to observe two points. Firstly, the supply voltage is specified by the expected worst case electrode load and the highest stimulation current.

In practice, electrode impedances largely vary over location and time, and stimulation currents vary over patients, physiological response, electrode location, etc. Thus, in the given example, for two different stimulation currents, the specified supply of 20V is wasted by 75% for the low stimulation current.

19 Constant Current Stimulation – Compliance (II)

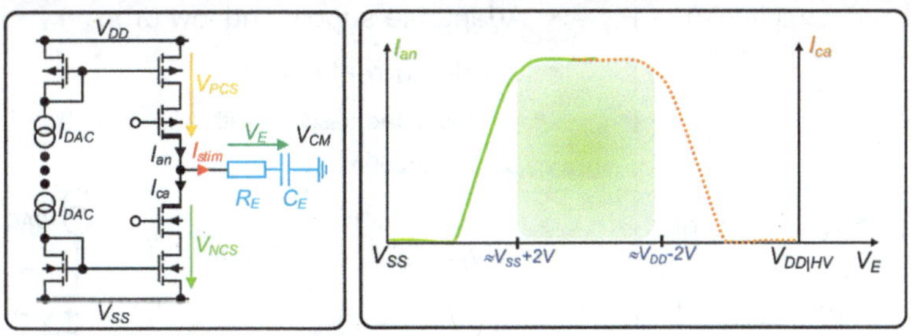

- Current source supply drop >V_{DSsat}
  ✗ Large current range (µA … mA) needs large V_{GS} needs large V_{DSsat}
- Example: V_{SS}+2V < V_E < V_{DD}-2V → **V_{DD}-V_{SS}=20V** 20% wasted
  → **V_{DD}-V_{SS}=5V** 80% wasted

Moreover, electronic current sources are based on transistors in saturation (MOS). Therefore, they reduce the output voltage compliance linearly with the number of transistors and their overdrive voltage V_{OV}. Since V_{OV} varies with the varying output current (e.g. 50dB output current DR in [Ortmanns 2007]), this yields a V_{OV} variation of more than x10, thus the drop can be significant. In the example, this drop might be acceptable, if the needed supply voltage is large. But in case of low load and thus low required output voltage, a major amount of the supply might be wasted over the current sources.

20 Constant Current Stimulation – Compliance (III)

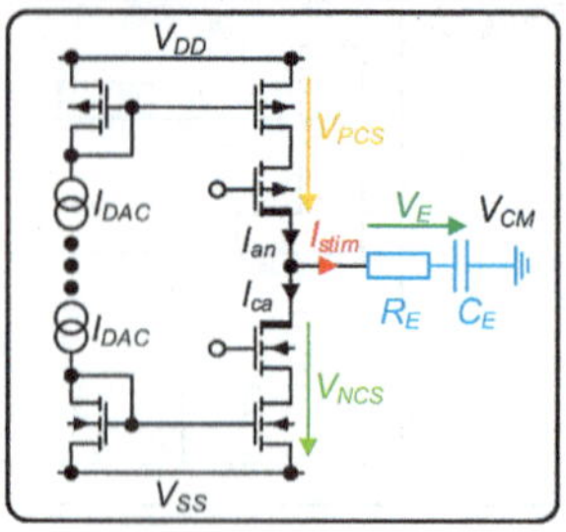

- Constant V_{DD} but variable stimulation current and "load"
  - ✗ Large fractions of power wasted
- Current source supply drop >V_{DSsat}
  - ✗ Large current range (µA … mA) needs large V_{GS} needs large V_{DSsat}
- Result:
  - ✗ CCS efficiency can drop to <10%

Challenge 1: Constant supply for different loads and stimulation currents
→ **Solution:** Adapt supply to needed compliance – need compliance monitor!

Challenge 2: Overhead voltage of constant current source reduces compliance
→ **Solution:** Make it as small as possible

A solution for both challenges can be found in the following ideas: The supply should be adapted to the needed voltage compliance. Besides the implementation of this adaptive supply in the power management, also a compliance monitor is needed, i.e. a simple circuit that detects if the CCS runs out of compliance and thus more supply would be needed. The second solution is to obviously realize the CCS with an as small as possible voltage drop over the current sources.

21 High Efficiency Constant Current Stimulator (I)

- Adaptable supply and low drop
- Triode current mirror
  - – Regulated cascode with $A_{1/2}$
  - ✓ $V_{DN|P}$=200mV
  - – $P_{A1,2}$ < 1..4µW
  - – V_{DDP}=20V, $I_{P/N}$=1-20mA

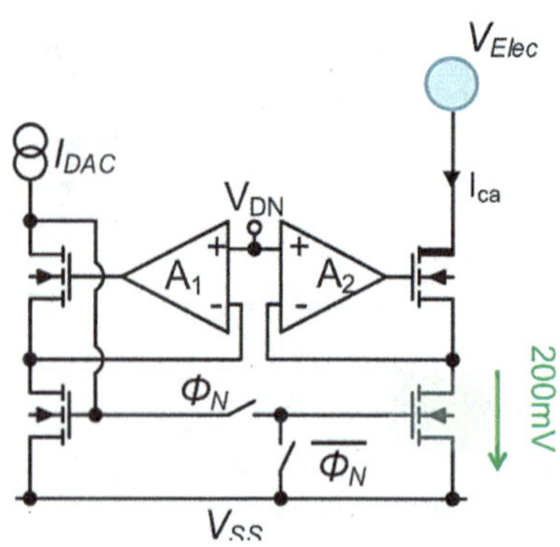

[c.f. Sooksood et al., ISSCC 11]

An implementation of both solutions is schematically illustrated [Noorsal 2012]. A low-voltage current steering DAC mirrors its current into the HV output of the CCS. The mirror is implemented using a linearly biased current source.

Adapted from the Widlar current source, the resistor is replaced by a linear transistor and the output impedance boosted by a regulated HV cascode. The bias potential of the linear transistor is generated from a similarly build mirror branch. Thereby, the V_{DS} of the current source can be fixed to e.g. 200mV, which is a compromise between matching and output compliance.

22

High Efficiency Constant Current Stimulator (II)

- Adaptable supply and low drop
- Triode current mirror
  - Regulated cascode with $A_{1/2}$
  - ✓ $V_{DN|P}$=200mV
  - $P_{A1,2}$ < 1..4µW
  - V_{DDP}=20V, $I_{ca/an}$=1-20mA
- Compliance monitor
  - ✓ Surveys control loop
  - ✓ Compliance info used to adapt supply V_{DDP}
  - ✗ HV mirror bias current still wasted!

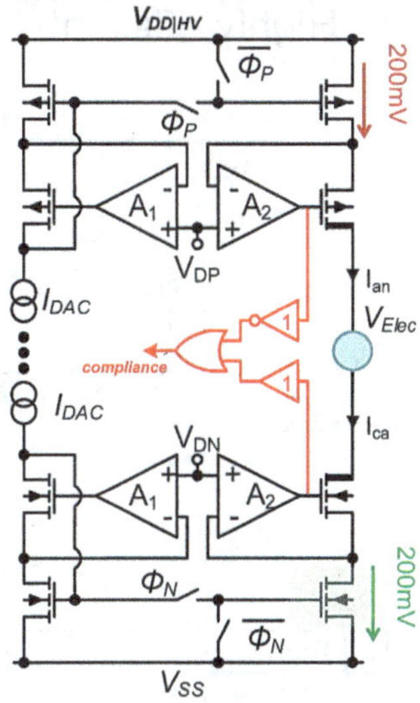

[c.f. Sooksood et al., ISSCC 11]

The bipolar output stage is shown here [Noorsal 2012]. As an additional advantage, this circuit provides a voltage compliance monitor by using the feedback mechanism of the regulated cascode. If there is a voltage compliance problem, the loop will try to adjust the output of the control amplifier towards its own positive supply to stabilize the current through the cascodes. This can be detected by using the shown inverters. The compliance signal is used to adjust the supply within the power management of the implantable system [Xu 2012].

23

Stimulator with Current Copying

- One DAC feeds N electrodes
  - ✓ V_G stored on C_{mem}
  - ✓ No static bias current
  - ✗ Regular update
- This can be merged with the circuit from the last slide
  - "Linear" biased M_1
  - Compliance monitor

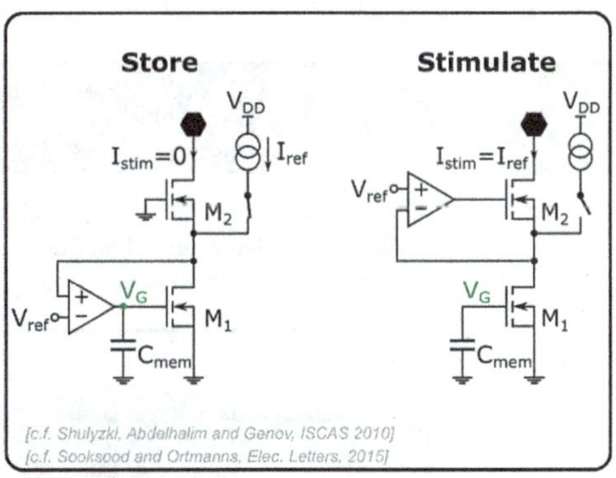

The previous output stage still suffers from significant power waste in the mirror branch. A solution to the wasting of a biasing current in the mirror paths has been shown in [Shulyzki 2010], which was named the current copier. The idea is to store the gate voltage V_G of the current source transistor M_1 after a short pulse in the current mirror path, in which a reference current I_{ref} is provided. This was originally used to use one DAC for several output stages. But it can also be advantageously used to save the power consumption in the mirror path and combined with the previous output stage [Sooksood 2012].

24 HV Stimulator Output Stage for Epiretinal Implant

- Highly efficient, high compliance output stage of a CCS

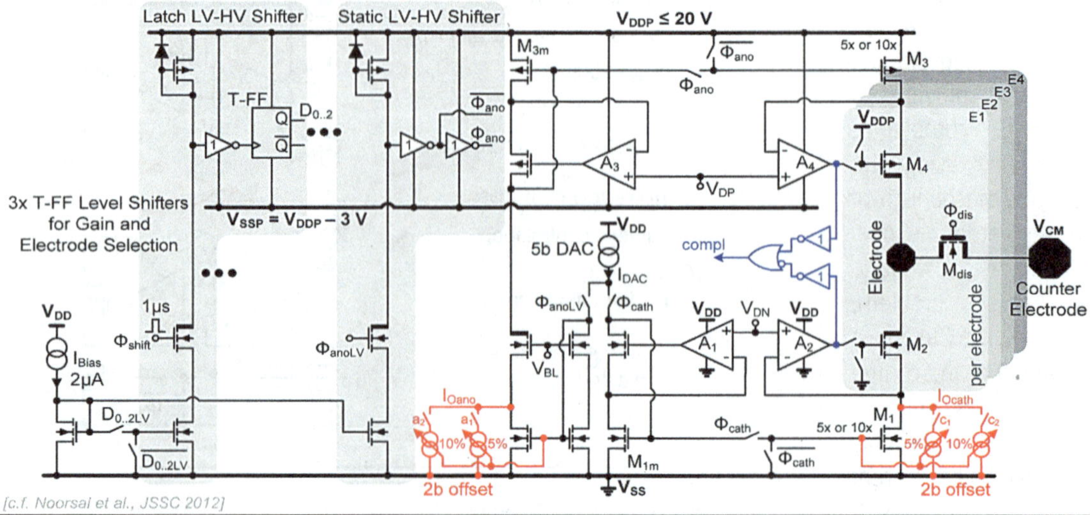

[c.f. Noorsal et al., JSSC 2012]

This figure shows a complete output stage schematic of a multichannel HV epiretinal stimulator [Norsaal 2012]. The HV control signals, which can have significant power consumption if permanently biased, are realized by pulsed bias currents and T-FF based control signal storage. The output stage is able to stimulate up to ±9V compliance from a 19V supply with up to 1mA output current with 50dB resolution. With a 1:4 multiplexed output stage, the design needed 0.2mm^2 die area, which was among the smallest available for the given functionality at the time with the potential to realize many hundreds of stimulation pixels on a few tens of mm^2 die area. The shown offset current sources were used to implement a dedicated charge balancing, a technique which will be looked at later in this chapter. The circuit was combined with a method to generate almost arbitrary stimulation waveforms, which made its usage highly flexible.

But what if the application requires a stimulator to be implemented on a fraction of that area? If several 1000s stimulation pixels have to be implemented on a few mm^2 only?

25 Stimulator for Highest Density Subretinal Implant

- Chip and system photo of a 1600 channel subretinal stimulator
- Implanted device seen in Eye Fundus
- Highest channel count in-vivo neurostimulator

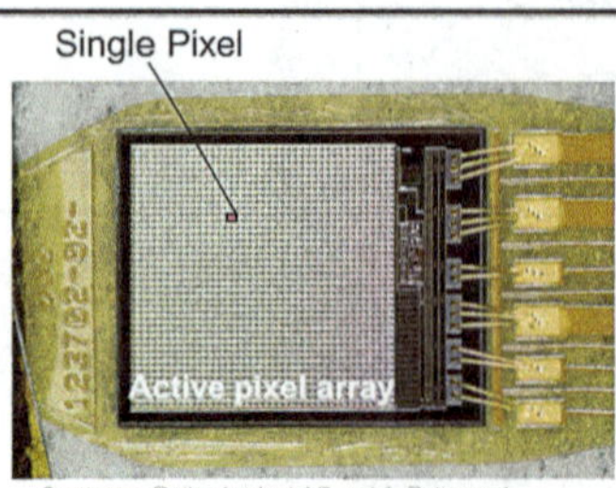

Courtesy: Retina Implant AG and A. Rothermel

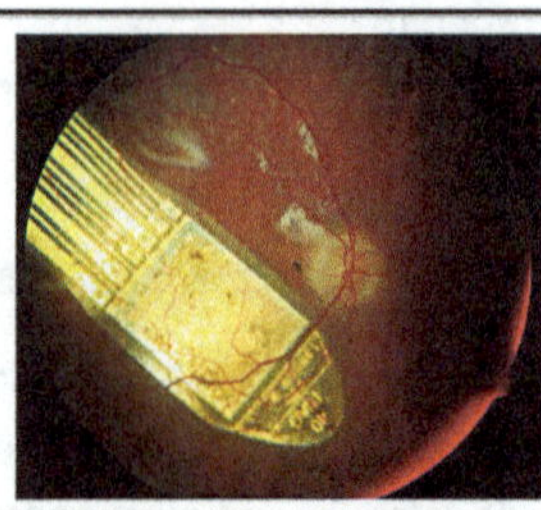

RETINA IMPLANT Alpha AMS. Source: Universitäts-Augenklinik Tübingen / Retina Implant AG

25 Stimulator for Highest Density Subretinal Implant

An application with such need is the subretinal stimulator, which was presented in [Rothermel 08].

The RETINA IMPLANT Alpha AMS chip is placed under the retina such that the function of the degenerated photoreceptors (rods and cones) can be replaced. The chip has a die area of only 3.2 x 4 mm^2. It features 1600 photodiodes, whose signals are „amplified" and passed through electrodes into the still functioning signal processing retinal layers. From there, following the natural optical path, the signal propagates via the optic nerve to the visual cortex where the incoming information is interpreted and visual impressions are displayed.

This system is the only imager chip, which has been implanted into the human eye to replace the natural photoreceptors, and it is still the in-vivo neurostimulator with the highest number of channels.

26 Subretinal Implant – Pixel Schematic

- Small stimulator with basic functionality
  - Anodic and cathodic output currents possible via M12
    - VH and VL exchange polarity
  - Passive electrode balancing (for "zero" DC current) via M15

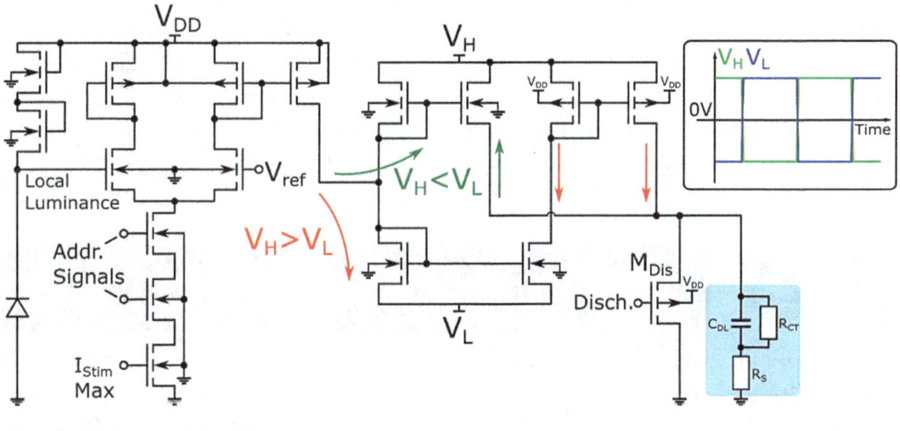

[c.f. A. Rothermel et al., ISSCC 2008]

Obviously, for such tiny area requirements, a fully equipped, flexible HV stimulator - as shown before - cannot be used. Instead, the stimulator is reduced to its basic functionality, i.e. a biphasic current source.

In the design of [Rothermel 2008], the overall stimulator ASIC was supplied by a rectangular AC voltage allowing DC free signaling, which is advantageous for isolation. The photodiode generates a photocurrent dependent on the local luminance. This current is being compared with a global luminance signal (VGL), and the result is the representation of the stimulation current. The maximum stimulation current per pixel is limited by a global signal. A part of this maximum stimulation current is mirrored into the output stage, where it is converted into cathodic or anodic current.

27 CVS with Impedance Adaptive Stimulation Control

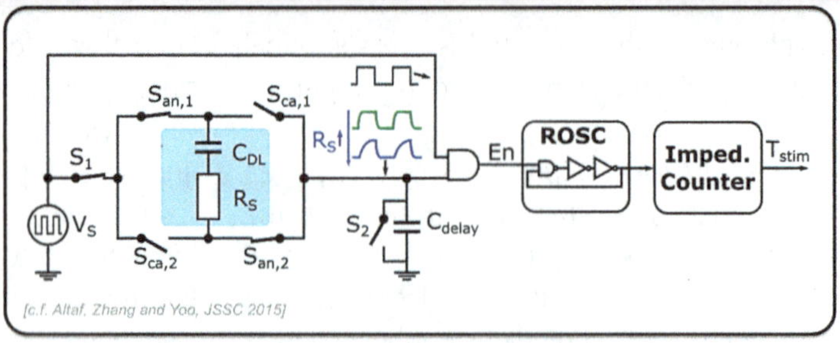

- 1 MHz source V_S used for stimulation and impedance estimation
- $C_{delay} << C_{DL} \rightarrow C_{delay}$ dominates RC constant during impedance estimation; S_2 shorts C_{delay} during stimulation
- Ring oscillator (ROSC) measures delay caused by $R_S C_{delay}$
- Stimulation time T_{stim} is adjusted based on electrode impedance

The main advantage of CVS architectures is their superior power efficiency. Also, large amounts of data, equipment and experience exist among neurologists from former work on CVS, making it still the method of choice in many cases. Even recent studies showed that individual patients respond better to CVS than CCS [Preda 2016].

A recent example for a CVS is shown in [Yoo 2015], where a pulsating voltage transcranial electrical stimulator was implemented. In this stimulation technique, square wave voltage pulses are applied as a stimulation source. This allows a very power efficient design, but comes with the drawback that the delivered charge per stimulation period is usually uncontrolled.

In [Yoo 2016] a highly digital stimulator circuit forms a time constant with the electrode impedance, which is measured by counting the periods of a ring oscillator. This determines the number of stimulation pulses that are applied. To avoid the accumulation of net-charge, the charge delivered in cathodic phase is retrieved back in anodic phasic with an inter-phase delay, by applying the same number of pulses.

A major drawback of the design lies in the realized charge balancing, which relies on applying the same amount of voltage pulses in anodic and cathodic phase, which will only result in an approximate charge balanced stimulation.

28 Reconfigurable CVS + CCS

In [Laotaveerungrueng 2010] a combined CCS/CVS stimulator has been proposed. It uses a HV differential amplifier to form a resistive feedback transimpedance amplifier (TIA) which converts the current mode input signal into a constant output voltage. By reconfiguring the HV output transistors of the CCS into the output stage of said TIA, the design achieves to reuse the largest area consumer within the reconfiguration. However, the implementation comes with two main drawbacks: First, the output stage operates in class-A during CVS, which requires a quiescent current from the HV supply. Secondly, the input of the TIA is directly connected to the working electrode. Especially capacitive loads at this point can make the amplifier unstable. Disadvantageously, the load at this point is the electrode itself, whose linear small signal model capacitor and resistor can easily vary over orders of magnitude for different electrode geometries, materials and aging.

Another CVS/CCS circuit was shown in [Guildvard 2012] using a current conveyor for the CCS as well as a HV TIA as CVS stimulator frontend. Disadvantageously, the amount of reuse is very low, which basically doubles the amount of HV circuitry.

28 Reconfigurable CVS + CCS

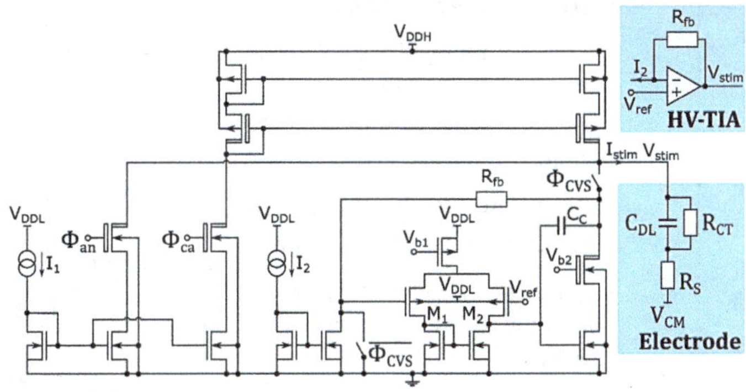

- Partwise reconfiguration of HV output stage into HV-TIA by using LV differential stage $M_{1/2}$

✓ Voltage and current stimulation
✓ Reuse of area consuming HV devices

X Electrode dependent stability
X Class-A output stage → static HV bias current

[c.f. N. Laotaveerungrueng et al., EMBC 2010]

29 Stimulator with Digital Feedback Loop

- Constant current by regulation of the electrode potential
- TIAs convert current to voltage
- Comparator converts difference between I_{stim} and I_{DAC} into the digital domain
- Integrated, digital signal is used to adjust V_{DD}
- Improved power efficiency
- Stability issues with variable electrode load

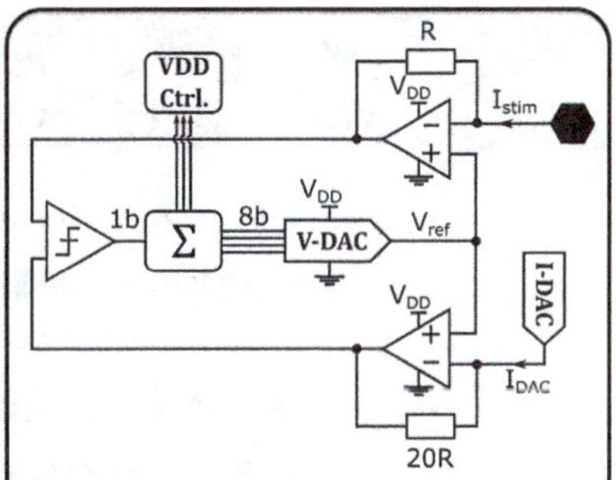

[c.f. Kassiri, Dutta, Soltani, Liu and Genov, ESSCIRC 2016]

In [Kassiri 2016] a digital feedback loop was presented to generate a CCS by adjusting the electrode voltage. A reference current (I-DAC) is used to synthesize the desired stimulation waveform. Both, reference and stimulation current, are converted into the voltage domain using two HV transimpedance amplifiers (TIA). The measured voltages are digitized using a quantizer, whose output is integrated in the digital domain. The 8-bit result of the integration adjusts the reference potential of the TIAs and the 4 most significant bits are used to adjust the stimulator supply voltage V_{DD}.

This design increases the power efficiency by minimizing the previously discussed headroom voltage. However, the input of the HV TIA, which is used to monitor the stimulation current, is directly connected to the working electrode leading to stability problems as the electrode represents a highly non-linear load.

30 Voltage-/Current-Stimulator with Digital Feedback

- Reconfigurable CCS/CVS
- Comparator digitizes difference between electrode potential and reference potential.
- Feedback loop controls CCS output to regulate output voltage
- 7b V-DAC for CVS waveform
- 5b I-DAC for CCS waveform
- Class-B operation during CVS and full reuse of HV output stage
  - High area efficiency
  - High power efficiency

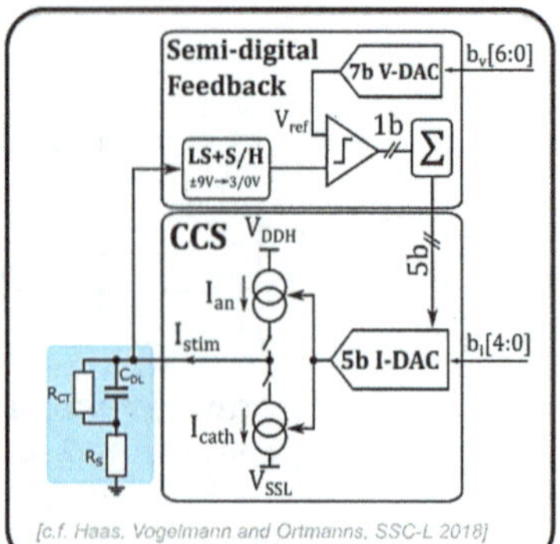

[c.f. Haas, Vogelmann and Ortmanns, SSC-L 2018]

Another architecture that allows CVS and CCS is presented in [Haas 2018]. The design uses a semi-digital feedback to control the output current of a constant current stimulator (CCS) and thereby generate an adjustable output voltage.

During CCS, an I-DAC is used for waveform synthesis, together with a HV output stage. When the stimulator is configured into CVS, the I-DAC input is generated by an on-chip control loop. A sample and hold, which includes a level shifter, converts the electrode potential from the HV domain into the LV domain, where the difference between the electrode potential and the output of a voltage DAC V_{ref} is digitized with a single bit quantizer and integrated in the digital domain to obtain the control signal for the CCS (5 bit I-DAC + 1 bit polarity).

This structure allows to completely reuse the CCS for CVS with very little area overhead. Together with its Class-B operation, this results in a high area and power efficiency.

31 Waveform Generation – Advantage of Flexibility

- **Selective activation of nerve fibers**
  - *Triangular* - peripheral nerve fiber *(Accornero 1977)*
  - *Quasitrapezoidal* - smaller nerve fiber *(Fang 1991)*
- **Asymmetric biphasic**
  - Avoid new excitation *(Ortmanns 2007)*
  - Limit channel interaction *(Machery 2006)*
- **Stimulation performance**
  - *Linear/exponential* decrease and Gaussian **stronger stimulation effect** *(Sahin 2007)*
  - *Exponential decay*

    reduce maximum electrode voltage *(Halpern 2010)*

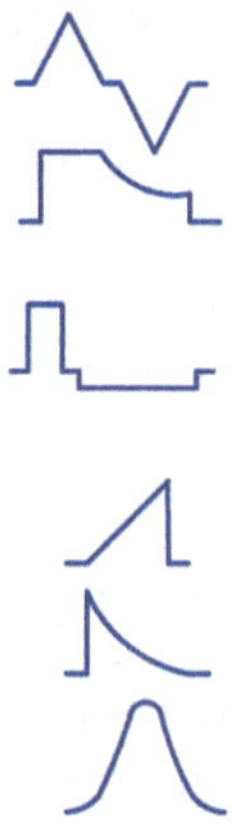

→ **Waveform should be "arbitrary" !**

31 Waveform Generation – Advantage of Flexibility

Most SotA stimulators offer biphasic stimulation pulses. In contrast, in the literature various advantages of the usage of more application specific stimulation waveforms have been reported, e.g. to limit channel interaction between adjacent stimulation sites and avoid new excitation, for selectivity, to minimize charge injection and to prevent electrode corrosion, and to enhance stimulation efficiency and safety.

32 Arbitrary Waveform Generation

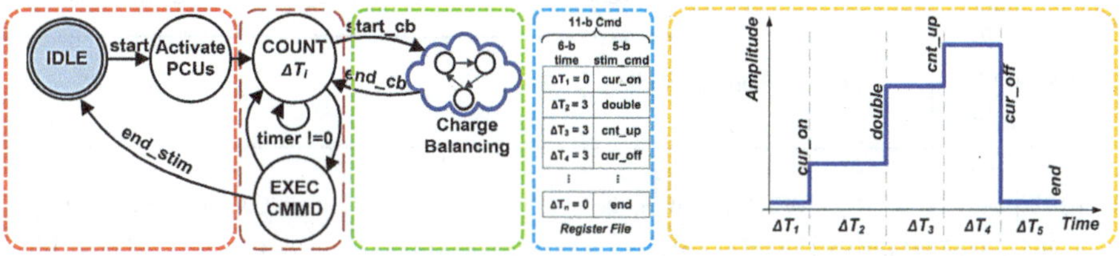

- Storage of 6-b time stamps ΔT and 5-b commands
- Sequence: count ΔT_i, then execute stimulation command Cmd_i
- Cmd changes stimulation amplitude, polarity, on/off
- Cmd can also
  - run loops
  - enter sub-state machines for charge balancing

[c.f. Noorsal, Ortmanns et al. 2012, JSSC]

Many authors responded to this need by allowing some kind of non-rectangular stimulation waveforms, but rarely the papers propose area and power efficient implementations. Many use microcontrollers [Stieglitz 1997], FPGA or DSP. Customized digital controllers, (RISC) have been used on integrated circuits [Boyer 2000, Chen 2012]. In addition, SRAM based designs have been used to store complete stimulation cycles a priori [Ba 2003].

An efficient way to realize arbitrary waveform control was introduced in [Noorsal 2012], where a number or time stamps together with commands is stored in a 11xN-bit SRAM, where N is the number of commands. The stimulation state machine counts the time ΔT_i and then executes the corresponding command. The commands directly modulate the stimulation DAC (current amplitude) or output stage on/off/polarity. Commands were also introduced to execute loops (for pulse trains), or to enter sub-state-machines for charge balancing, etc. By one more bit in the commands, it was possible to simultaneously execute two different waveshapes at various electrodes.

33 Arbitrary Waveform Generation – Examples

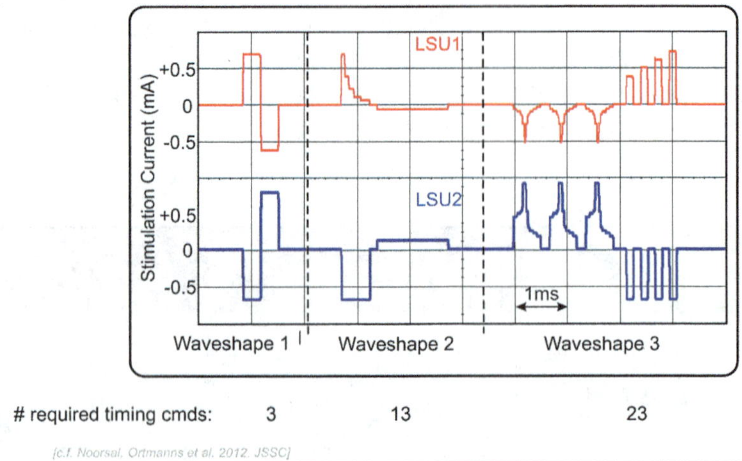

This figure shows three subsequently executed stimulations, where two electrodes have been stimulated with two different arbitrary waveshapes. The measurement shows the current waveform. Examples on ø=200µm Pt-black electrodes in PBS can be found in [Noorsal 2012]. First is a simple rectangular biphasic pulse, but different polarity of the two channels, is shown. It only needs 3 commands (11x3-bits SRAM). Next a more complex exponential decaying pulse is generated, which needs 13 commands. The last shown example is a customized wavepattern using a pulse train up/down with some arbitrary shapes, which in total needed 23 control commands.

IN VIVO IMPEDANCE ESTIMATION

Based on the previously discussed electrode model, techniques have been implemented into SoCs, which allow the in-vivo estimation of the electrode impedance. This information can be used to detect electrode degradation or failure in the implanted state, allowing the constant monitoring of the electrode/tissue interface.

34 Electrode Impedance Estimation (I)

In [Lo 2016] a system is presented that injects a small, bipolar current pulse and records the electrode response. Three measurement points are used to obtain the values of the electrode model.

In order to acquire the electrode response, the neural recorder is bypassed by a HV multiplexer, as it would saturate due to its high gain.

As the HV devices in this multiplexer are quite area consuming, this allows to monitor the electrode impedance only at a subset of all stimulation sites. Additionally the evaluation of only 3 measurement points can be subject to large variations in the presence of noise.

34 Electrode Impedance Estimation (I)

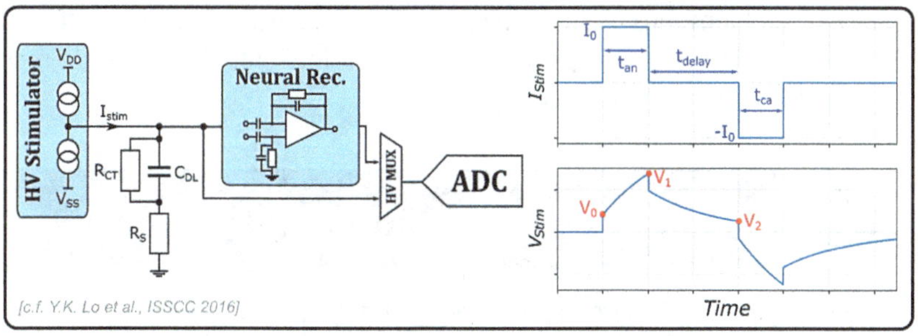

- Biphasic current pulse is used to estimate electrode impedance
- Neural recorder is bypassed to avoid input saturation
- Evaluation based on 3 measurement points $V_0 - V_2$ and adjusted stimulation amplitude + duration

$$R_S = {}^{V_0}/_{I_0} \qquad C_{DL} = \frac{I_0 \cdot t_{an}}{V_1 - V_0} \qquad R_{CT} = \frac{t_{delay}}{C_{DL} \ln\left(\frac{V_2}{V_1 - I_0 \cdot R_S}\right)}$$

35 Electrode Impedance Estimation (II)

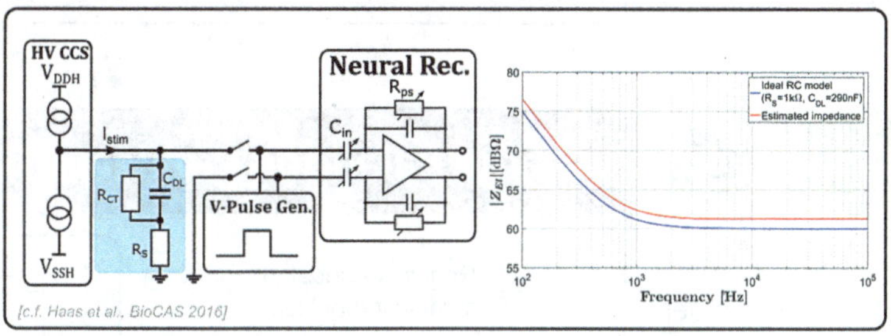

- Variable gain neural recorder with A_{rec} = 0 dB/40 dB
- Voltage pulse generator for estimation of recorder TF $H_{rec}(f)$
- Current pulse I_{stim} from CCS to record electrode response
- Estimation of electrode impedance in frequency domain, based on $H_{ref}(f)$ and recorded impulse response.

Another impedance estimation technique [Haas 2016] implements an adjustable LNA gain, which can be set to 0 dB during the estimation procedure. Thereby the saturation of the amplifier is avoided during the estimation. In order to account for the influence of the recorders transfer function (TF) on the impulse response of the electrode estimation, first the impulse response of the recorder is measured by disconnecting the recorder from the working electrode over the recorder protection switches.

This implementation yields the advantage, that no additional, area consuming HV circuitry is required to bypass the neural recorder. Thereby it allows to monitor the electrode impedance at each working electrode with little circuit overhead. Additionally the estimation in the frequency domain evaluates more data points, compared to the previously presented method, making it more robust.

CHARGE BALANCING

The charge balancing problem arises from our discussion of electrode polarization. In order to avoid electrolysis, strong faradaic currents, pH change and thus destruction of device and tissue, any DC current flow onto the electrode must be avoided.

36

Charge Balancing Problem (I)

- Electrode potential V_E during bipolar current stimulation
- Charge balanced stimulation if: $Q_c = Q_a$
- No harmful DC voltage due to charge accumulation

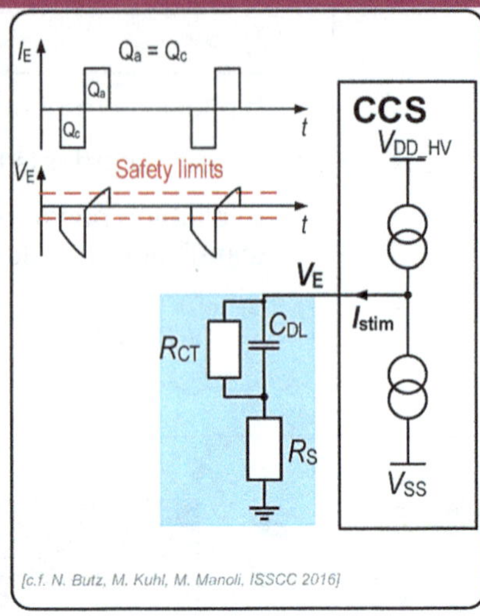

[c.f. N. Butz, M. Kuhl, M. Manoli, ISSCC 2016]

For this, very early in the application of neural stimulation, the application of bipolar stimulation was proposed. By stimulating an anodic pulse to compensate the charge of a leading cathodic pulse, and by making $Q_A=Q_C$, the net DC current to the electrode should be kept at zero.

37

Charge Balancing Problem (II)

- Results of unbalanced stimulation pulses
  - → excess voltage
  - → irreversible & toxic reactions
  - → tissue damage
  - → electrode corrosion

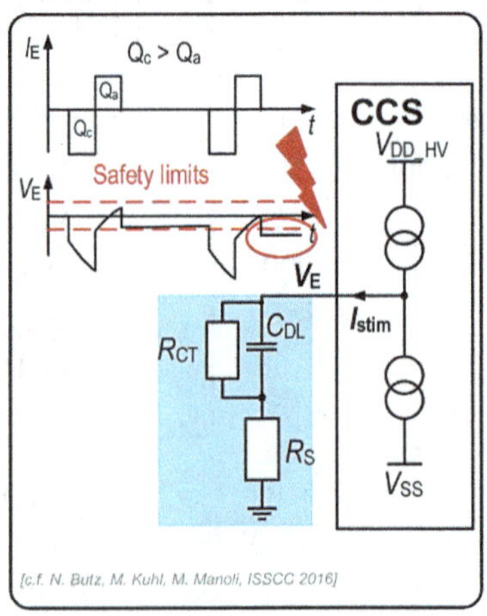

[c.f. N. Butz, M. Kuhl, M. Manoli, ISSCC 2016]

But this is only true in first order approximation. Firstly, after the leading pulse, charge could leak from the electrode and even a perfectly balanced counterpulse might yield a net current. Most importantly, the two stimulation current sources will never be perfectly balanced. For stimulation currents in the range of 1-10mA, the matching should be in the range of 1e-6, which is obviously not possible [Ortmanns 2007]. The result is illustrated here. Any mismatch of the bipolar charges leads to an excess electrode voltage over time. If this electrode voltage raises too high, this will yield strong faradaic currents.

38

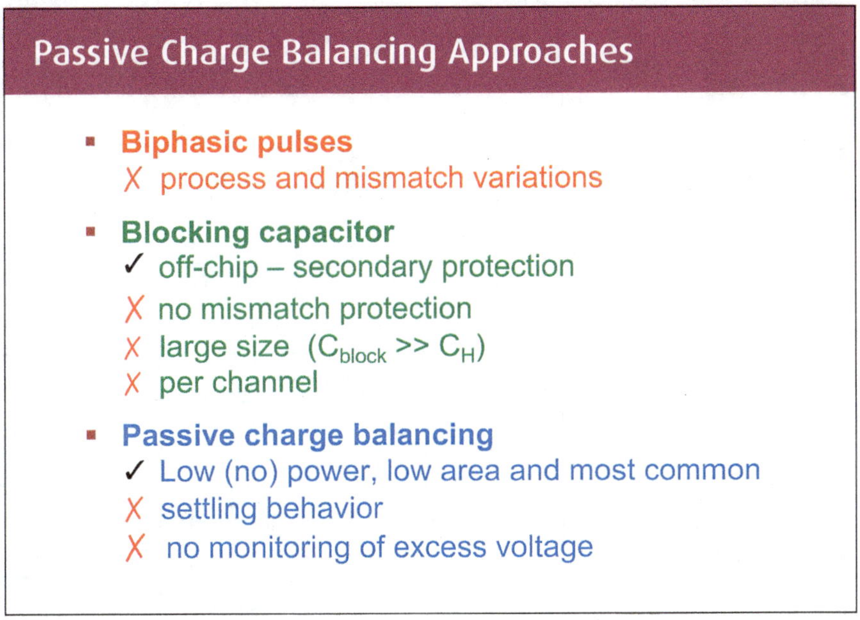

Various methods have been proposed in the past to counteract this. The first is to use blocking capacitors. By AC-coupling of the electrode to the stimulator, DC current shall be prevented. In order to have a negligible effect on the stimulation efficiency, the blocking capacitor must be much larger than the Helmholtz double-layer capacitor. Such capacitors in the μF range are passive devices, which obviously are not useable in highly multichannel applications. Moreover, it has been variously shown that using blocking capacitors can still lead to harmful conditions to the electrode [Dongen 2016].

Passive charge balancing simply uses a discharge switch, which – after every stimulation – shorts the stimulation electrode to a common mode body potential. Such passive discharge has two disadvantages. Firstly, it is unmonitored. Thus, we don't know if the electrode has been balanced or not. Secondly, the mismatch is only statistically known and the discharging time has to be short in highly multichannel devices [Sooksood 2010].

39

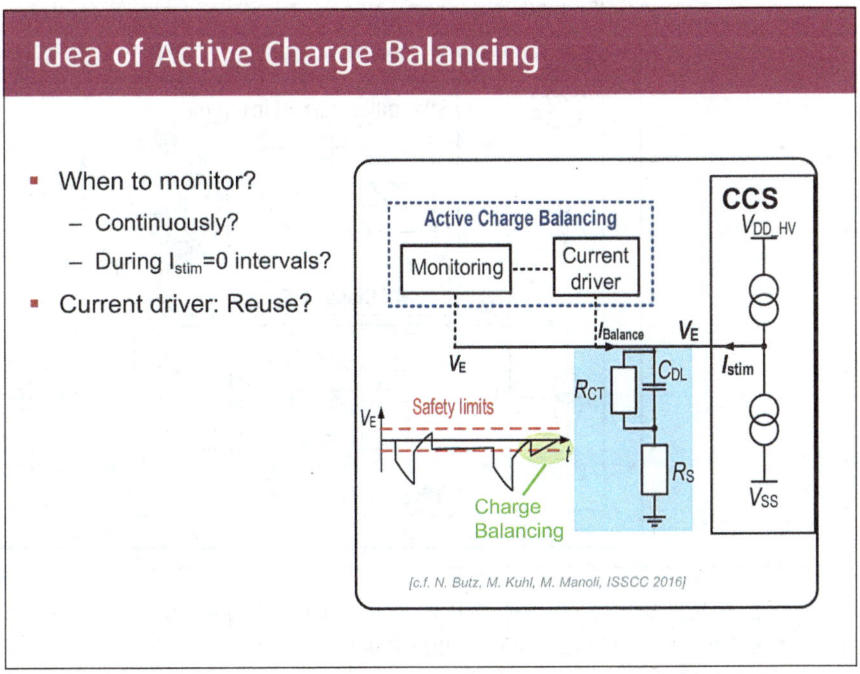

From these problem, the idea of active charge balancing was born. There, the idea is to monitor the electrode voltage subsequent to every stimulation or in regular time intervals. If the excess voltage exceeds a predefined window, then a countermeasure can be initiated [Ortmanns 2007b].

40 Active Charge Balancing – Excess-Voltage Nullor

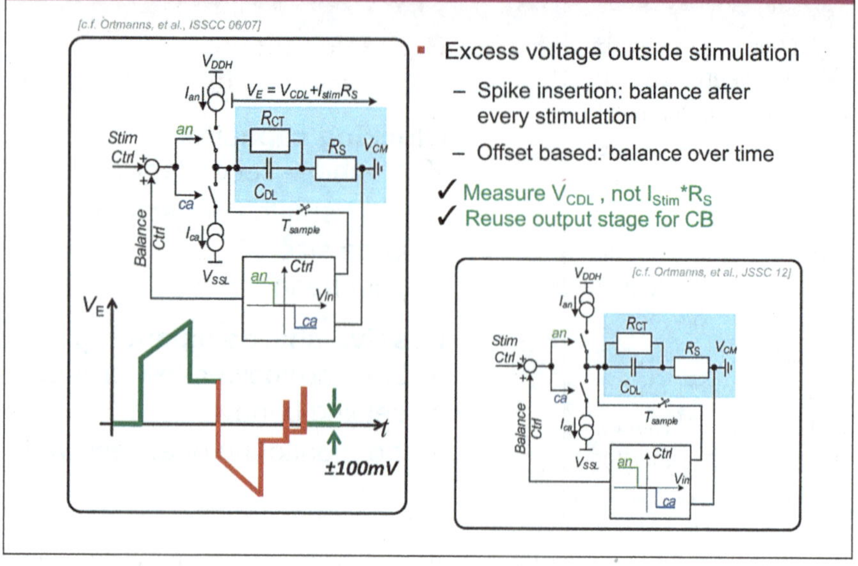

In its original implementation, two countermeasures were proposed. Firstly, short charge packets were delivered from the anyhow implemented CCS in order to balance the electrode from an excess voltage [Ortmanns 2007]. This was called the "voltage nullor", since it nulls the electrode voltage instantaneously after every stimulation. Secondly, it was proposed to use a slower reactions by using an integrator as controller [Sooksood 2010]. Thus, monitoring an excess voltage would explicitly imbalance the CCS, such that the next stimulation cycle is less mismatched.

41 Active Charge Balancing – All-Analog Charge Control

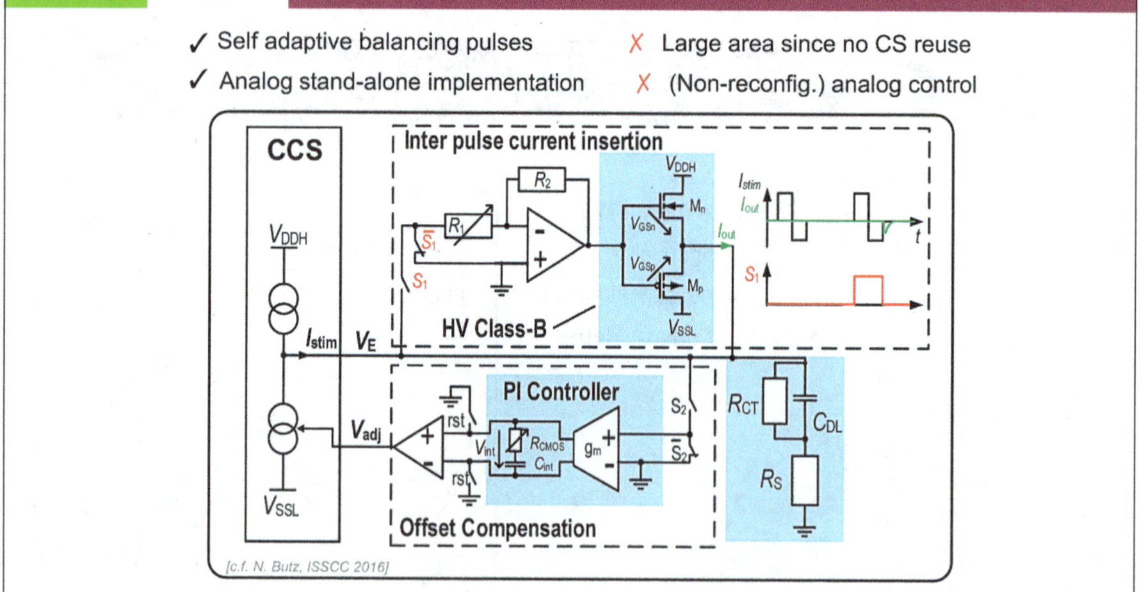

In [Butz 2016], the same two active techniques were realized as a stand alone charge balancing ASIC, implementing the spike-based balancer as "Consequence-based Inter-Pulse Charge Control," whereas the offset based balancer was named the "Cause-based Offset Compensation."

42 Charge Balancing – A Few More References

- Blocking capacitor resulting in electrode DC offset

  Marijn N. van Dongen and Wouter A. Serdijn, Does a coupling capacitor enhance the charge balance during neural stimulation? An empirical study. Med Biol Eng Comput. 2016; 54: 93–101.

- Max. charge error and excessive voltage from passive balancing

  Luis E. Rueda G., Marco Ballini, Nick Van Hellepute, Srinjoy Mitra Analysis of Passive Charge Balancing for Safe Current-Mode Neural Stimulation ISCAS 2017

  K. Sooksood et al., "An active approach for charge balancing in functional electrical stimulation," IEEE Trans. Biomed. Circuits Syst., 2010.

- Some "new" techniques
  - Don't modulate anodic vs. cathodic amplitude by pulse length

    E. Maghsoudloo *et al.*, "A new charge balancing scheme for electrical microstimulators based on modulated anodic stimulation pulse width," *IEEE ISCAS*, 2016.

  - Single current source, bipolar stimulation with inherent matching

    S. Nag *et al.*, "Flexible charge balanced stimulator with 5.6 fc accuracy for 140 nc injections," *IEEE Trans. Biomed. Circuits Syst.*, 2013

This topic of charge balancing is still actively researched, where this list of recent publications can serve as a basis of further reading.

43 Measured Charge Balancing

- 200μA pulses with t_{cath}=1ms, 200Hz repetition, +8% mismatch, Ø200μm Pt Black electrode (R_E = 3.4kΩ, C_E = 110nF)

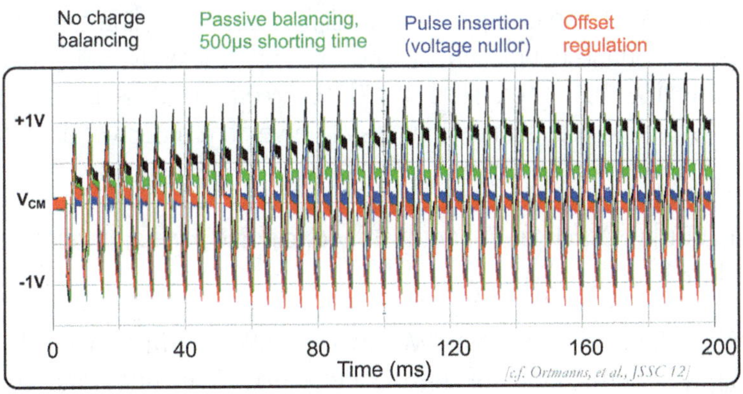

This slide, shows the effectiveness of charge balancing applied on Pt-black electrodes with 200μA at 200Hz.

By having an intrinsic 8% current mismatch without any charge balancing, the electrode potential increases over time which can destroy the electrode. With passive discharge for 500μs, the voltage is reduced but still high. Next, when the pulse insertion method is used, the electrode potential is balanced instantaneously to reference potential at every stimulation cycle. If the offset regulation charge balancing is activated, after approx. 40ms the resting electrode potential is regulated around the common voltage level.

44 References

1. [Accornero 1977] N. Accornero, G. Bini, G. L. Lenzi, and M. Manfredi, "Selective activation of peripheral nerve fibre groups of diferent diameter by triangular shaped stimulus pulses." J Physiol., vol. 273(3), pp. 539 – 560, 1977.

2. [Aquilina 2006] O. Aquilina , A brief history of cardiac pacing, Images Paediatr Cardiol 2006;27:17-81 [Ba 2003] A. Ba and M. Sawan, "Multiwaveforms generator dedicated to selective and continuous stimulations of the bladder," Proc. 25th Annu. Int. Conf. IEEE EMBS, vol. 2, pp. 1569–1572, 2003.

3. [Boyer 2000] S. Boyer, M. Sawan, M. Abdel-Gawad, S. Robin, and M. M. Elhilali, "Im- plantable selective stimulator to improve bladder voiding: design and chronic experiments in dogs," IEEE Trans. Rehabil. Eng., vol. 8, no. 4, pp. 464–470, Dec. 2000.

4. [Butz 2016] N. Butz, A. Taschwer, Y. Manoli and M. Kuhl, "22.6 A 22V compliant 56μW active charge balancer enabling 100% charge compensation even in monophasic and 36% amplitude correction in biphasic neural stimulators," 2016 IEEE International Solid-State Circuits Conference (ISSCC), San Francisco, CA, 2016, pp. 390-391.

5. [Chen 2012] K. Chen, Y.-K. Lo, Z. Yang, J. D. Weiland, M. S. Humayun, and W. Liu, "A system verification platform for high-density epiretinal prostheses," IEEE Trans. Biomed. Circuits Syst., vol. PP, no. 99, p. 1, 2012.

6. [Dongen 2016] M.N. van Dongen, W.A. Serdijn, "Does a coupling capacitor enhance the charge balance during neural stimulation? An empirical study.", Med Biol Eng Comput. 2016 Jan;54(1):93-101. [Accornero 1977] N. Accornero, G. Bini, G. L. Lenzi, and M. Manfredi, "Selective activation of peripheral nerve fibre groups of diferent diameter by triangular shaped stimulus pulses." J Physiol., vol. 273(3), pp. 539 – 560, 1977.

7. [Aquilina 2006] O. Aquilina , A brief history of cardiac pacing, Images Paediatr Cardiol 2006;27:17-81

8. [Ba 2003] A. Ba and M. Sawan, "Multiwaveforms generator dedicated to selective and continuous stimulations of the bladder," Proc. 25th Annu. Int. Conf. IEEE EMBS, vol. 2, pp. 1569–1572, 2003.

9. [Boyer 2000] S. Boyer, M. Sawan, M. Abdel-Gawad, S. Robin, and M. M. Elhilali, "Im- plantable selective stimulator to improve bladder voiding: design and chronic experiments in dogs," IEEE Trans. Rehabil. Eng., vol. 8, no. 4, pp. 464–470, Dec. 2000.

10. [Butz 2016] N. Butz, A. Taschwer, Y. Manoli and M. Kuhl, "22.6 A 22V compliant 56µW active charge balancer enabling 100% charge compensation even in monophasic and 36% amplitude correction in biphasic neural stimulators," 2016 IEEE International Solid-State Circuits Conference (ISSCC), San Francisco, CA, 2016, pp. 390-391.

11. [Chen 2012] K. Chen, Y.-K. Lo, Z. Yang, J. D. Weiland, M. S. Humayun, and W. Liu, "A system verification platform for high-density epiretinal prostheses," IEEE Trans. Biomed. Circuits Syst., vol. PP, no. 99, p. 1, 2012.

12. [Dongen 2016] M.N. van Dongen, W.A. Serdijn, "Does a coupling capacitor enhance the charge balance during neural stimulation? An empirical study.", Med Biol Eng Comput. 2016 Jan;54(1):93-101.

13. [Fang 1991] Z. P. Fang and J. T. Mortimer, "Selective activation of small motor axons by quasitrapezoidal current pulses," IEEE Trans. Biomed. Eng., vol. 38, no. 2, pp. 168–174, 1991. [Grill 2004] W.M.Grill, Electrical Stimulation of the Peripheral Nervous System: Biophysics and Excitation Properties. World Scientific, 2004, ch. 2.1, pp. 319–341.

14. [Guilvard 2012] A. Guilvard, A. Eftekhar, S. Luan, C. Toumazou and T. G. Constandinou, "A fullyprogrammable neural interface for multi-polar, multi-channel stimulation strategies," 2012 IEEE International Symposium on Circuits and Systems, Seoul, 2012, pp. 2235-2238.

15. [Haddad 2006] S. Haddad, R. Houben, W. Serdijn, The history of cardiac pacemakers: an electronics perspective. IEEE Engineering in Medicine and Biology Magazine. 25. 38-48, 2006.

16. [Haas 2016] M. Haas, J. Anders and M. Ortmanns, "A bidirectional neural interface featuring a tunable recorder and electrode impedance estimation," 2016 IEEE Biomedical Circuits and Systems Conference (BioCAS), Shanghai, 2016, pp. 372-375.

17. [Haas 2018] M. Haas, P. Vogelmann, M. Ortmanns, A Neuromodulator Frontend with Reconfigurable Class-B Current and Voltage Controlled Stimulator, IEEE Solid-State Circuits Letters, 2018. Accepted for publication.

18. [Halpern 2010] M. E. Halpern and J. Fallon, "Current waveforms for neural stimulation-charge delivery with reduced maximum electrode voltage," IEEE Trans. Biomed. Eng., vol. 57, no. 9, pp. 2304–2312, Sep. 2010.

19. [Kassiri 2016] H Kassiri, G Dutta, N Soltani, C Liu, Y Hu, R Genov, An impedance-tracking batteryless arbitrary-waveform neurostimulator with load-adaptive 20V voltage compliance European Solid-State Circuits Conference, ESSCIRC Conference 2016: 42nd, 225-228

20. [Laotaveerungrueng 2010] N. Laotaveerungrueng, et al., "A high-voltage, high-current CMOS pulse generator ASIC for deep brain stimulation," 2010 Annual International Conference of the IEEE Engineering in Medicine and Biology, Buenos Aires, 2010, pp. 1519-1522.

21. [Lee 2015] H.M. Lee, et al., A Power-Efficient Switched-Capacitor Stimulating System for Electrical/Optical Deep Brain Stimulation, IEEE JSSC, 50(1), Jan. 2015

22. [Lo 2016] Y. K. Lo, Chih-Wei Chang, Yen-Cheng Kuan, Stanislav Culaclii, Brian Kim, Kuanfu Chen, Parag Gad, V. Reggie Edgerton, Wentai Liu., "22.2 A 176-channel 0.5cm3 0.7g wireless implant for motor function recovery after spinal cord injury," 2016 IEEE International Solid-State Circuits Conference (ISSCC), San Francisco, CA, 2016, pp. 382-383.

23. [Macherey 2006] O. Macherey, A. van Wieringen, R. Carlyon, J. Deeks, and J. Wouters, "Asymmetric pulses in cochlear implants: Efects of pulse shape, polarity, and rate," J. Assoc. Res. Otolaryngol., vol. 7, no. 3, pp. 253–266, 2006.

24. [Merrill 2005] D.R. Merrill, et al: "Electrical stimulation of excitable tissue: design of efficacious and safe protocols". J. of Neuroscience Meth. 141 (2): 171–198, 2005.

25. [Ortmanns 2007] M. Ortmanns, A. Rocke, M. Gehrke and H. J. Tiedtke, "A 232-Channel Epiretinal Stimulator ASIC," in IEEE Journal of Solid-State Circuits, vol. 42, no. 12, pp. 2946-2959, Dec. 2007.

26. [Ortmanns 2007b] M. Ortmanns, "Charge Balancing in Functional Electrical Stimulators: A Comparative Study," 2007 IEEE International Symposium on Circuits and Systems, New Orleans, LA, 2007, pp. 573-576.

27. [Noorsal 2012] E. Noorsal, K. Sooksood, H. Xu, R. Hornig, J. Becker and M. Ortmanns, "A Neural Stimulator Frontend With High-Voltage Compliance and Programmable Pulse Shape for Epiretinal Implants," in IEEE Journal of Solid-State Circuits, vol. 47, no. 1, pp. 244-256, Jan. 2012.

28. [Preda 2016] F. Preda et al., "Switching from constant voltage to constant current in deep brainstimulation: a multicenter experience of mixed implants for movement disorders," European Journal of Neurology, vol. 23, no. 1, pp. 190–195, 2016. [Randles 1947] Randles, J. E. B. (1947). "Kinetics of rapid electrode reactions". Discussions of the Faraday Society. 1: 11

29. [Rothermel 2008] A. Rothermel, V. Wieczorek, L. Liu, A. Stett, M. Gerhardt, A. Harscher, S. Kibbel, "A 1600-pixel Subretinal Chip with DC-free Terminals and ±2V Supply Optimized for Long Lifetime and High Stimulation Efficiency," 2008 IEEE International Solid-State Circuits Conference - Digest of Technical Papers, San Francisco, CA, 2008, pp. 144-602.

30. [Sahin 2007] M. Sahin and Y. Tie, "Non-rectangular waveforms for neural stimulation with practical electrodes," J. Neural Eng., vol. 4, no. 3, pp. 227–233, Sep. 2007. [Simpson 2007] Simpson, Jim & Ghovanloo, Maysam. (2007). An Experimental Study of Voltage, Current, and Charge Controlled Stimulation Front-End Circuitry. Proceedings - IEEE International Symposium on Circuits and Systems. 325 - 328. 10.1109/ISCAS.2007.378401. [Shannon 1992] Shannon, R.V. (April 1992). IEEE TBME, 39 (4): 424–426.

31. [Shulyzki 2010] Shulyzki, R., Abdelhalim, K., and Genov, R.: 'CMOS current-copying neural stimulator with OTA-sharing', Proc. IEEE ISCAS 2010, Paris, France, May 2010, pp. 1232–1235

32. [Sooksood 2010] Sooksood K, Stieglitz T, Ortmanns M., An active approach for charge balancing in functional electrical stimulation, IEEE Trans Biomed Circuits Syst. 2010 Jun;4(3)

33. [Sooksood 2012] K. Sooksood and M. Ortmanns, "Power efficient output stage for high density implantable stimulators," in Electronics Letters, vol. 48, no. 10, pp. 551-552, May 10 2012.

34. [Stieglitz 1997] T. Stieglitz, T. Matal, and M. Staemmler, "A modular multichannel stimulator for arbitrary shaped current pulses for experimental and clinical use in FES," Proc. 19th Annu. Int. Conf. IEEE EMBS, vol. 4, pp. 1777 –1780, Oct - Nov 1997.

35. [Tsoucalas 2014] Gregory Tsoucalas, Marianna Karamanou, Maria Lymperi, Vassiliki Gennimata, George Androutsos, The "torpedo" effect in medicine, Int Marit Health 2014; 64, 2: 65–67.

36. [Xu 2012] H. Xu, E. Noorsal, K. Sooksood, J. Becker and M. Ortmanns, "A multichannel neurostimulator with transcutaneous closed-loop power control and self-adaptive supply," 2012 Proceedings of the ESSCIRC (ESSCIRC), Bordeaux, 2012, pp. 309-312.

37. [Yoo 2015] J. Yoo et al., "A 16-Channel Patient-Specific Seizure Onset and Termination Detection SoC With Impedance-Adaptive Transcranial Electrical Stimulator", Journal of Solid-State Circuits, vol. 50, no. 11, pp. 2728-2740, 2015

CHAPTER 04

Wireless Power Transfer Circuits for Biomedical Microsystems

Maysam Ghovanloo

Bionic Sciences Inc., USA

Wireless power transmission (WPT) is rapidly emerging in several areas that demand various levels of power requirement from nanowattsin some internet of things (IoT) wireless sensors and radio frequency identification (RFID) tags to watts and kilowatts in mobile computing and electric vehicles (EV), respectively. While use of WPT in some of these applications is a matter of convenience, in a considerable group of implantable medical devices (IMDs) with power consumption in the milliwatts to watts range, it is necessity and a matter of feasibility of a therapy.

1 Introduction

Several methods have been proposed for WPT in IMD applications: 1) Electromagnetic (EM) coupling in the near-field, and 2) mid-field, 3) far-field electromagnetic radiation, 4) capacitive coupling, 5) ultrasonic coupling, 6) mechanical vibration, and 7) optical power transmission. So far, an inductive link, which operates based on near-field magnetic coupling (1), offers the most viable and popular method to wirelessly transfer energy with high power transmission efficiency (PTE) at the frequency range of 0.1 to 100's of MHz, depending on the size of the IMD and amount of power delivered to the load (PDL). This method, which includes many successful examples, such as cochlear implants, retinal prostheses, pain management systems, brain-computer interfaces (BCI), and artificial hearts, is the focus of this chapter.

We will take a closer look at a wide range of IMDs both in clinical use and those that are under development. We cover the fundamentals of inductive coupling among coils based on the Faraday law and the theory behind coupled-mode magnetic resonance to construct circuit models that allow us to devise a method for designing and optimizing coil geometries and electrical characteristics based on the key requirements and constraints of a certain IMD application. We will look at some examples, including the specific case of mm-sized IMDs that are recently proposed for distributed wireless neural interfaces.

2 Outline

- ***Introduction***
- ***Near-Field Wireless Power Transmission***
- ***Dynamic Impedance Matching via Q-Modulation***
- ***Data Telemetry Using Multiple Carriers***
- ***Pulse Harmonic Modulation (PHM)***
- ***Pulse Delay Modulation (PDM)***
- ***Pulse-Width Modulation Impulse-Radio Ultra-Wideband (PWM-IR-UWB)***
- ***Efficient AC-to-DC Conversion and Power Management Circuits***
- ***Conclusions***

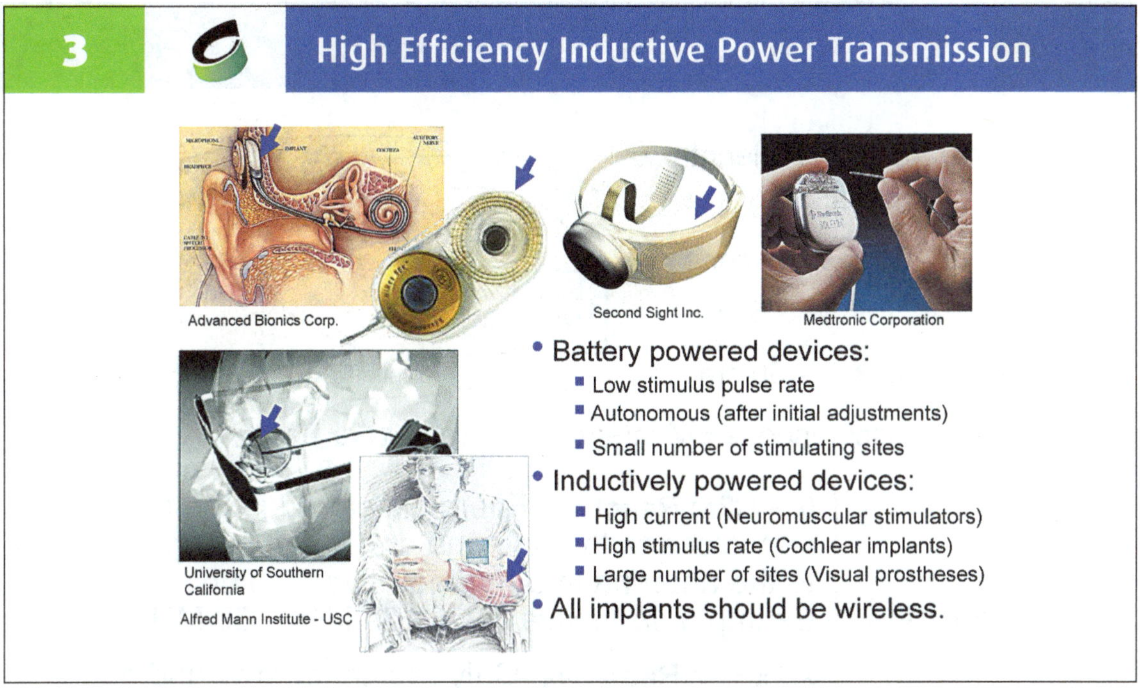

Powering mechanism is one of the most important aspects of designing an implantable medical device (IMD) and heavily depends on its medical application and anatomical location, where it is being implanted. IMDs that have reached the clinical stage and become commercially available for patients after receiving the necessary approval from regulatory bodies, such as the Food and Drug Administration (FDA) in the USA, can be divided into three categories:

- IMDs such as pacemakers and deep brain stimulators that have a relatively small average power consumption in the range of 10s to 100s of microwatts (mW) and a small number of stimulation sites (1~4). These IMDs are often powered by primary (non-rechargeable) batteries that have sufficient energy storage capacity to operate the IMD for 2~10 years, following which period they are explanted and replaced via a routine surgery. As a result, these IMDs are larger, often the size of a match box, more than half of which is occupied by the battery, and implanted in the body where there is enough space, such as in the chest or abdominal area.
- IMDs such as cochlear and retinal implants that have considerably larger power consumption in the range of 10s to 100s of milliwatts (mW), a large number of stimulation sites (10s~100s), and implanted in anatomical locations, where the space it very limited, e.g. in the mastoid bone behind the ears or next to the eyeball. These IMDs need to be powered through an inductive link across the skin (a.k.a. transcutaneous link) between an implanted receiver (Rx) coil within the IMD and an external transmitter (Tx) coil, which is often held in place with a pair of magnets or on a wearable device, e.g. eyeglass.
- There are also IMDs that have high power consumption but placed in anatomical locations where there is sufficient space, such as pain management systems implanted in the back or abdomen, in which case the IMD is equipped with both a rechargeable battery and an inductive link to charge it on a regular basis.

A common theme among all these IMDs is being wireless, requiring them to have access to power and data without breaching the skin barrier. The near-field inductive links to be discussed in the rest of this chapter are usable in the 2nd and 3rd categories, as well as those that are being developed for a variety of BCI applications in the cranial space, which fall in the 2nd category.

4 Reactive vs. Radiative

Carrier signal wavelength: λ

- $0 \sim \lambda$: Near field → $1/r^3$
- $\lambda \sim 2\lambda$: Transition zone
- $2\lambda \sim \infty$: Far field → $1/r$

f = 1 GHz → λ = 30 cm

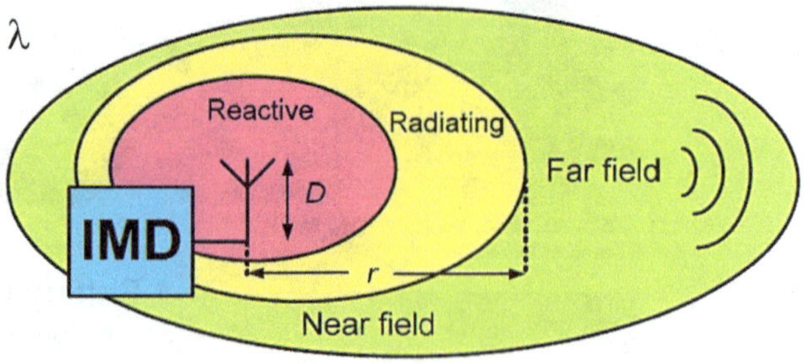

→ All transcutaneous EM interactions are in **near field**.

Near field

- **$0 \sim \lambda/2\pi$: Reactive**: Energy is stored in the field very close to the antenna and it can return back to the antenna in a regenerative fashion; Example: *Inductive links* (125 kHz ~ 50 MHz)
- **$\lambda/2\pi \sim \lambda$: Radiative**: Only radiant energy and no storage; Example: *MICS* Band* (402~405 MHz)

* Medical Implant Communication Service (MICS)

P.V. Nikitin et al., *Intl. Conf. on RFID*, 2007

The area around any radio frequency (RF) transmitter (Tx) can be divided into three regions: near-field (0 ~ λ), transition zone or mid-field (λ ~ 2), and far-field (2λ ~ ∞, where λ is the carrier signal wavelength [Ref1]. The near-field region itself can be divided into two sub-regions, one (0 ~ l/2π) is called the reactive zone, where energy is stored in the field very close to the antenna and it can return back to the antenna in a regenerative fashion, for instance, Inductive links that operate within 125 kHz ~ 100 MHz. The other one (l/2p ~ l) is called radiative zone, in which we have radiant energy and minimal storage, for instance, the Medical Implant Communication Service (MICS) band that operates within 402~405 MHz range [Ref2]. Considering limitations in the choice of carrier frequencies in WPT through inductive coupling (< 100 MHz, λ > 3 m), and the proximity of the Tx and Rx coils in IMD applications, it can be concluded that inductive links used in IMDs generally operate within the reactive near-field region.

[Ref1] C. A. Balanis, *Antenna Theory: Analysis and Design*, 3rd Ed., Wiley, 1997.

[Ref2] P.V. Nikitin, K.V.S. Rao and S. Lazar, "An overview of near field UHF RFID," *Intl. Conf. on RFID*, Apr. 2007.

5 Transcutaneous Link Power Losses

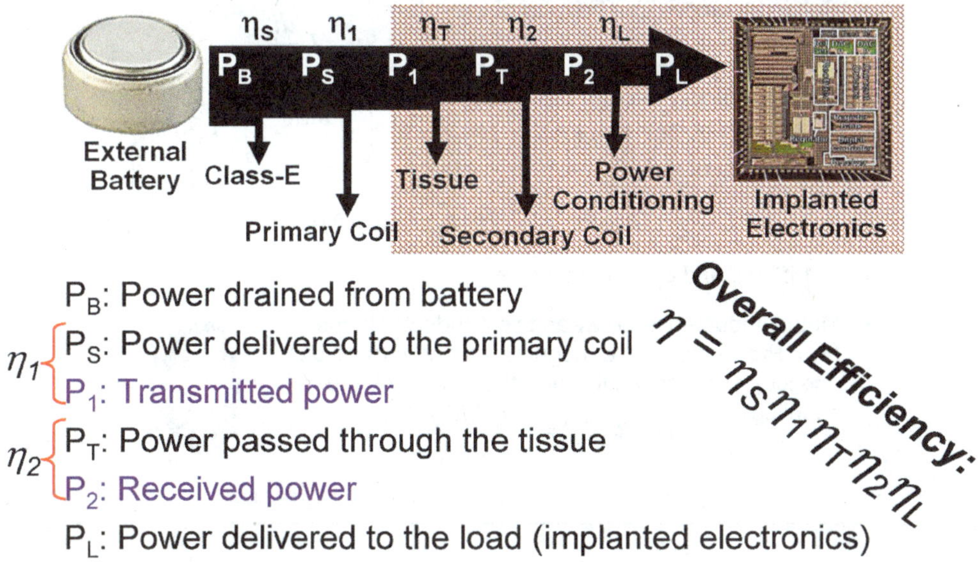

Another common requirement among all IMDs, besides being wireless, is low power dissipation to the extent that the dissipated power as heat either from the IMD or through the process of WPT from the external energy source (battery) to the implant electronics does not increase the steady state temperature within the tissue environment around the IMD by more than 1°C, even though occasional temperature rise up to 5°C over short periods might also be tolerated. This would require a careful design of the entire chain of power transmission and management, particularly the steps that are the least efficient.

Generally the key steps involved in WPT from the external battery to IMD electronics, considering the fact that only AC signals pass through an inductive link, while both energy source and electronics have DC supplies, involve:

1. DC to AC power conversion, at a certain carrier frequency, *f*, often through an oscillator followed by a power amplifier (PA), both of which can be combined in certain PA topologies, such as class-E PA.

2. The PA driving the Tx coil through an impedance matching circuit and generating an AC magnetic field by the Tx coil (a.k.a. primary coil), part of which is absorbed in the surrounding tissue, which consists of >90% water, and part of which passes through the Rx coil (a.k.a. secondary coil) and induces current in it.

3. AC carrier signal induced in the Rx coil is delivered to the power conditioning block within the IMD through a matching circuit to be rectified and regulated to create a stable DC power supply for the rest of the IMD electronics.

Every one of these steps involves power dissipation as heat, with P_1 and P_2, the power dissipated in the Tx and Rx coils due to their parasitic resistance, R_{S1} and R_{S2}, respectively, being the most important sources of power dissipation.

We should point out that the overall PTE would be the multiplication of the partial efficiencies of all these steps, which we would like to push towards 100%. Therefore, it is the inductive link that creates a bottleneck in this particular near-field WPT process, signifying the importance of the proper design and optimization of the Tx-Rx coil geometries and their associated impedance matching circuits.

6 Self and Mutual Inductance

- Self inductance (L): The ratio of the magnetic flux generated in an area enclosed by a conductor loop to the current passing through the loop. $v = L\frac{di}{dt}$
- If $r/R << 1$ → One turn loop: $L(R,r) \approx \mu_0 R\left(\ln\left(\frac{8R}{r}\right) - 2\right)$
- For N turns of radii R_i (i = 1, 2, … N),

$$L = \sum_{i=1}^{N} L(R_i, r) + \sum_{i=1}^{i=N}\sum_{j=1}^{j=N} M_{ij}(R_i, R_j, d_r = 0)(1 - \alpha_{i,j})$$

$\alpha_{i,j} = 1$ if i = j, and $\alpha_{i,j} = 0$ otherwise

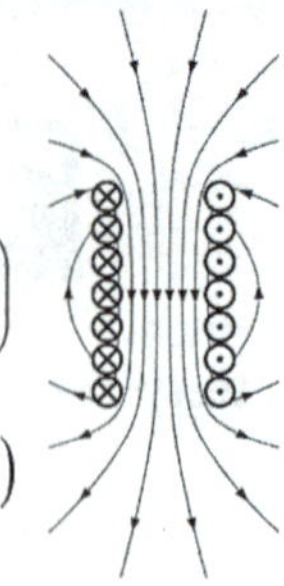

- Mutual inductance between two conductor loops (M_{ij}): Proportion of the magnetic flux generated by one loop that passes through the other loop (flux coupling). $M_{12}(R_1, R_2, d_{12}) = \frac{\pi\mu_0 N_1 R_1^2 N_2 R_2^2}{2\sqrt{(R_1^2 + {d_{12}}^2)^3}}$

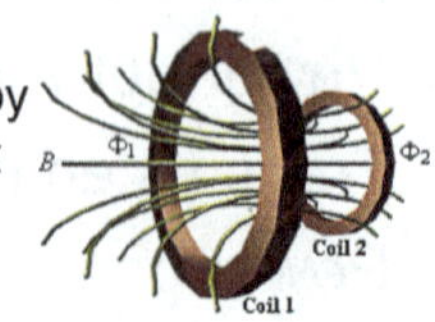

C.M. Zierhofer and E.S. Hochmair, *TBME* 1996

There are numerous models for deriving the self and mutual inductance of different types of coils made of wires or lithographically-defined planar spiral patterns, which can be used to derive the inductance as well as parasitic capacitance and resistance of each coil with various levels of accuracy [Ref 1]. In order to more accurately calculate self and mutual inductances of coils with various geometries, one should either use tabulated parameterized equations that are derived empirically or use finite element electromagnetic software, such as FastHenry, SONET, or HFSS (Ansoft, Pittsburgh, PA).

[Ref 1] C.M. Zierhofer, and E.S. Hochmair, "Geometric approach for coupling enhancement of magnetically coupled coils," *IEEE Trans. Biomedical Engineering,* vol. 43, no. 7, pp.708-714, Jul. 1996.

7 Magnetic Coupling Between Two Coils

- Coupling coefficient, k_{ij}, between two coils is the most important factor in determining the PTE of the inductive link.

$$k_{ij} = M_{ij}/\sqrt{L_i L_j}$$

- k_{ij} can be defined as the portion of the flux generated by the Tx coil that pass through the Rx coil (flux linkage)

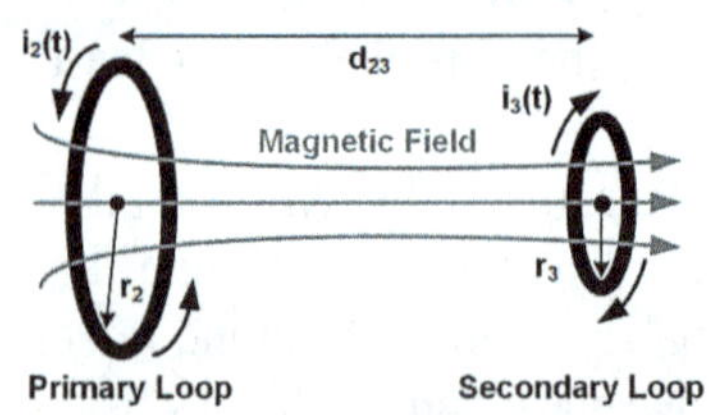

inductive coupling between a primary loop with the radius of r_2 and a secondary loop with the radius of r_3, which are separated by d_{23}.

- Inductive coupling between two loops would be stronger if a higher portion of the magnetic field generated by the primary loop passes through the secondary loop.

7 Magnetic Coupling Between Two Coils

The fundamental physics behind near-field WPT between two or more coils is expressed in the Faraday's law of induction, which states that when the total magnetic flux through a conductive loop – defined as the integral over the surface enclosed by the loop of the magnetic flux – varies with time, a current is induced in the loop itself. This, in turn, results in an electromotive force (EMF) induced in the loop [Ref 1]. Thus, a primary loop which is driven by the PA will generate a varying magnetic field, part of which passes through the secondary loop, depending on their relative size, proximity, and orientation with respect to one another, resulting in an induced current in the secondary loop.

In addition to the distance and geometry, the coils' alignment has a significant effect on their mutual inductance. For example, it can be shown that if one of the coils is tilted by an angle φ, their mutual inductance reduces by a factor of *cos(φ)* [Ref 2].

[Ref 1] M. Sadiku, *Elements of Electromagnetics*, 4th ed. Oxford University Press, 2007.

[Ref 2] M. Soma, D.G. Galbraith, and R.L. White, "Radio-frequency coils in implantable devices: misalignment analysis and design procedure," *IEEE Trans. Biomed. Eng.* vol. 34, pp. 276-282, Apr. 1987.

8 Simple Circuit Model of Coupled Coils

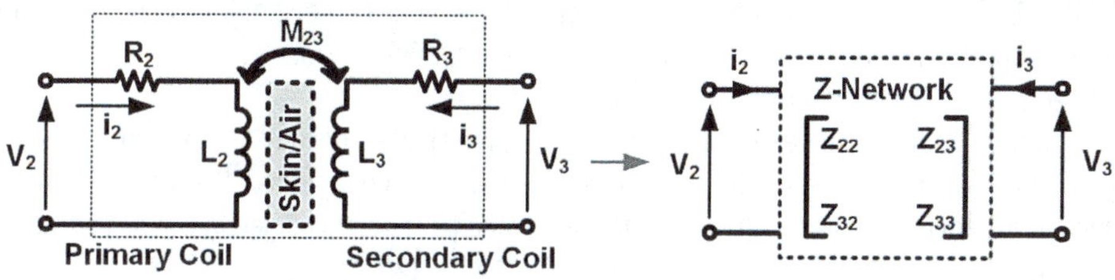

- In time domain:

$$V_2(t) = R_2 \cdot i_2(t) + L_2 \cdot di_2(t)/dt + M_{23} \cdot di_3(t)/dt$$

$$V_3(t) = M_{23} \cdot di_2(t)/dt + R_3 \cdot i_3(t) + L_3 \cdot di_3(t)/dt$$

- In Laplace domain:

$$\begin{bmatrix} V_2(s) \\ V_3(s) \end{bmatrix} = \begin{bmatrix} R_2 + L_2 s & M_{23}s \\ M_{23}s & R_3 + L_3 s \end{bmatrix} \cdot \begin{bmatrix} i_2(s) \\ i_3(s) \end{bmatrix} = Z \cdot \begin{bmatrix} i_2(s) \\ i_3(s) \end{bmatrix}$$

In a simple 2-coil link, a time-variant current $i_2(t)$ in the primary coil, L_2, generates a time-variant magnetic field, part of which passes through the secondary coil, L_3. This part of the time-varying magnetic field generates voltage $V_3(t)$ across L_3 and current $i_3(t)$ through the secondary loop due to its mutual inductance, M_{23}, with L_2. R_2 and R_3 are the ohmic losses of L_2 and L_3, respectively.

9 Resonant vs. Non-Resonant WPT

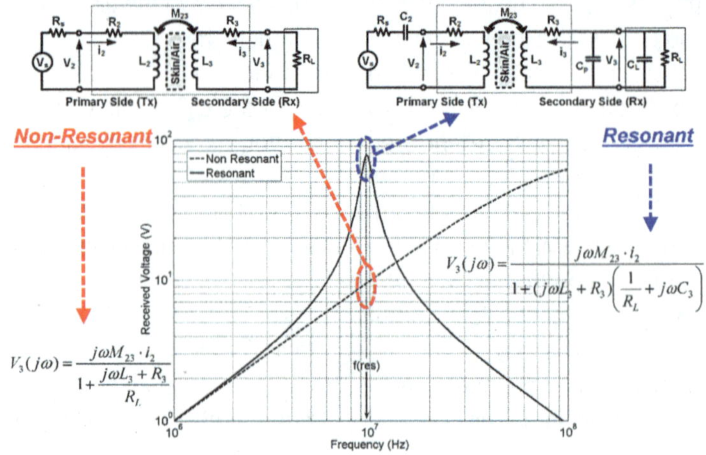

Even though WPT occurs between two coils that are simply coupled, in order to significantly increase V_3 at a certain frequency of interest, f_0, we should maximize the impedance across R_L at f_0. The best way to do this is to add a capacitor, C_3, in parallel with L_3 to form a parallel resonant LC-tank circuit. The resonance frequency of the L_3C_3-tank circuit should be tuned to match that of the external Tx carrier frequency. In choosing C_3, one should also consider the parasitic capacitance of L_3, interconnects, and the input capacitance of the stage following the L_3C_3-tank circuit, which is usually an AC-to-DC converter (e.g. a rectifier). If we represent the sum of all these parasitic capacitors by Cp, then $C_3 = C_p + C_L$, where C_L is the capacitance that is physically added across the load. If the load also has a capacitive component, that capacitance should also be taken into account in calculating C_3 [Ref 1].

[Ref 1] M. Kiani and M. Ghovanloo, "Inductive Power Transmission Systems," Wiley Encyclopedia of Electrical and Electronics Engineering, DOI: 10.1002/047134608X.W8306 (2016).

10 Calculating Power Transfer Efficiency [PTE] (I)

PTE is highly dependent on: $\mathbf{k_{23}}$, $\mathbf{Q_2}$, and $\mathbf{Q_3}$

$Q_2 = \omega_0 L_2/R_2$, $Q_3 = \omega_0 L_3/R_3$

Secondary $R_3L_3C_3$ can be reflected onto the primary:

$Q_{3L} = R_P/\omega_0 L_3$

At resonance ($\omega = \omega_0$):
Series L_2C_2 shorts
Parallel $L_{ref}C_{ref}$ opens

We are only interested in that portion of the power that is delivered **from V_s to R_L**

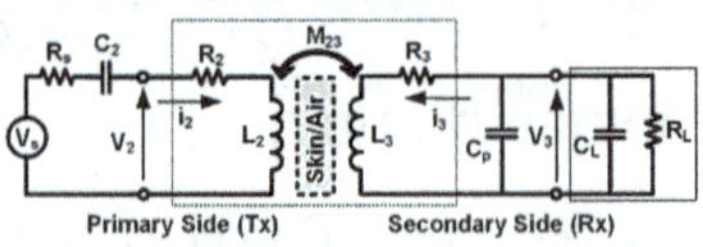

$$R_{P3} = Q_3^2 R_3 \qquad R_P = R_{P3} \,||\, R_L$$

$$\begin{cases} R_{ref} = k_{23}^2 (L_2/L_3) R_P = k_{23}^2 \omega_0 L_2 Q_{3L} \\ C_{ref} = (L_3/L_2)(C_3/k_{23}^2) = 1/(\omega_0^2 L_2 k_{23}^2) \end{cases}$$

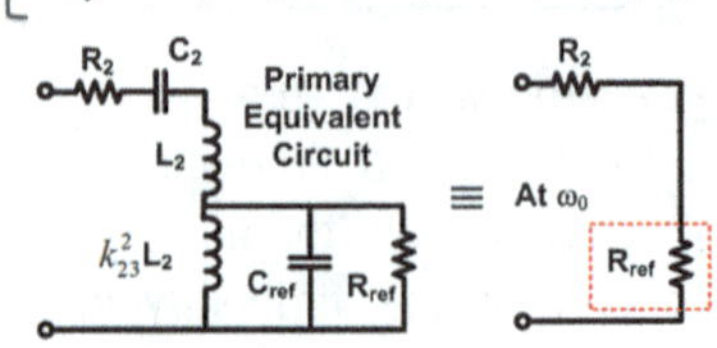

R.R. Harrison, ISCAS 2007

10 Calculating Power Transfer Efficiency [PTE] (I)

Power transfer efficiency (PTE) is a key parameter in the design of WPT links for IMDs by identifying source specifications, heat dissipation, power transmission capacity, and possibility of interference with other devices. The PTE is defined as the ratio between the power delivered to the load (PDL) and the power drained from the energy source, *Vs*, and delivered to the primary loop. To better understand the effects of inductive link circuit parameters on the PTE, we define the Tx and Rx coils' quality factors (Q-factor) as $Q_2 = \omega_0 L_2/(R_2 + R_s)$, and $Q_3 = \omega_0 L_3/R_3$, respectively, where R_s represents the power amplifier (PA) loss.

To further simplify the circuit analysis, the secondary loop, including R_L, can be reflected on to the primary side with the reflected impedance, $L_{ref} \parallel C_{ref} \parallel R_{ref}$. To find R_{ref}, the secondary side is modeled with a parallel load resistance. The series loss resistance of L_3 i.e. R_3 can be transformed to a parallel resistance[Ref 1]. Therefore, the equivalent parallel resistance in the secondary loop is $R_p = R_{P3} \parallel R_L$.

Due to the mutual coupling of L_2 and L_3, the secondary load resistance and capacitance can be transferred to the primary side. $Q_{3L} = R_p/\omega_0 L_3$ is referred to as the loaded quality factor of the secondary coil [Ref 2]. It should be noted that C_{ref} still resonates out with $k_{23}^2 L_2$ at f_0, i.e. becomes an open-circuit, leaving behind only a resistance, R_{ref}, in the primary loop. It should be noted that L_2 and C_2, which form a series LC-tank, are chosen such that they also resonate at f_0, ideally turning into short-circuit.In the simplified circuit at resonance, the input power provided by the source is simply divides between R_s+R_2 and R_{ref}. The power absorbed by R_s+R_2 is dissipated as heat in the PA and the primary coil, while the power delivered to R_{ref} is that portion of the source power that is transferred across the link to the secondary loop. This power further divides between R_3 and R_L, which are the only power consuming components on the Rx side.

[Ref 1] T. Lee, *The design of CMOS radio-frequency integrated circuits*, 2nd Ed., Cambridge University Press, New York, NY, 2004.

[Ref 2] R. Harrison, "Designing efficient inductive power links for implantable devices," *IEEE Int. Symp. Circuits Syst.*, pp. 2080–2083, May 2007.

11 Calculating Power Transfer Efficiency [PTE] (II)

Key factors: **k_{23}, Q_2, Q_3**

$$\eta_{2-coil} = \frac{R_{ref}}{R_s + R_2 + R_{ref}} \frac{R_{P3}}{R_{P3} + R_L} = \frac{k_{23}^2 Q_2 Q_{3L}}{1 + k_{23}^2 Q_2 Q_{3L}} \cdot \frac{Q_{3L}}{Q_L}$$

$$Q_{3L} = Q_3 Q_L / (Q_3 + Q_L) \qquad Q_L = R_L / \omega_0 L_3$$

For a given set of Q_2, Q_3 and k_{23} values, there is an optimal load, $R_{L,PTE}$, which can maximize the PTE at that particular arrangement, such as the coupling distance, d_{23}.

$$R_{L,PTE} = \omega_0 L_3 Q_{L,\mathrm{PTE}} \qquad Q_{L,PTE} = \frac{Q_3}{(1 + k_{23}^2 Q_2 Q_3)^{1/2}}$$

For a given set of Q_1, Q_2 and k_{12} values, the power **delivered to the load (PDL)** will be maximized if $R_1 = R_{ref}$.

$$Q_{L,PDL} = \frac{Q_3}{1 + k_{23}^2 Q_2 Q_3}$$

M.W. Baker & R. Sarpeshkar, *TBioCAS* 2007

11 Calculating Power Transfer Efficiency [PTE] (II)

Q_{3L} is often referred to as the load quality factor. It can be seen that k_{23}, Q_2, and Q_3 are the key factors that should be increased to maximize the PTE [Ref 1].

By comparing optimal Q_Ls for PTE and PDL, one can conclude that maximum PTE and PDL cannot be achieved simultaneously for a given R_L.

[Ref 1] M. Baker and R. Sarpeshkar, "Feedback analysis and design of RF power links for low-power bionic systems," *IEEE Trans. Biomed. Circuits Syst.*, vol. 1, pp. 28–38, Mar. 2007.

12 Why Printed Spiral Coils?

- PSCs can be fabricated on **flexible** substrates to **conform** to the outer body or brain **surface curvature**.
- PSCs occupy a **small volume** → Planar shape suitable for implantation under the skin or within the epidural space.
- Wire-wound coils cannot be **batch-fabricated** or reduced in size without the use of sophisticated machinery.
- PSCs can be **lithographically defined** in one or more layers on rigid or flexible substrates.
- PSCs offer more flexibility in **optimizing** their **geometries**.
- PSC optimization requires accurate models that consider the **PSC environment**.

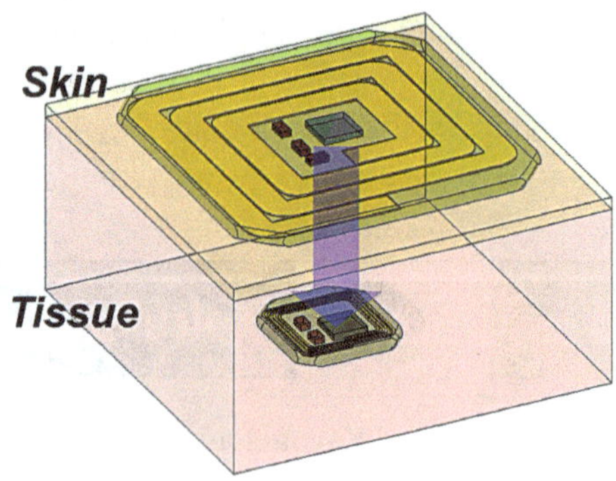

Considering the key factors in determining the PTE, the quality factors of both Tx and Rx coils need to be maximized. This is challenging particularly on the Rx side because it needs to be embedded in the IMD, where the size is extremely constrained. As a result using wire-wound coils (WWC) with high enough wire diameter to achieve low parasitic resistance may not be feasible. This is why for a variety of reasons, planar spiral coils (PSC) are quite popular in IMD applications. But how should we maximize the Q-factor of a PSC?

13 Optimizing PSC Geometries

Iterative procedure to achieve the optimal PSC geometries:

Parameter	Symbol	Design Value
Implanted PSC outer diameter	d_{o2}	10 mm
Minimum PSC inner diameter*	d_{imin}	0 or 8 mm
PSC relative distance	D	10 mm
Link operating frequency	f	13.56 MHz
Secondary nominal loading	R_L	500 Ω
Minimum conductor width	w_{min}	150 μm
Minimum conductor spacing	s_{min}	150 μm
Conductor thickness	t_c	38 μm**
Conductor material properties	ρ, μ_{rc}	~17 nΩm, ~1**
Substrate thickness	t_s	1.5 mm
Substrate dielectric constant	ε_{rs}	4.4 (FR4)

*Depending on whether a chip or magnet is going to be placed in the center of the PSC or not.
**1-oz copper on FR4 printed circuit board.

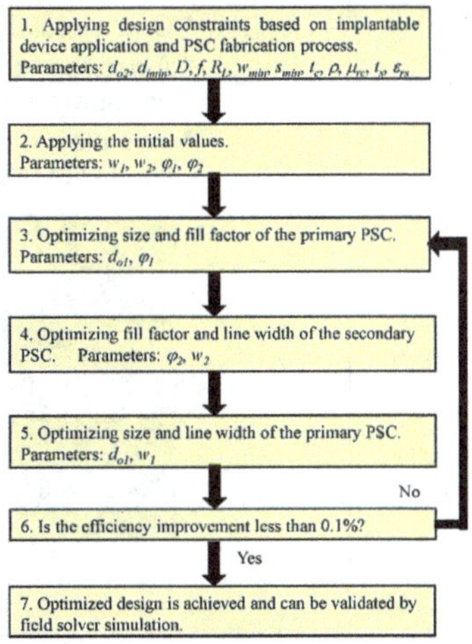

Jow and Ghovanloo, IEEE TBioCAS 2007

We have devised a theoretical model to calculate the inductance, parasitic capacitance, and parasitic resistance of a PSC based on its geometry and conductor characteristics in Ref [1], while considering the coil's surrounding environment. Then we used this model to devise the steps and details of an iterative PSC optimization to achieve the best Q-factor and maximize the PTE.

[Ref 1] U. M. Jow and M. Ghovanloo, "Design and optimization of printed spiral coils for efficient transcutaneous inductive power transmission," *IEEE Trans. Biomed. Circuits Syst.*, vol. 1, pp. 193–202, Sep. 2007.

14 Iterative Procedure for PSC Optimization

Step 1: Applying design constraints based on the implantable device application and PSC fabrication technology.
Step 2: Initial values for PSC geometries.
Step 3: Size and fill-factor of the implanted PSC.
Step 4: Fill-factor and conductor width of external PSC.
Step 5: Size and conductor width of external PSC.
Step 6 & 7: Check whether PCE is maximized, if not go to Step 3.

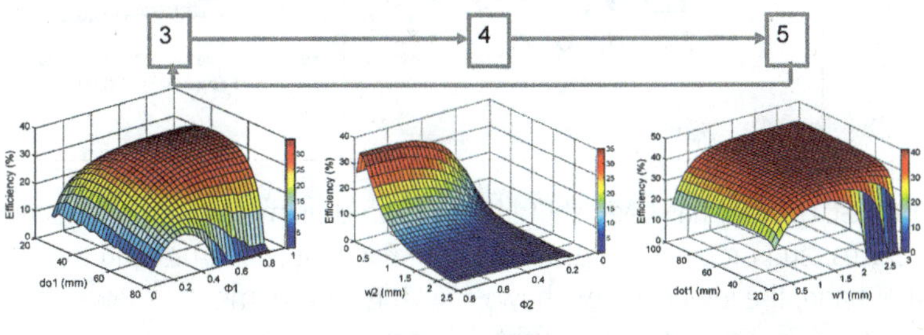

Jow and Ghovanloo, *IEEE TBioCAS* 2007

The iterative procedure starts by applying the design constraints based on the IMD application and PSC fabrication technology and continues until the PTE reaches its maximum level.

15

Measurement Setup

- Commercial PSC fabrication process, coated with silicone.
- A network analyzer (R&S ZVB4) was used → S-parameters → Z-parameters → calculate k and Q → Calculate η
- Two plastic bags (~50 μm thick) hanged from a clamp, and filled with beef to emulate implantation environment.

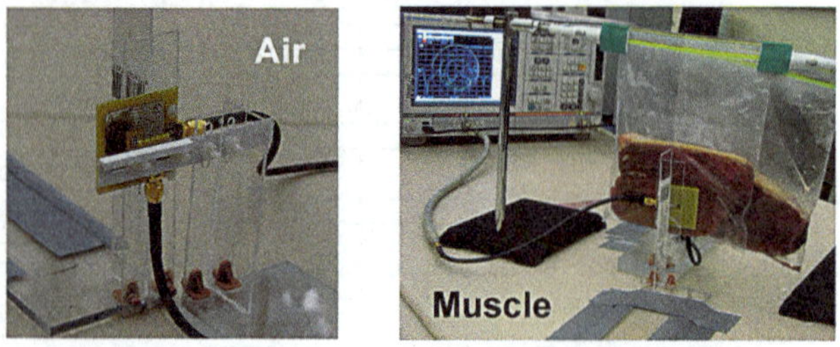

Jow and Ghovanloo, *IEEE TBioCAS* 2009

Results from the PSC optimization procedure are then compared with a finite element model (FEM) in HFSS as well as measurements in three different surrounding environments: 1) air, 2) saline, and 3) muscle tissue [Ref 1].

The implantable Rx PSC was sandwiched between the two bags of saline while the external PSC was aligned with it, touching the outer surface of one of the bags.

[Ref 1] U. Jow and M. Ghovanloo, "Modeling and optimization of printed spiral coils in air, saline, and muscle tissue environments," *IEEE Trans. Biomed. Cir. Syst.*, vol. 3, no. 5, pp. 339-347, Oct. 2009.

16

Calculated, Simulated, and Measured Q

PSC11 (external) optimized for **air**
128 → 28
78% drop!

PSC21 (external) optimized for **muscle**
122 → 92
25% drop

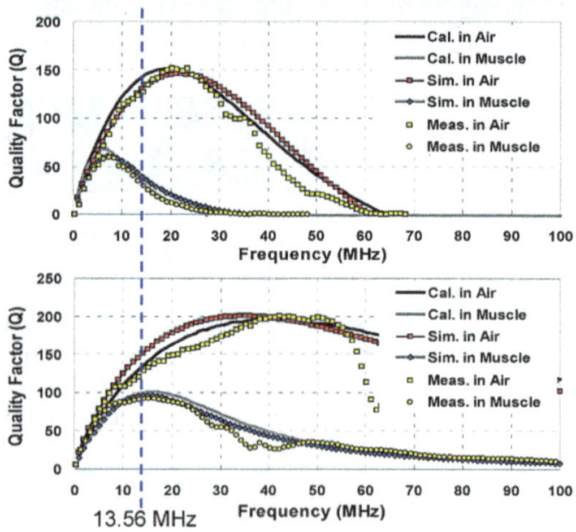

The first observation from this experiment is that the theoretical model, HFSS simulations, and measurement results are well in agreement. The second observation is the huge drop in the Q of the PSC, which geometry has been optimized in the air at the carrier operating frequency of 13.56 MHz from 128 down to only 28. In this design, the peak of the PSC Q-factor has been shifted towards lower frequencies, and drops at the desired frequency. On the other hand, the lower curves show that if the PSC geometry is optimized in the right environment, i.e. tissue surroundings, the peak of the Q occurs exactly at the right frequency, thanks to the coil optimization procedure, Q only drops from 122 to 92, which is more than 3 times higher than the Q of the PSC that is optimized in the air.

This experiment shows the importance of considering the surrounding environment and its effect on the coil parasitic components during the design and optimization process.

17

Calculated, Simulated, and Measured η%

Air Environment
72.2% vs. 55.6%

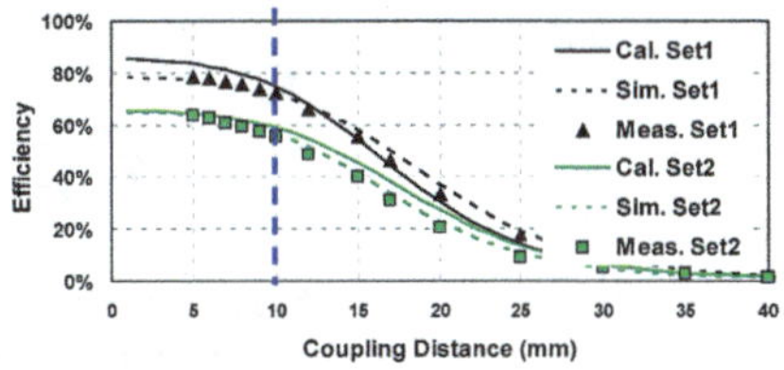

Muscle Environment
21.8% vs. 30.8%

41.3% Improvement!

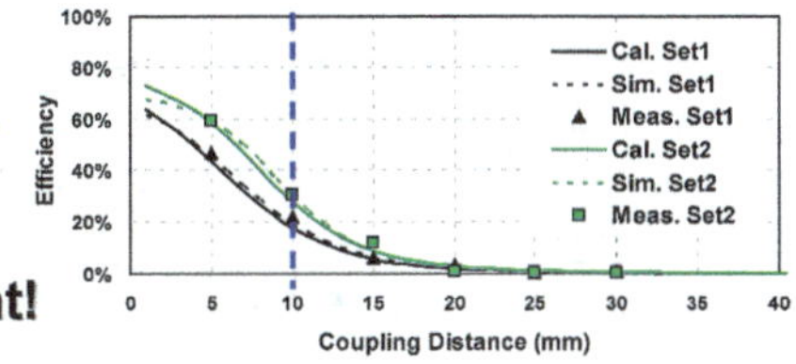

Jow and Ghovanloo, *IEEE TBioCAS* 2009

Similarly, there is very good agreement between the PTE of the inductive link in a pair of optimized PSCs in theoretical calculations, HFSS simulations, and measurements. It can be seen that even though the PSCs that are optimized for operation in the tissue environment achieve lower PTE in air, they provide up to 41.3% higher PTE when they operate in the environment that they are designed for [Ref 1].

[Ref 1] U. Jow and M. Ghovanloo, "Modeling and optimization of printed spiral coils in air, saline, and muscle tissue environments," *IEEE Trans. Biomed. Cir. Syst.*, vol. 3, no. 5, pp. 339-347, Oct. 2009.

18

PTE in 2-Coil Inductive Links

$$\eta_{2-coil} = \frac{k_{23}^2 Q_2 Q_{3L}}{1 + k_{23}^2 Q_2 Q_{3L}} \cdot \frac{Q_{3L}}{Q_L}$$

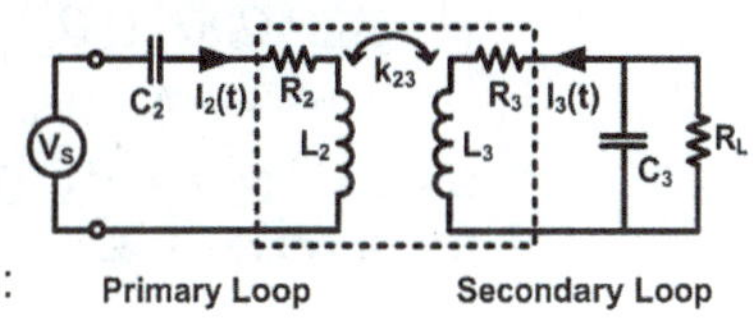

- PTE is highly dependent on: k_{23}, Q_2, Q_3, Q_L (load Q)

$Q_{3L} = Q_3 Q_L / (Q_3 + Q_L)$

$Q_L = R_L / \omega_0 L$

- Large R_L (Q_L): Low efficiency in the secondary loop
- Small R_L (Q_L): Low efficiency in the primary loop

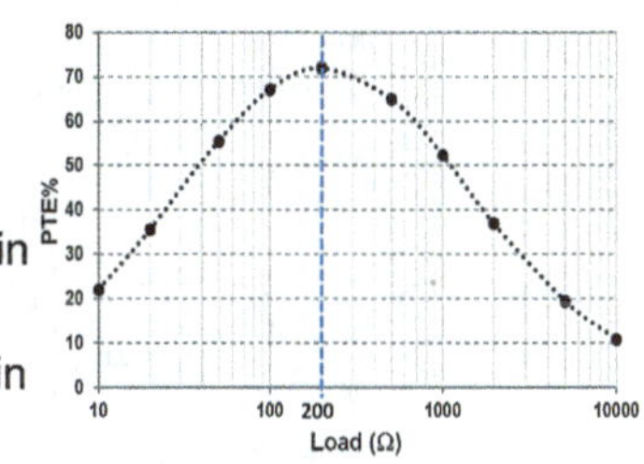

R.R. Harrison, *ISCAS* 2007

So far we learned that in a traditional 2-coil link, coupling coefficient and quality factor of the coils have the highest impact on the PTE. It should be noted, however, that in this coil configuration, Q_2 of the Tx coil (L_2) is affected by the source (often a PA) output impedance, and the Q_3 of the Rx coil is affected by the loading, R_L. As a result, PTE is heavily dependent on R_L also, and there is a particular R_L that would maximize the PTE.

PTE dependence on R_L is undesired. Because in some implantable devices that are mainly stimulating the neural tissue, such as cochlear or retinal implants, depending on the loudness of the incoming sound or brightness of the viewed scene, the loading of the Rx coil varies, and the average PTE could be low [Ref1]. Therefore, it is desired to maintain high PTE for a range of R_L values.

[Ref1] R. Harrison, "Designing efficient inductive power links for implantable devices," *IEEE Int. Symp. Circuits Syst.*, pp. 2080–2083, May 2007.

19

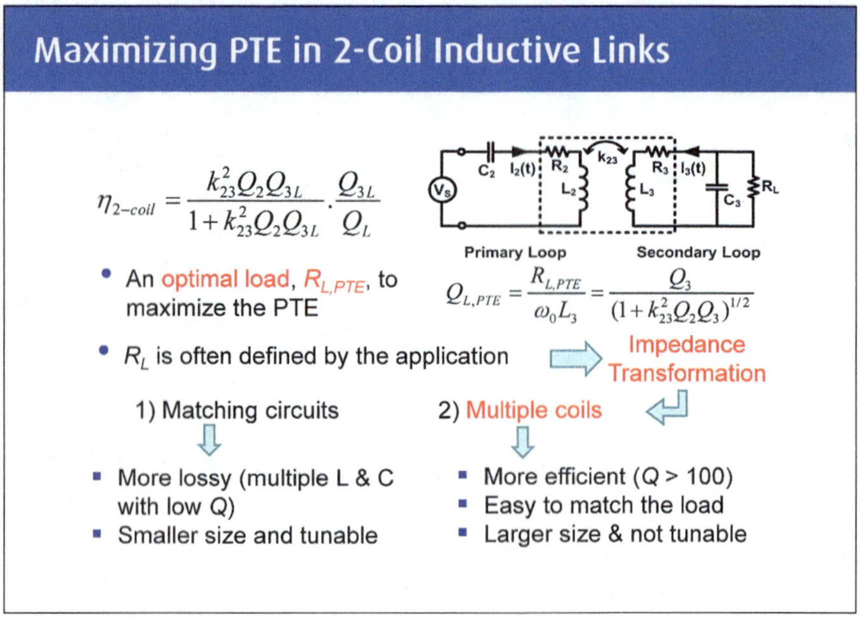

Two possible solutions for the aforementioned problem, aside from designing the inductive link such that the nominal loading would be close to the optimal R_L value, are using matching circuits or multiple inductors for impedance transformation. Each method has its pros and cons. For instance, matching circuit components would be difficult to integrate on chip because of the relatively low operating frequency of the inductive link. Using multiple coils also add to the size of the IMD and the design process could be more complex.

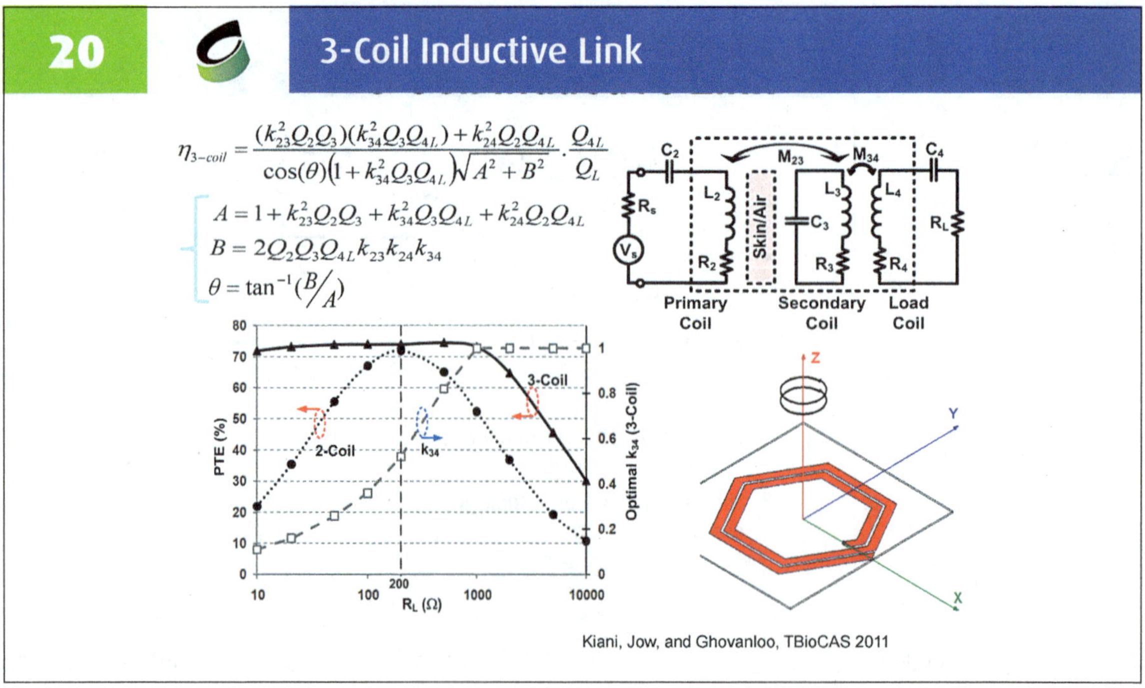

A possible solution for maintaining high PTE for a wide range of R_L values, or transforming the actual R_L to the optimal R_L is adding a resonator, L_3C_3, on the Rx side. Since unlike L_4, this resonator is not loaded by R_L, it can maintain its high Q-factor.

Here we have derived the expression for PTE in a 3-coil link, and it can be seen that similar to 2-coil links, to achieve high PTE, Q of the coils need to be

20 3-Coil Inductive Link

high. However, there is an optimal value for M_{34}, the mutual coupling between the added resonator and the Rx coil, and consequently their coupling coefficient, L_{34}, which maximizes the PTE. As shown in this comparison between 2-coil and 3-coil link PTE, k_{34} can convert almost any value of R_L to the optimally matched R_L, looking through L_3, as long as it is not saturated (k_{34} = 1) [Ref1].

It should be noted that since k_{34} depends on the size and relative positioning of L_3 and L_4, changing k_{34} with R_L once the implant is fabricated is not feasible. Nonetheless, the added resonator give designers more degrees of freedom to convert any nominal R_L to its optimal value.

[Ref1] M. Kiani, U. Jow, and M. Ghovanloo, "Design and optimization of a 3-coil inductive link for efficient wireless power transmission," *IEEE Trans. Biomed. Cir. Syst.*, vol. 5, pp. 579-591, Dec. 2011.

21 PTE and PDL in 3-Coil Link

Maximizing PTE should not be at the cost of decreasing PDL. ☺

$$P_{L,3\text{-}coil} = \frac{V_s^2}{2R_2}\frac{(k_{23}^2Q_2Q_3)(k_{34}^2Q_3Q_{4L})}{(1+k_{23}^2Q_2Q_3+k_{34}^2Q_3Q_{4L})^2}\cdot\frac{Q_{4L}}{Q_L}$$

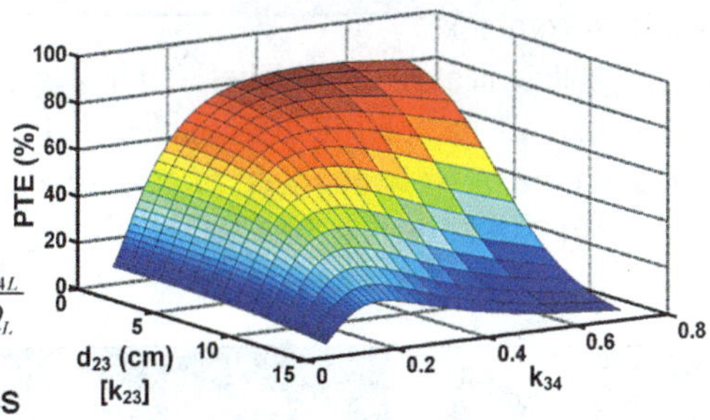

The optimal design maximizes both PTE and PDL.

$$k_{23,PDL} = \left(\frac{1+k_{34}^2Q_3Q_{4L}}{Q_2Q_3}\right)^{1/2}$$

$$k_{34,PDL} = \left(\frac{1+k_{23}^2Q_2Q_3}{Q_3Q_{4L}}\right)^{1/2}$$

☺

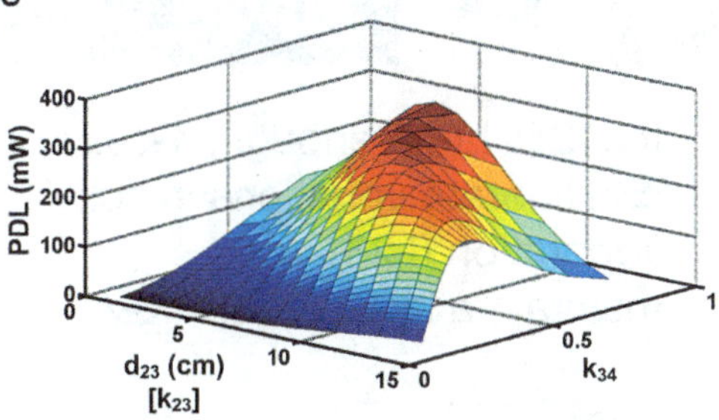

Kiani, Jow, and Ghovanloo, TBioCAS 2011

One should note that when designing inductive links for biomedical or any other applications, in addition to PTE the amount of power delivered to the load (PDL) is also important.

PDL expression for a 3-coil link shows that it is proportional to the source voltage squared, V_s^2, which can be adjusted for a desired PDL. However, high levels of V_s may impose risk or increase the size and cost of components needed in the PA, or reduce its efficiency. Therefore, in the design of an inductive link, when the coupling coefficients between 3 coils are being selected, it is better to move to areas of the 3D surface where both PTE and PDL are high. Fortunately in 3-coil inductive links these regions have considerable overlap [Ref1].

[Ref1] M. Kiani, U. Jow, and M. Ghovanloo, "Design and optimization of a 3-coil inductive link for efficient wireless power transmission," *IEEE Trans. Biomed. Cir. Syst.*, vol. 5, pp. 579-591, Dec. 2011.

22 4-Coil Inductive Link

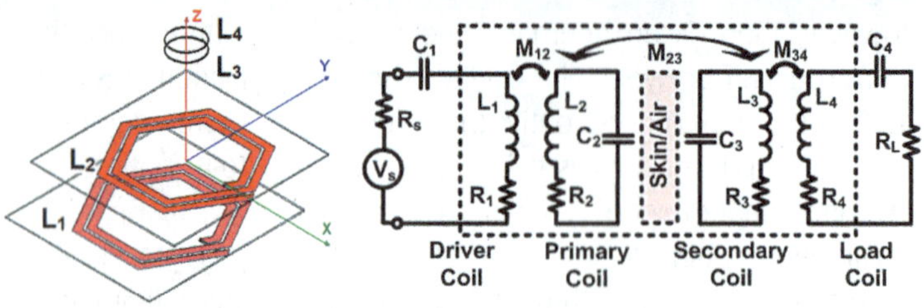

$$\eta_{4\text{-}coil} = \frac{(k_{12}^2Q_1Q_2)(k_{23}^2Q_2Q_3)(k_{34}^2Q_3Q_{4L})}{[(1+k_{12}^2Q_1Q_2).(1+k_{34}^2Q_3Q_{4L})+k_{23}^2Q_2Q_3].[1+k_{23}^2Q_2Q_3+k_{34}^2Q_3Q_{4L}]}\cdot\frac{Q_{4L}}{Q_L}$$

- 4-Coil link adds an additional DoF for impedance matching on the source side.
- If k_{12} is large, the reflected load onto L_1 increases dramatically, which helps maximize the PTE at the cost of reducing PDL.

Kiani and Ghovanloo, TCAS-I 2012

It is possible to add an additional resonator, L_2C_2, to the Tx side, and create a 4-coil inductive link, in which all resonators are tuned at the carrier frequency. This resonator adds yet another DoF for impedance matching on the source side. The complete expression for the PTE in a 4-coil link is presented here and it is rather complex. However, it still indicates that higher Q for every resonator results in higher PTE. The values of the k_{34} and k_{12}, on the other hand, need to be carefully optimized for the best possible PTE [Ref1].

[Ref1] M. Kiani and M. Ghovanloo, "The circuit theory behind coupled-mode magnetic resonance based wireless power transmission," *IEEE Trans. Cir. Syst.-I*, vol. 59, Sept. 2012.

23 PTE and PDL in 4-Coil Link

If k12 is large enough, 4-coil can tolerate variations in coil separation (k23) and maintain a large PTE. ☺

$$P_{L,4\text{-}coil} = \frac{V_s^2}{2R_1}\times \frac{(k_{12}^2Q_1Q_2)(k_{23}^2Q_2Q_3)(k_{34}^2Q_3Q_{4L})}{[(1+k_{12}^2Q_1Q_2).(1+k_{34}^2Q_3Q_{4L})+k_{23}^2Q_2Q_3]^2}\cdot\frac{Q_{4L}}{Q_L}$$

$$k_{23,PTE} = \left(\frac{\sqrt{1+k_{12}^2Q_1Q_2.(1+k_{34}^2Q_3Q_{4L})}}{Q_2Q_3}\right)^{1/2}$$

$$k_{23,PDL} = \left(\frac{(1+k_{12}^2Q_1Q_2).(1+k_{34}^2Q_3Q_{4L})}{Q_2Q_3}\right)^{1/2}$$

Small overlap between high PTE and PDL areas. ☹

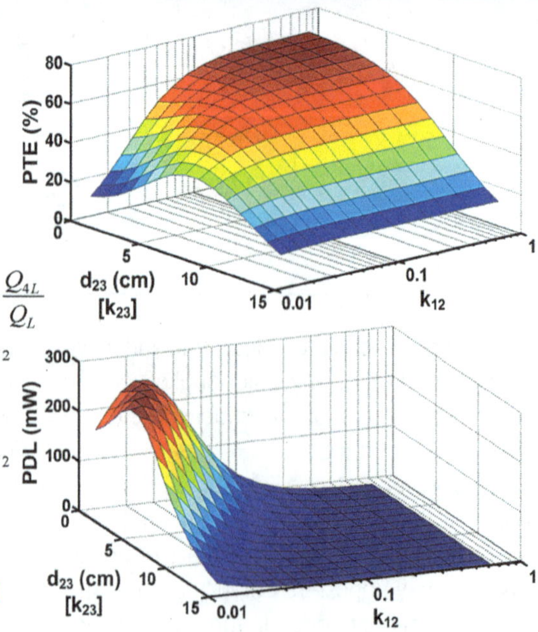

Kiani and Ghovanloo, TCAS-I 2012

23 PTE and PDL in 4-Coil Link

Similar to 2-coil and 3-coil inductive links, here we look at the effect of changing coil couplings on the PTE and PDL of the 4-coil link, particularly the impact of the new resonator and its couplings with the adjacent coils, k_{12} and k_{23}. [Ref1]

PDL expression for a 4-coil link shows that it is also proportional to the source voltage squared, V_s^2, which can be adjusted for a desired PDL. However, high levels of V_s would be undesired. Looking at the areas of the 3D surfaces, where both PTE and PDL are high, unfortunately in 4-coil inductive links these regions do not have considerable overlap [Ref2]. As a result, the designer should select between high PTE or high PDL, depending on the application.

[Ref1] M. Kiani, U. Jow, and M. Ghovanloo, "Design and optimization of a 3-coil inductive link for efficient wireless power transmission," *IEEE Trans. Biomed. Cir. Syst.*, vol. 5, pp. 579-591, Dec. 2011.

[Ref2] M. Kiani and M. Ghovanloo, "The circuit theory behind coupled-mode magnetic resonance based wireless power transmission," *IEEE Trans. Cir. Syst.-I*, vol. 59, Sept. 2012.

24 3-Coil vs. 4-Coil Link Measurements

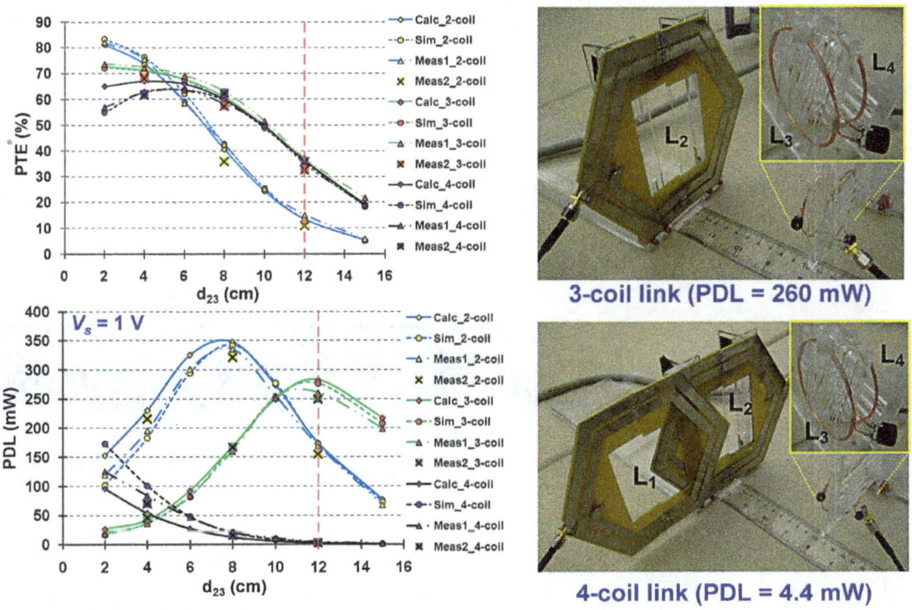

Kiani, Jow, and Ghovanloo, TBioCAS 2011

A detailed comparison between 2-coil, 3-coil, and 4-coil links, including theoretical analysis, as well as simulation and measurement results can be found in [Ref1]. This comparison clearly shows the pros and cons of each approach at different coil separations and power levels.

[Ref1] M. Kiani, U. Jow, and M. Ghovanloo, "Design and optimization of a 3-coil inductive link for efficient wireless power transmission," *IEEE Trans. Biomed. Cir. Syst.*, vol. 5, pp. 579-591, Dec. 2011.

25

New Figure of Merit (FoM) for Inductive Power Transmission Links

- Trade-offs between high **PTE** and sufficient **PDL**
- High PTE to reduce:
  - Reduce heat dissipation
  - Tissue exposure to magnetic field (safety)
  - Interference with nearby electronics (FCC)
- Sufficient PDL to:
  - Small PA transistors
  - Reducing Vs for safety and cost reduction

$$FoM = \frac{\eta_{m-coil}^{n} \times P_{L,m-coil}}{V_s^2}$$

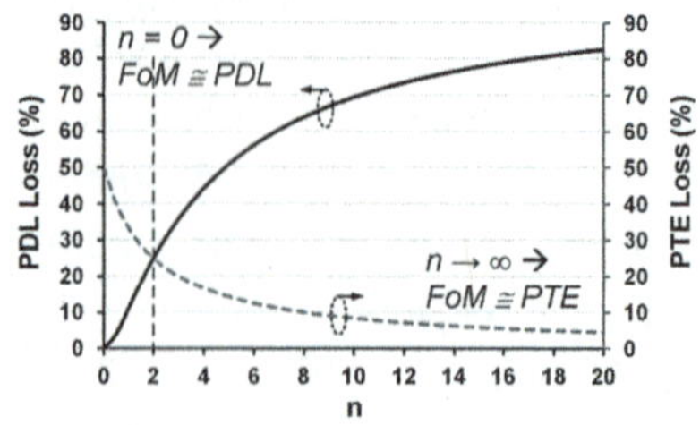

Kiani and Ghovanloo, *TIE* 2013

In order to choose the right configuration among 2-, 3-, and 4-coil links, and optimize a design, one should consider the trade-offs between high PTE and sufficiently high PDL based on the specific needs of any particular WPT application from ultra low power implantable microelectronic devices (IMD) to high power electric vehicles (EV). This will impact many practical aspects, such as safety, size, and voltage/current tolerance of various components [Ref1].

In order to facilitate this choice and clarify the design and optimization procedure, we have proposed a figure-of-merit (FoM) that include the key parameters in WPT, allowing the designer to adjust the importance of PTE vs. PDL by setting a factor, n. If n = 0, all the emphasis will be on the PDL. Selecting n = 2 gives equal weights to PDL and PTE. Any larger value of n > 2 gives PTE more emphasis over PDL to the extent that n= ∞ only considers PTE in the inductive link optimization.

[Ref1] M. Kiani and M. Ghovanloo, "A figure-of-merit for designing high performance inductive power transmission links," *IEEE Trans. Indus. Elect.*, vol. 60, pp. 5292-5305, Nov. 2013.

26

Optimal Multi-coil Link for Charging Handheld Mobile Devices based on FoM

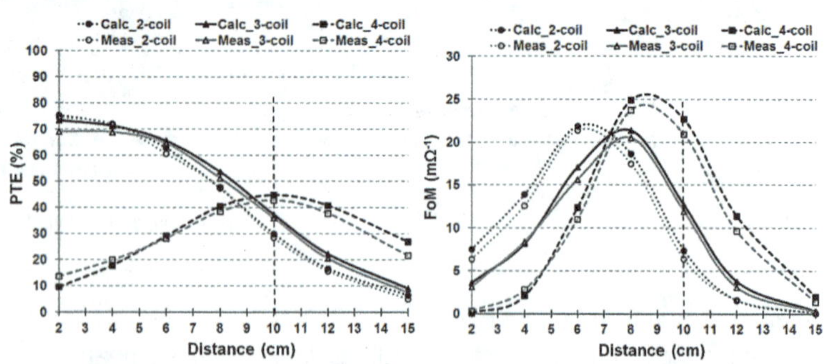

The 4-coil link has superior FoM at d_{23} = 10 cm at the cost of much lower PTE, and consequently the FoM, at shorter coupling distances.

Rx coil dia. = 4 cm, R_s = 0.5 Ω, R_L = 5 Ω, f_0 = 13.56 MHz, d_{23} = 10 cm

Kiani and Ghovanloo, *TIE* 2013

It can be seen that in a set of 2-, 3-, and 4-coil links that are all optimized based on FoM, the 4-coil link offers the highest PTE at longer coil separations in the order of > 9cm. However, at shorter distances, it cannot offer high PTE, and falls well below 2- and 3-coil links.

27 Optimal Multi-Coil Link Based on FoM

	2-Coil Link	3-Coil Link	4-Coil Link
Strong coupling (k)	☺	☺	☹
Weak coupling (k)	☹	☺	☺
Large PDL (small R_s)	☺	☺	☹
Small PDL (large R_s)	☹☹	☹	☺
Coupling variations & small R_s	☺	☺☺	☹
Coupling variations & large R_s	☹	☺	☺☺
Size constraints	☺	☹	☹☹

Kiani and Ghovanloo, *TIE* 2013

This table summarizes the pros and cons of multi-coil links from various practical aspects, particularly those that are defined by the application or fabrication process.

28 3-Coil Inductive Links for Load Matching

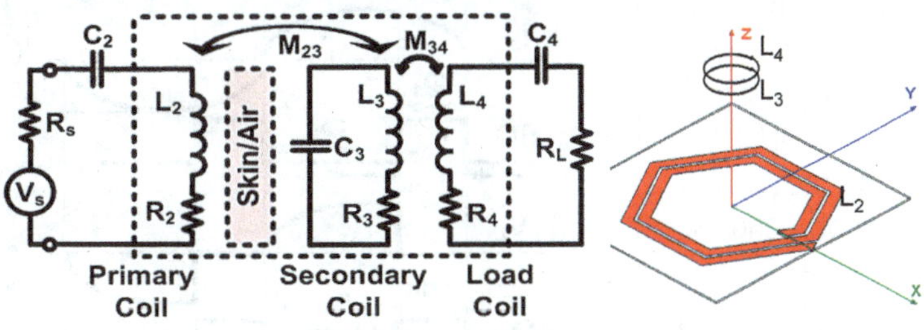

$$\eta_{3\text{-}coil} = \eta_{23}\eta_{34} = \frac{(k_{23}^2 Q_2 Q_3)(k_{34}^2 Q_3 Q_{4L})}{[(1+k_{23}^2 Q_2 Q_3 + k_{34}^2 Q_3 Q_{4L})(1+k_{34}^2 Q_3 Q_{4L})]} \cdot \frac{Q_{4L}}{Q_L}$$

Problem: k_{23} and R_L change but k_{34} is not adjustable during operation!

Earlier we mentioned that one of the benefits of 3-coil and 4-coil links is allowing impedance matching at the load and source by adjusting the mutual coupling between L_3 and L_4, M_{34}, and L_1 and L_2, M_{12}, respectively. While these parameters can be carefully adjusted during the design procedure for nominal values of R_L and M_{23}, the mutual coupling between the Tx and Rx resonators, they cannot be adjusted during implant operation, when both R_L and M_{23} are likely to change, when, for instance the stimulus current or the distance or alignment between the Tx and Rx coils change, resulting in the original design parameters to be sub-optimal.

29

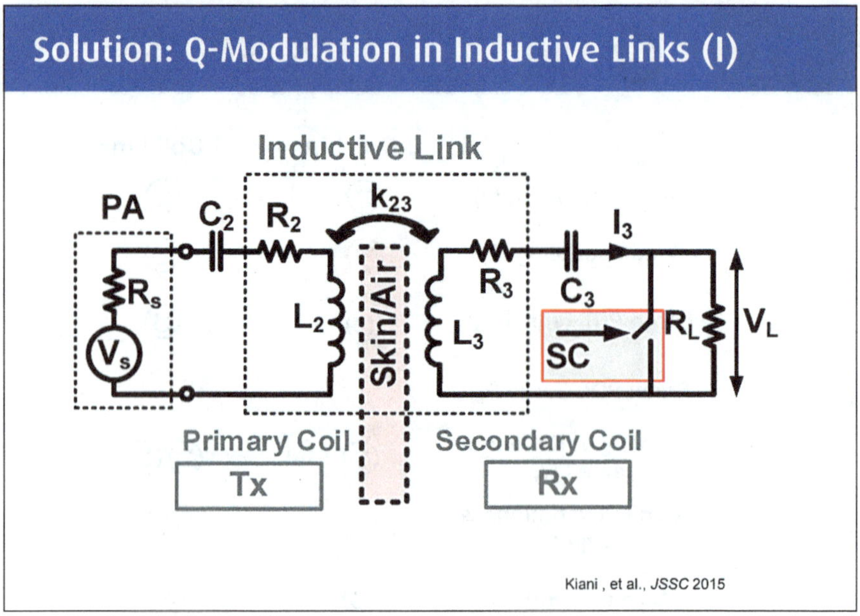

To address this problem, we have proposed adding a switch, SC, across R_L in order to change the effective value of R_L by adjusting the duty cycle of SC. Unlike M_{34} and M_{12}, which cannot be modified on-the-fly during IMD operation, duty cycle of the SC can be adjusted electronically in a closed-loop to continuously ensure the optimal effective R_L is reflected onto the primary side, and consequently the highest possible PTE is achieved [Ref1].

[Ref1] M. Kiani, B. Lee, P. Yeon, and M. Ghovanloo, "A Q-modulation technique for efficient inductive power transmission," IEEE Journal of Solid-State Circuits, vol. 50, no. 12, pp. 2839 - 2848, Dec. 2015.

30

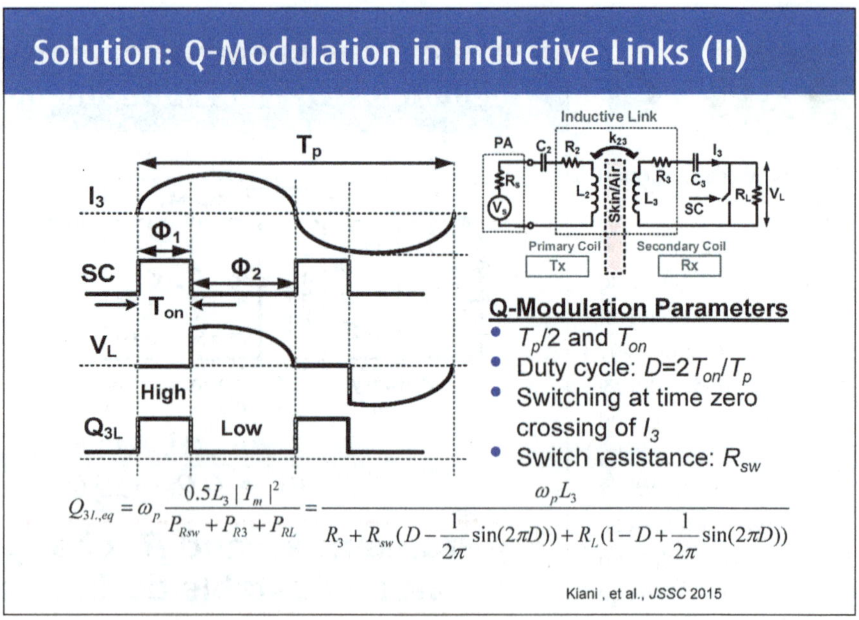

It is possible to synchronize the SC operation with the power carrier signal. In this case, if T_p is considered the power carrier period, and SC is closed for a duration of T_{on} at the beginning of each half cycle, then duty cycle, D, can be defined as $2T_{on}/T_p$. If the ON resistance of the SC is considered to be R_{SW}, then the effective loaded quality factor of the Rx LC tank, $Q_{3L,eq}$ can be found as a function of D and R_L, and adjusted to be equal to the optimal Q_{3L}, which maximizes the PTE. This is why we have called this method, "Q-modulation."

31 Dynamic Load Matching in Q-Modulation

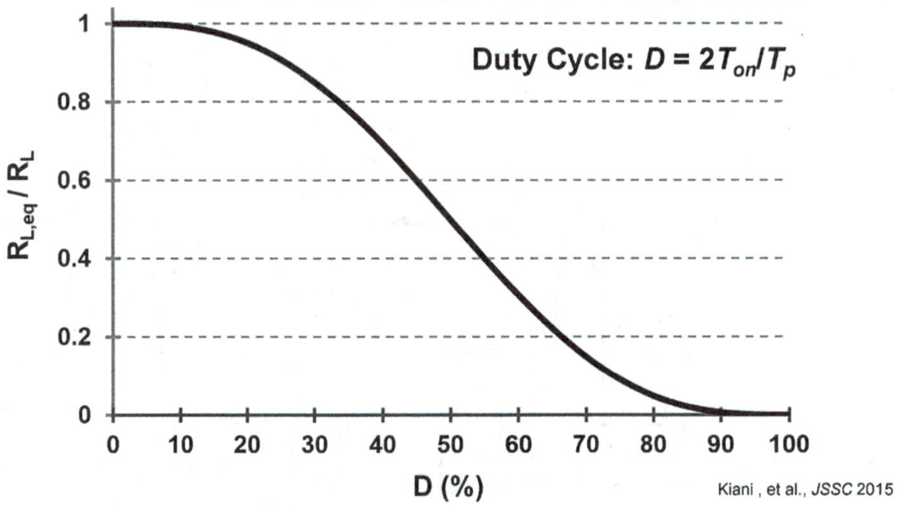

Kiani , et al., *JSSC* 2015

Here is how $R_{L,eq}$, the effective value of R_L, changes with D according to the equation in the previous slide [Ref1].

[Ref1] M. Kiani, B. Lee, P. Yeon, and M. Ghovanloo, "A Q-modulation technique for efficient inductive power transmission," IEEE Journal of Solid-State Circuits, vol. 50, no. 12, pp. 2839 - 2848, Dec. 2015.

32 PTE Comparison between 3-Coil vs.Q-Modulation Inductive Links

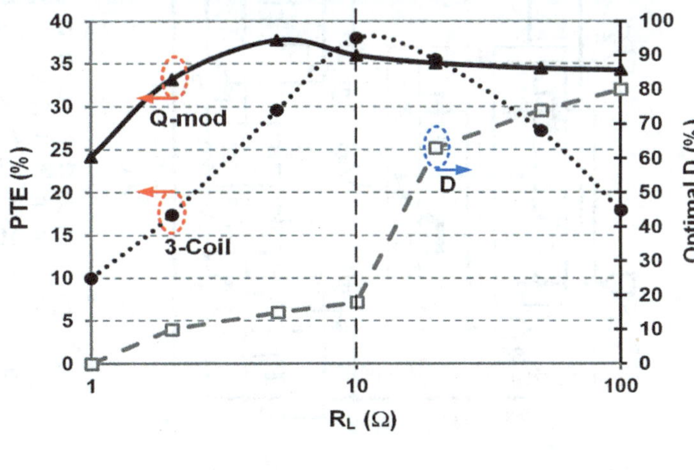

Kiani , et al., *JSSC* 2015

Here we have compared the PTE of a 3-coil inductive link with that of a 2-coil link, equipped with Q-modulation switch, when R_L varies over a wide range by two orders of magnitude, from 1 Ω to 100 Ω.

It can be seen that the 3-coil link, which is optimized for R_L = 10 Ω, reaches PTE > 35% at the optimal R_L value, however, shows much lower PTE when R_L is not optimal and k_{34} is not adjusted for the highest PTE. On the other hand, the 2-coil link has been able to maintain its high PTE > 25% across the wide range of R_L variations by changing D from 0% (no modulation) for very low R_L to 80% for high R_L. In this experiment D has been adjusted manually.

33 Q-Modulation Power Management (QMPM) Block Diagram

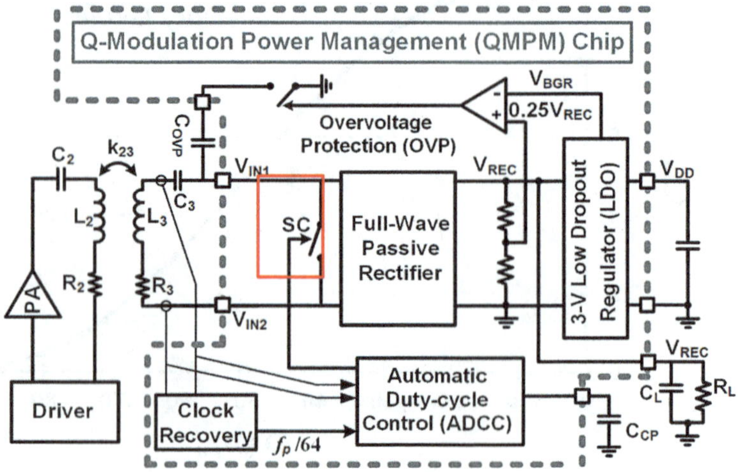

Kiani , et al., *JSSC* 2015

It is possible to adjust D automatically in a closed-loop based on any measurable parameter of interest in the inductive link. Here the block diagram of a 2-coil inductive link is shown, in which the SC duty cycle is changed based on the power carrier signal amplitude, V_{IN1} and V_{IN2}. The Automatic Duty-Cycle Control (ADCC) block in this case ensures to keep V_{IN1} and V_{IN2} at the highest possible level, corresponding to the highest possible PDL, by dynamically changing D.

34 Full-WavePassive Rectifier

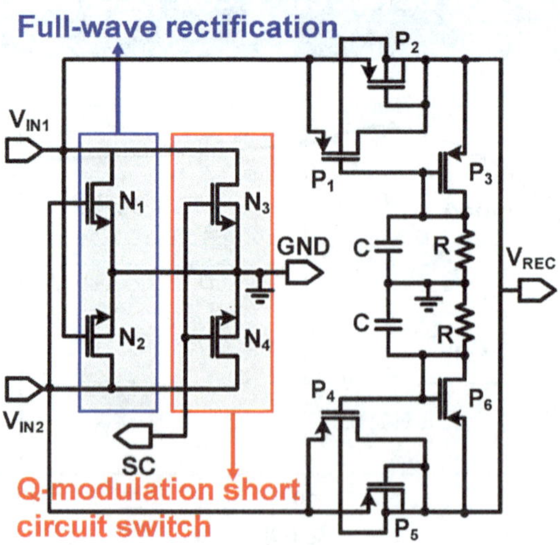

- **Device Sizing:**
  $N_{1\text{-}4}$: W = 38.4 mm
  $P_{1,3,4,6}$: W = 51.2 mm
  $P_{2,5}$: W = 6.4 mm
  L = 0.6 µm for all transistors
  R = 1 MΩ
  C = 10 pF
- **f_P = 2 MHz**
- **P_{out} = 1.45 W**

Kiani , et al., *JSSC* 2015

Schematic diagram of the full-wave rectifier used in the previous slide circuit is shown here, including Q-modulation switch, which consists of transistors N_3 and N_4. Here P_1-P_6 are diode-connected PMOS transistors with threshold voltage adjustment, and N_1 and N_2 are diode-connected NMOS transistors that provide the return current path in the full-wave rectifier.

35 Automatic Duty Cycle Control (ADCC)

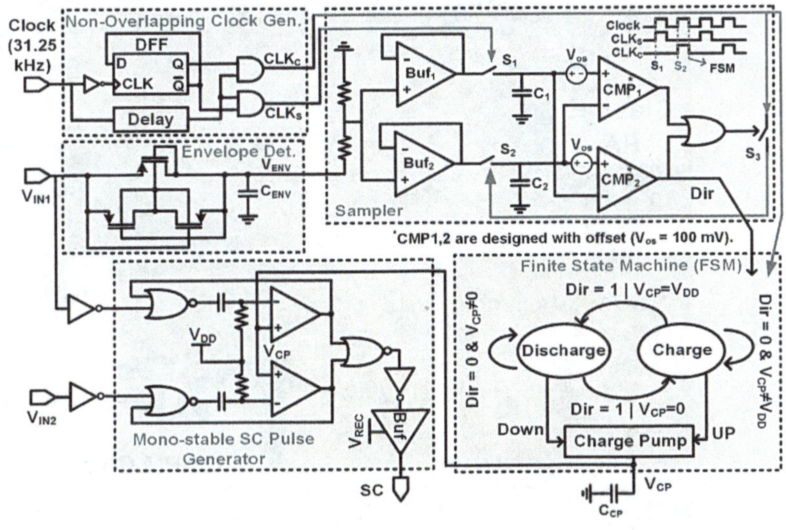

Kiani , et al., *JSSC* 2015

A more detailed schematic diagram of the ADCC block is shown here. V_{IN1} is passed through an envelope detector and sampled with a pair of non-overlapping clocks. Then comparators with 100 mV offset indicate whether V_{IN1} is increasing or decreasing by changing the direction flag, Dir. Dir is fed into a finite state machine (FSM), which changes a control voltage, V_{CP}, which in turn controls the duty cycle of the SC switch. V_{CP} is fed into a mono-stable circuit that is synchronized with the carrier signal, and generates SC pulses at the beginning of each carrier half-cycle.

36 Q-Modulation Power Management ASIC

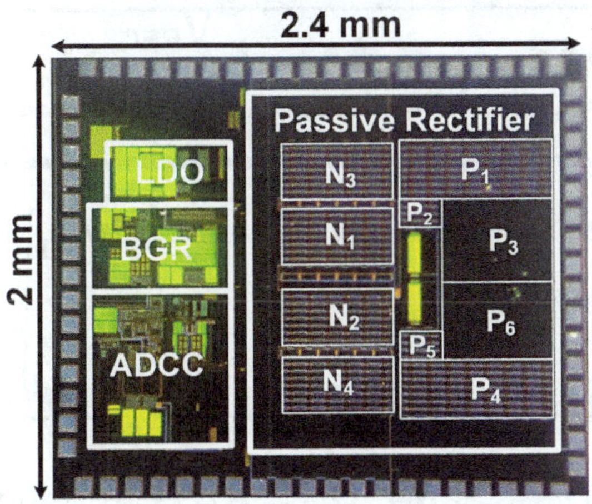

- Process: TSMC 0.35 µm
- f_P = 2 MHz
- ADCC Freq. = 31.25 kHz
- *D* Range: 10% – 70%
- P_{out} = 1.45 W

Kiani , et al., *JSSC* 2015

To evaluate the functionality of the Q-modulation technique, a Q-modulation based power management (QMPM) ASIC including a full-wave rectifier, regulator, ADCC, and Q-modulation circuitry has been fabricated in the TSMC 0.35-mm process. The carrier frequency is chosen at 2 MHz, nominal loading is R_L = 15 Ω, and the ADCC checks V_{IN1} envelope at 31.25 kHz.

37 Measurement Setup

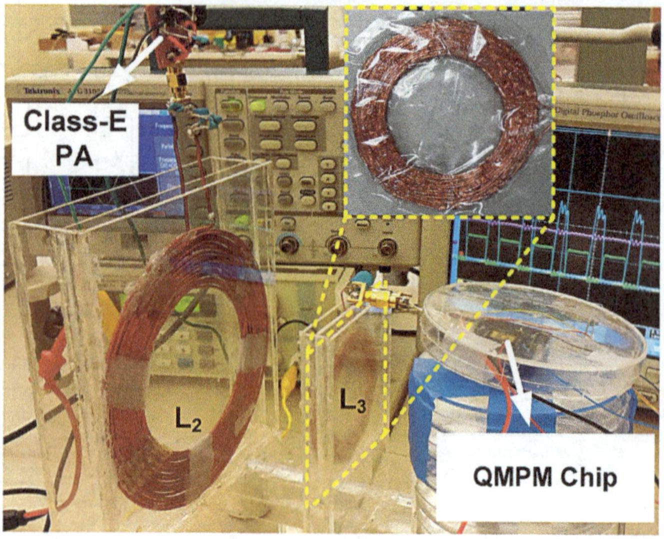

- L_2 Dia. = 14 cm
- L_2 = 40 μH
- L_3 Dia. = 6.5 cm
- L_3 = 20 μH
- C_L = 4 μF
- f_P = 2 MHz

The measurement setup consists of a 2-coil link. On the Tx side, L_2 is driven by a class-E power amplifier at a constant amplitude. On the Rx side, QMPM follows L_3, and delivers the rectified and regulated power to R_L in parallel with a 4 mF capacitor.

38 Rectifier Measurement Results without Q-Modulation

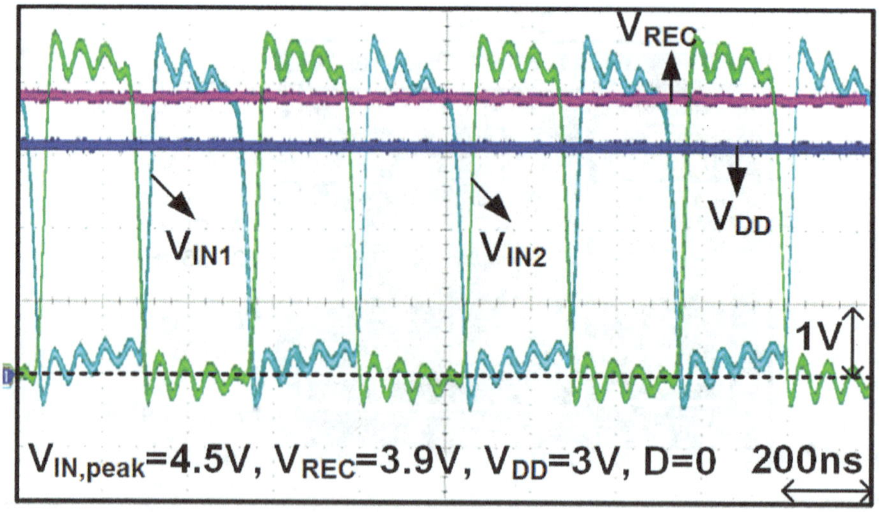

Without Q-modulation (D = 0), an input carrier signal with 4.5 V peak generates V_{REC} = 3.9 V.

39 Rectifier Measurement Results with Q-Modulation

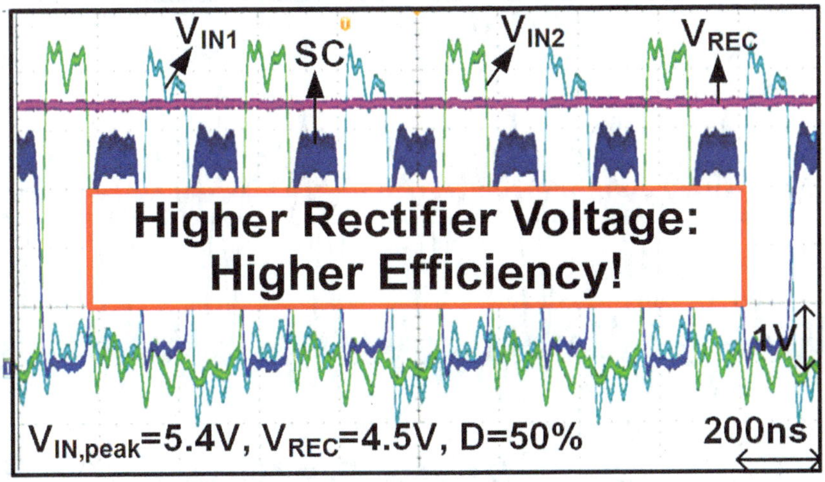

Without changing the transmitter output power/amplitude, when Q-modulation is activated with D = 50% duty cycle, it results in the carrier signal to peak at 5.4 V on the Rx side and generate V_{REC} = 4.5 V. This shows that the efficiency of the inductive link has increased considerably and both PTE and PDL have gone up compared to D = 0 in the last slide.

40 Measurement Results with Q-Modulation (I)

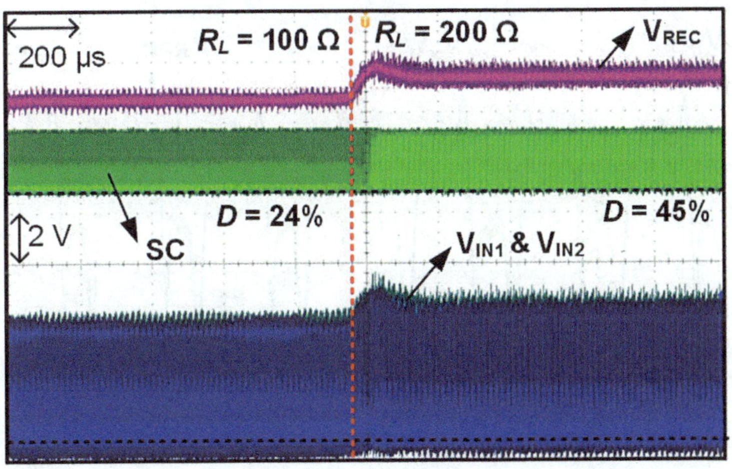

Kiani , et al., *JSSC* 2015

QMPM can also react to changes in R_L automatically by changing D in a way that V_{IN1} and V_{IN2} are maximized. In the following two slides, the changes in the key parameters and switching waveforms are demonstrated for a sudden change from R_L = 100 Ω to R_L = 200 Ω.

41 Measurement Results with Q-Modulation (II)

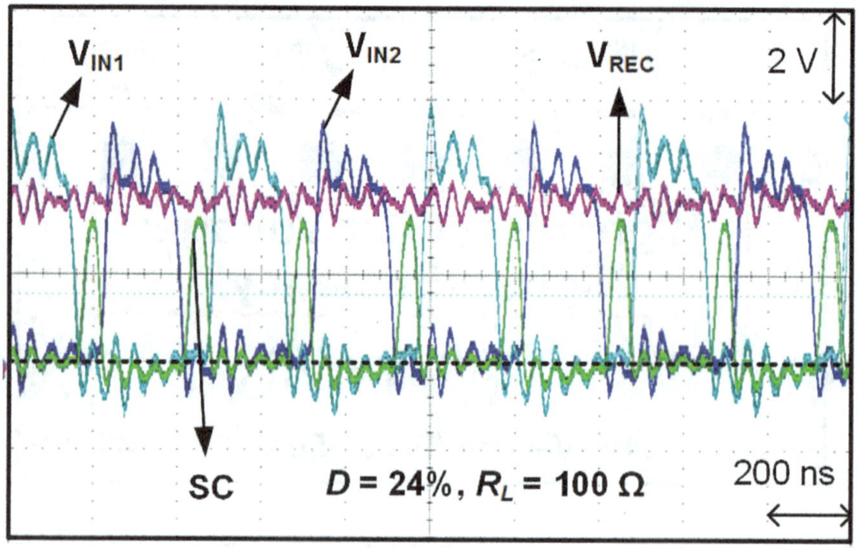

The effective R_L is reached with D = 24% for the actual R_L = 100 Ω.

42 Measurement Results with Q-Modulation (III)

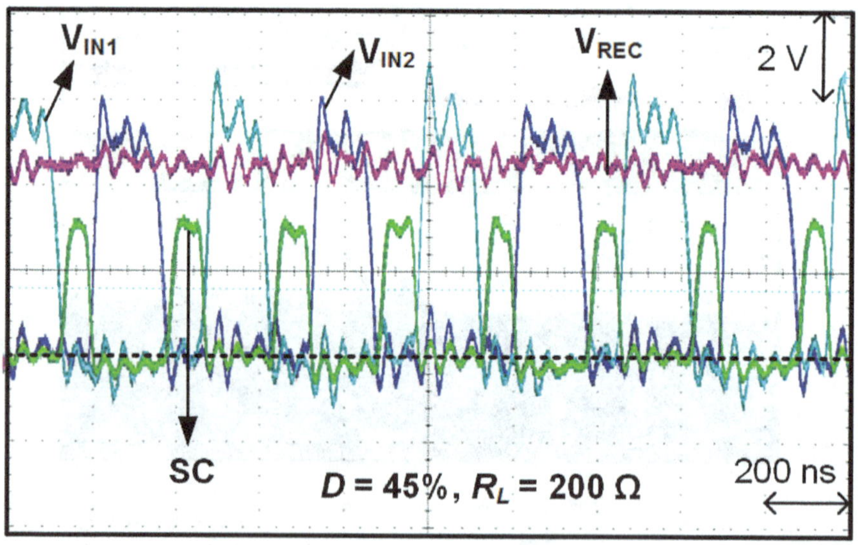

The effective R_L is reached with D = 45% for the actual R_L = 200 Ω.

43 Load Variation with Q-Mod.

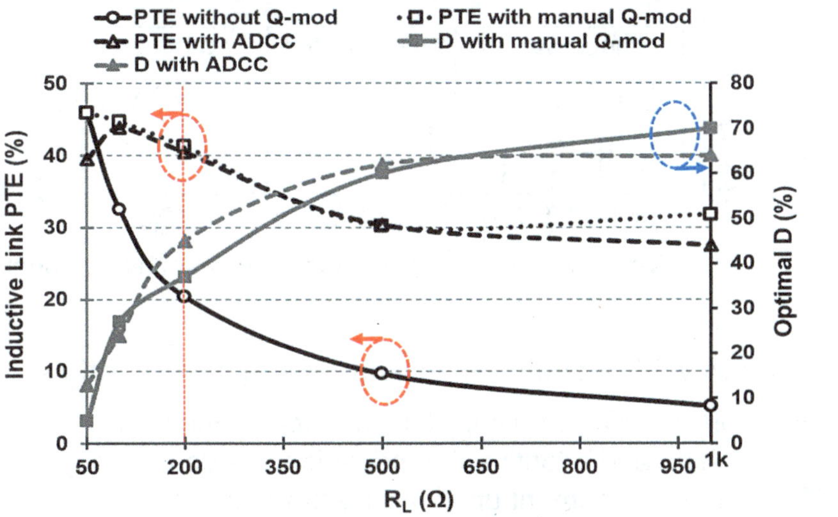

By changing R_L in a wide range from 50 Ω to 1 kΩ, and measuring the inductive link PTE, functionality of the Q-modulation method and the QMPM prototype ASIC have been verified. It can be seen that without Q-modulation, PTE rapidly degrades by increasing R_L from more than 40% down to less than 10%. By activating Q-modulation, high PTE above 30% has been maintained across the wide range of R_L values.

44 Multi-Cycle Q-Modulation

It should be noted that SC switching in Q-modulation does not necessarily need to happen at every carrier half cycle. The switching can be at a much lower rate and include multiple carrier cycles, during which time SC is short and others during which SC is open, as explained in [Ref1].

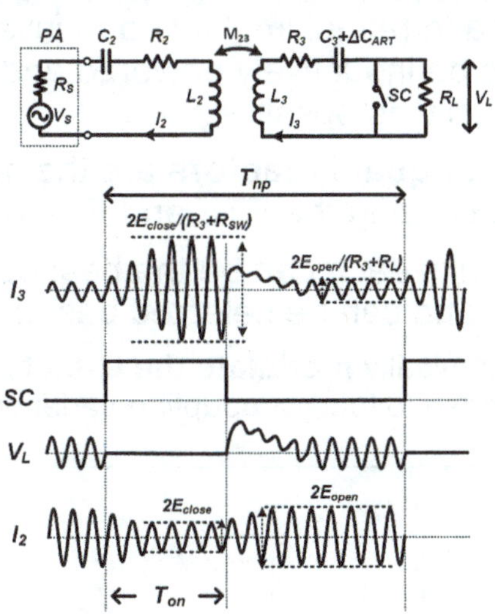

$$\eta_{2-coil} = \frac{k_{23}^2 Q_2 Q_{3L}}{1 + k_{23}^2 Q_2 Q_{3L}} \cdot \frac{Q_{3L}}{Q_L}$$

- Modulate Q_{3L} by multiple periods of carrier frequency (T_{np}, T_{on}).
- Modeling & Theoretical Analysis to achieve highest PTE.

Lee , et al., *IEEE TIE* 2016

[Ref1] B. Lee, P. Yeon, and M. Ghovanloo, "A multi-cycle Q-modulation for dynamic optimization of inductive links," *IEEE Transactions on Industrial Electronics*, vol. 63, no. 8, pp. 5091 - 5100, Aug. 2016.

45 Wirelessly-Powered IMD Structure

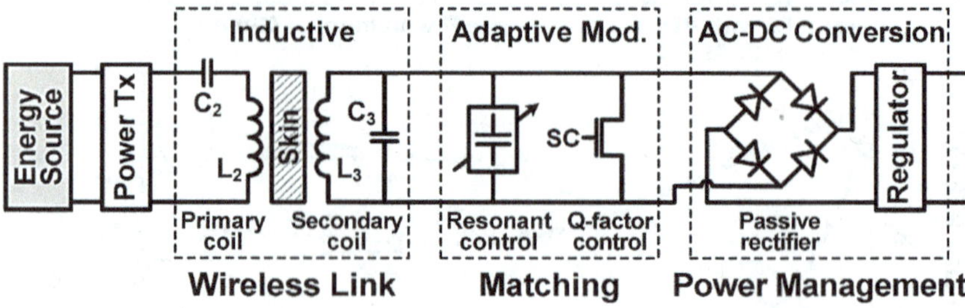

Various WPT techniques in every stage of power flow:

- Wireless power transmission link (across the skin)
- Matching and adaptive Q-modulation (inside the body)
- Power management units (inside the body)

To conclude, here we present a chain of power management blocks before and after the inductive link, which are needed to maintain high PTE in the presence of coil misalignments and load variations all the way from an external energy source, such as the battery in the behind-the-ear (BTE) unit of a cochlear implant, all the way to the implantable device (IMD).

46 Conclusions

- High performance **implantable microelectronic devices** (IMD), such as cochlear and retinal implants and invasive brain-computer interfaces need to be inductively powered and communicated as efficiently and reliably as possible.
- Coils' **coupling and quality factors** are the most important parameters in maximizing the PTE and PDL of an inductive link.
- 2-, 3-, and 4-coil links each can be the best options depending on the design specs, and can be selected based on a FoM.
- It is possible to dynamically **modulate the Q-factor** of the Rx coil by switching it in response to load or coupling variations.

CHAPTER 05

Artificial Intelligence Processors for Biomedical Circuits and Systems

Kwonjoon Lee and Hoi-Jun Yoo

KAIST, Korea

In this chapter, we will discuss the topic of AI processors in the bio-medical circuit and systems with four subsections. Firstly, we will clearly define the definition and scope of artificial intelligence (AI), machine learning, and deep learning with a detailed explanation of basic deep learning operations. Secondly, we will introduce various bio-medical circuit and systems applying (deep) neural network techniques. Thirdly, we will cover some case studies for hardware implementation of deep learning processor. Lastly, as a future direction of the AI (deep learning) processor, we will introduce a neuromorphic computing with emerging nonvolatile memory (eNVM), which shows a promising potential for more efficient deep learning operation than traditional Von-Neumann architecture-based deep learning processor.

1 Outline

- ❑ **Introduction :**
  Artificial Intelligence, Machine Learning, and Deep Learning
- ❑ **(Deep) Neural Network Applications for Bio-medical Circuit and Systems**
- ❑ **H/W Implementation of Deep Learning Inference Processors :** 1) DNPU, 2) UNPU
- ❑ **Future Direction :**
  Neuromorphic Computing with Nonvolatile Memory

2 Introduction

❑ AI, Machine Learning, and Deep Learning

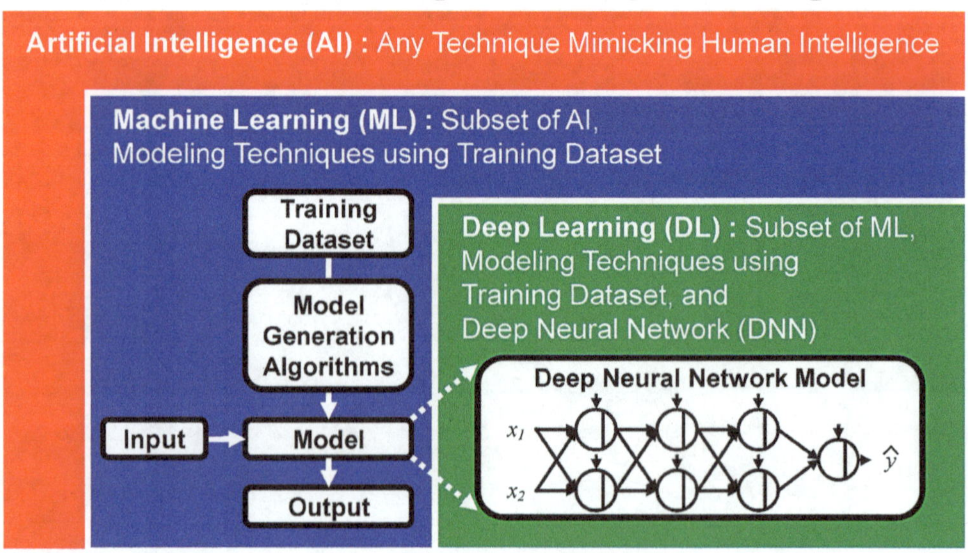

2 Introduction

Recently, deep learning-based techniques have achieved impressive results in various applications such as image/face recognition, while showing higher accuracy compared to the conventional techniques. With the increased interests on deep learning, there are some misuse and abuse of the terms of AI, machine learning, and deep learning without clear definition. However, It is important to be aware of the difference between AI, machine learning, and deep learning.

Firstly, AI is a general term for all kinds of techniques mimicking human intelligence, thus including machine learning and deep learning. Machine learning is a modeling technique based on the training dataset which consists of input data or input data & labeled answer in order to represent real life system with all kinds of mathematical model. The name of machine learning is derived from its spontaneous operation finding a proper model based on the training dataset without human intervention. Especially, this spontaneous operation to find a proper model is called by 'Training' or "Learning". Lastly, deep learning, the subset of machine learning is also a modeling technique based on the training dataset in order to represent real life system with a deep neural network model.

3 Machine Learning

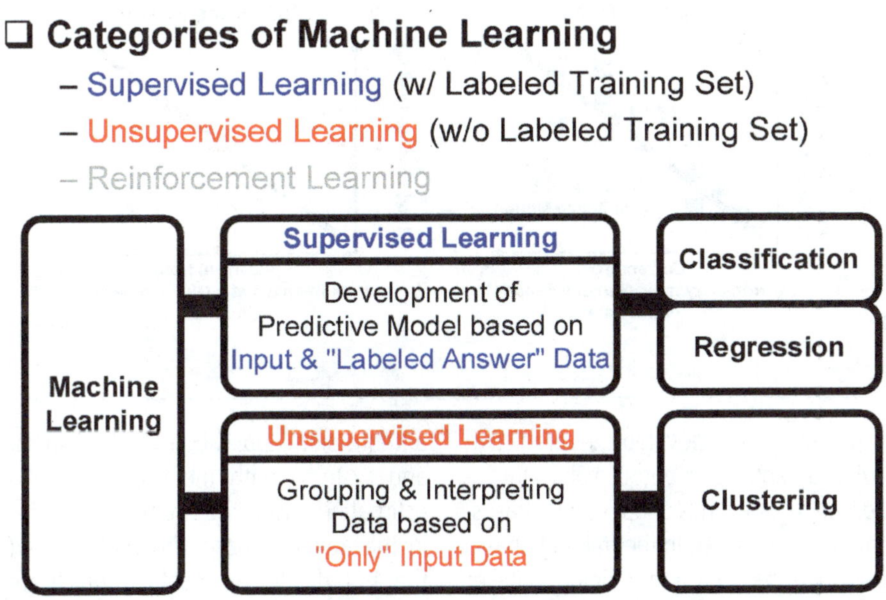

The range of machine learning is extensive. Machine learning can be categorized into sub-categories which include supervised learning, unsupervised learning, and reinforcement learning. In this tutorial, we will not cover the reinforcement learning due to its unique but inappropriate characteristics for biomedical circuit and systems.

Supervised learning is a machine learning technique to develop a model based on training dataset, which contains input data and labeled

3 Machine Learning

answer. Supervised learning can be categorized into regression application and classification application. Unsupervised learning is a machine learning technique to develop a model based on training dataset, which contains only input data. Therefore, there are no human intervention in the process of training and development of training dataset. Clustering is widely used as a typical application of the unsupervised learning. Detailed explanation about regression, classification, and clustering will be covered at the next page.

4 Supervised Learning (I)

❑ **Regression**

- Development of Predictive Model (Continuous Output)
- Training Phase and Inference Phase
- Training Set : Input, Labeled Answer (Continuous Value)

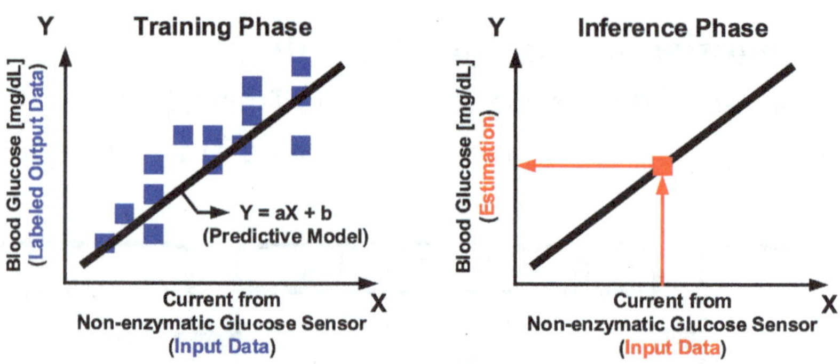

Regression is used to develop a predictive model having a continuous output value. In the regression application, two kinds of operation phases, training and inference are exist. In the training phase, model is developed based on a training dataset which contains input data and labeled answer. In the inference phase, another input data, which is different from the input data in the training dataset is inserted into the developed model in order to predict continuous output value. The following figures show a regression application in the bio-medical circuit and systems with the example of blood glucose estimation using non-enzymatic glucose sensor. In the training phase, the coefficients (a and b) of the 1st order linear equation, which are the model, are determined based on the training dataset. In the inference phase, we can get the blood glucose estimation by inserting the input data that is not included in the training dataset into the developed model in the training phase.

5 Supervised Learning (II)

❑ **Classification**

- Development of Predictive Model (Categorical Output)
- Training Phase and Inference Phase
- Training Set : Input, Labeled Answer (Categorical Value)

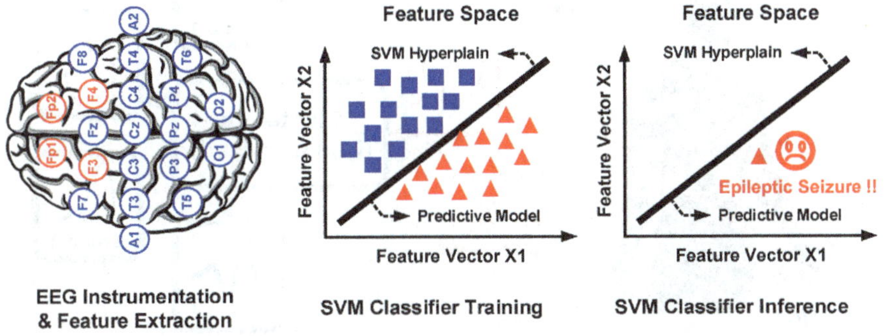

Classification is used to develop a predictive model having a categorical output value. In the case of classification application, two kinds of operation phases, training and inference, are exist like a regression application. In the training phase, model is developed based on a training dataset which contains input data and labeled answer. Note that the labeled answer in the training dataset has a categorical value, which is different from the labeled answer having a continuous value in the regression application. In the inference phase, another input data, which is different from the input data in the training dataset is inserted into the developed model in order to predict categorical output value from the input data. The following figures show a classification application in the bio-medical circuit and systems with the example of epileptic seizure classification using electroencephalogram (EEG) feature extraction and support vector machine (SVM). In the training phase, predictive model which is a maximum margin hyperplain to classify normal state and epileptic seizure state from patient's eeg signal is generated base on training dataset and extracted features. In inference phase, data point, which is expressed features in feature space by extracting features from another input data, is inserted into feature space and SVM classify whether the data point is in normal state or epileptic seizure state.

6 Unsupervised Learning

Clustering is a typical application of the unsupervised learning. In the clustering application, input data having high correlation is analyzed and grouped by clustering algorithms. The results of clustering and classification can be similar however, it is important to be aware of the difference between clustering and classification. In the case of classification, predictive model is developed by training and training dataset, which contains input data and labeled categorical answer. On the other hand, clustering application uses only unlabeled training dataset in order to analyze or group high correlated data. The following figures show a clustering application in the bio-medical circuit and system with the example of photoplethysmography (PPG) signal grouping and signal quality labeling.

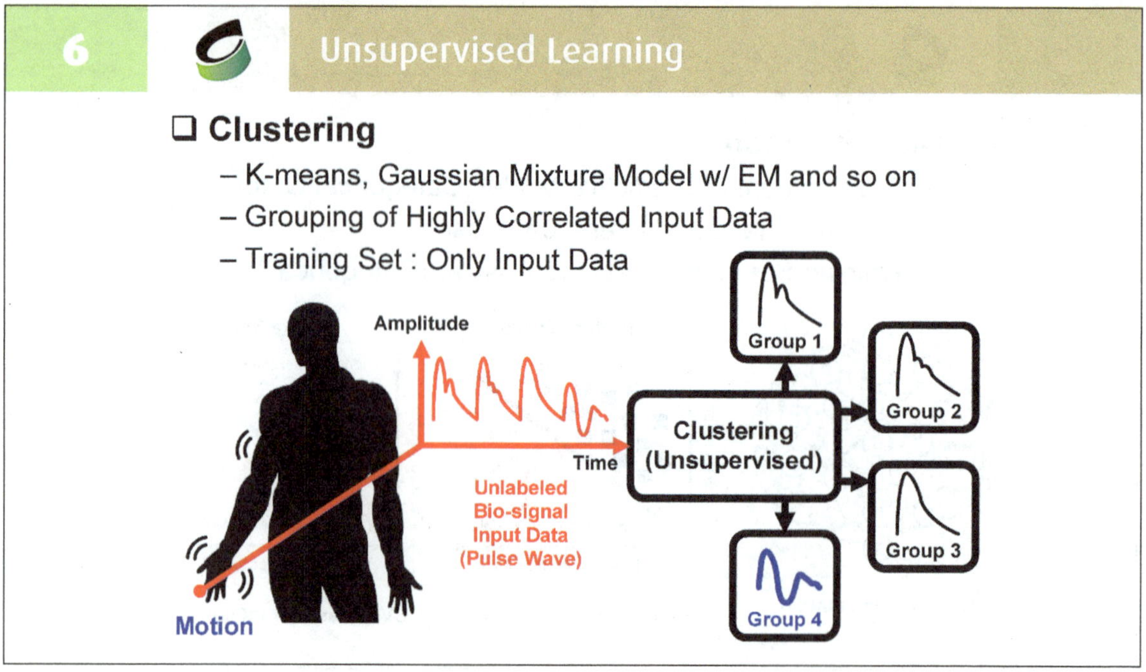

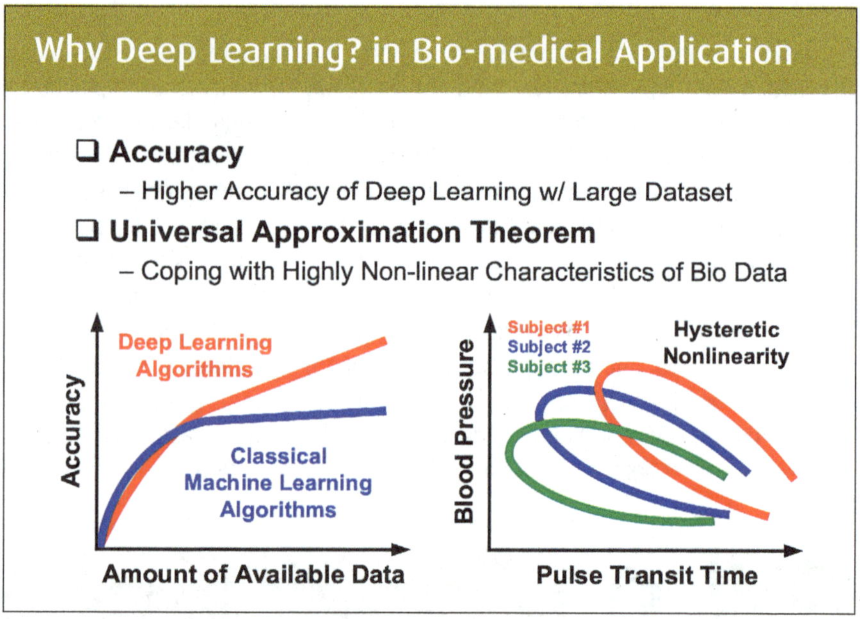

Before delving into the deep learning and various deep neural networks, we need to check an appropriateness of adopting deep learning and various deep neural networks. Recently, deep learning and various deep neural networks are getting spotlight due to its high accuracy compared to the conventional machine learning algorithms as shown in the left figure. Especially, with large amount of available dataset, accuracy performance of deep learning algorithms outperforms that of conventional machine learning algorithms. Also, bio-data which is collected from human body frequently shows highly non-linear characteristics. For example, as shown in the right figure, relationship between the human systolic blood pressure (SBP) and pulse transit time (PTT), the time it take for the pulse wave generated by left ventricular contraction to propagate between two points of arterial tree, shows hysteretic non-linearity as well as subject-dependency. However, universal approximation theorem enables the deep learning-based techniques to cope with highly non-linear characteristics of bio-data, thus encouraging vigorous adoption of deep learning in bio-medical application.

8 Deep Learning

❑ **Deep Neural Network (DNN)**

– Aggregation of Multiple Neurons & Multiple Layers

❑ **Definition of Deep Learning**

– Development of Model (Weight, Bias) w/ Training Data Set

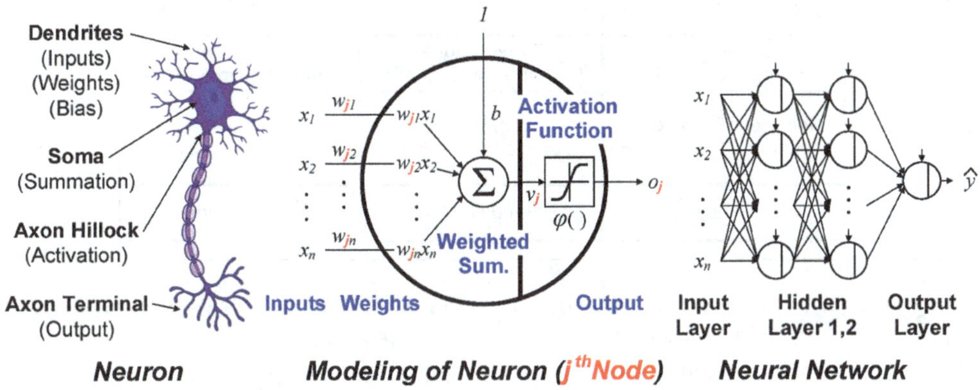

Next, we will talk about deep learning, which is the main topic of this chapter. Deep learning uses training dataset and deep neural network model which mimics biological neuron of the human, while enabling a function of regression and classification with the training and inference operation. In the neuron, electrochemical input signals are inserted into the dendrites and summed at the soma. In response to the summed electrochemical signals, the axon hillock generates activation signal, and then, the axon terminal sends the transmitted activation signal to the dendrites of the next neurons. Node, a basic element of deep neural network mimics abovementioned human neuron's operation.

The input vector of the node can be identically matched to abovementioned electrochemical input signals, and the weight & bias of the node mimics the dendrites of the neuron, while reflecting different strength of weight connections. The weighted summation and non-linear activation function such as sigmoid, tangent hyperbolic (tanh), and rectified linear unit (ReLU) in the node reflect the axon hillock's operation. Finally, the output of the node has a same function of the axon terminal.

Deep neural network is composed of input layer, output layer, and multiple (more than two) hidden layers which are located between the input and output layer as shown in the bottom-right figure. Also, each layer (Input, hidden, and output) can include one or more nodes. Note that the nodes in the hidden and output layer have internal operation such as weighted summation and non-linear activation, whereas the nodes in the input layer do not include any operation. Definition of deep learning is a development of deep neural network model, which is to obtain weight & bias values minimizing the error between labeled answer in the training dataset and predicted output value from the deep neural network model. Deep learning supports regression and classification application with training & inference operation. The detailed training & inference operation of deep neural network will be covered later.

9 Typical DNNs

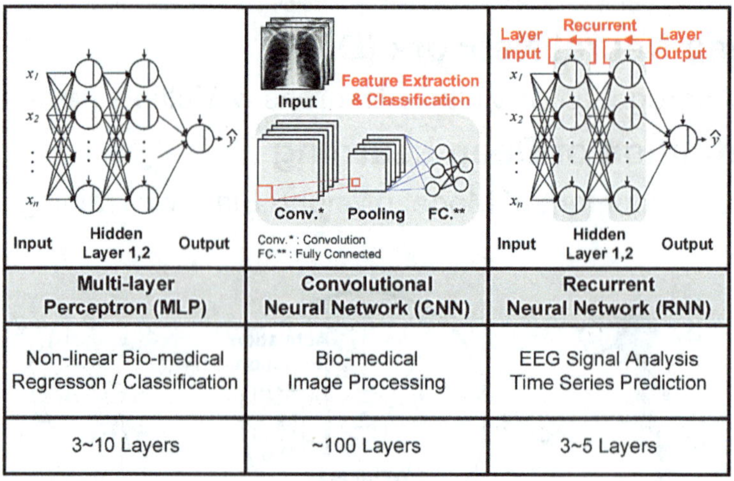

Next, three kinds of typical and basic deep neural networks, which are widely used in the bio-medical applications, will be introduced.

Firstly, multi-layer perceptron (MLP) is aggregation of single-layer perceptron, which is composed of one node (operation of weighted summation, and activation), inputs into the node, and weight & bias connections between the inputs and node. In the MLP neural network, there are input layer, output layer, and hidden layers between the input layer and output layer. One layer (hidden layer, output layer) in the MLP can include multiple single-layer perceptrons, and there are fully connected weight & bias connections between two different single-layered perceptrons in adjacent two layers. In this regard, MLP can be called fully connected layers (FCL), which is a part of convolutional neural network. The MLP (generally includes 3~10 layers) is widely used in regression or classification of the bio-medical application which contains non-linear characteristics.

Secondly, convolutional neural network (CNN) is specialized deep neural network for image classification application, thus widely used in the bio-medical image classification such as anomaly (cancer) detection in the X-ray image. CNN is composed of two kinds of different neural networks, feature extraction neural networks and classification neural networks. Feature extraction neural network includes convolution layers and pooling layers, while classification neural network consists of fully connected layers. In the convolution layers, feature maps are generated by convolution operation between input image and convolution filter (automatically trained based on training dataset and training operation) with activation, thus emphasizing unique feature of input image for classification. In the pooling layers, size (or dimension) of feature maps is reduced by (max or average) pooling operation, thus inserting the extracted and reduced features into the fully connected layers. In the fully connected layers, the extracted and reduced features are used for classification operation in the CNN. There are many kinds of CNNs, and some CNNs consist of more than 100 layers.

Lastly, basic recurrent neural network (RNN) is considerably different from abovementioned MLP and CNN. This is because RNN has feedback path, which is called recurrent path. Recurrent path can be generated by getting one layer's outputs and inserting into the (or previous) layer's inputs. Due to the recurrent path, one layer including recurrent path simultaneously gets current inputs and previous time step's outputs (the function of previous time step's inputs), thus obtaining characteristics of 'memory'. Therefore, RNN is used to predict current or future event (or value) based on the previous time step's data and context. There are many kinds of RNNs but basic RNN as shown in the figure generally consists of 3~5 layers. The RNN can be used in bio-medical application such as time series prediction and eeg signal analysis.

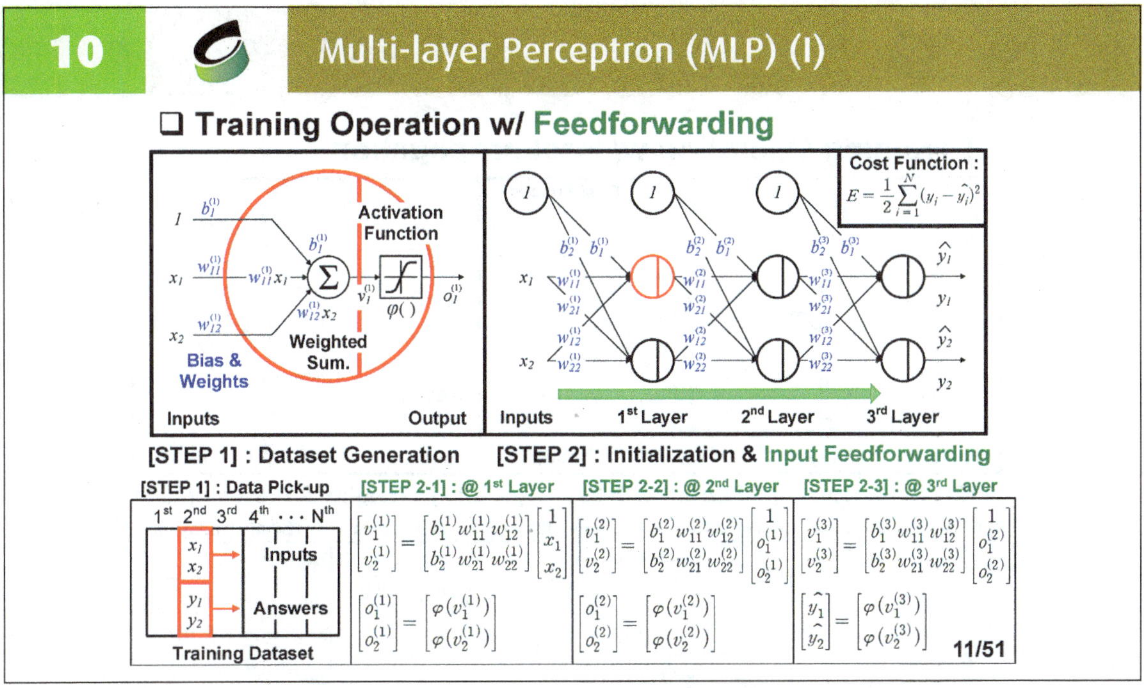

Next, we will explain detailed training operation of multi-layer perceptron (MLP), a typical and basic deep neural network. Firstly, we will check the model structure of MLP in the upper-right figure. The MLP is composed of input layer, two hidden layers (1st layer, 2nd layer), and output layer (3rd layer). Each layer (Input, 1st, 2nd, and 3rd layer) has two nodes, and there are fully connected weight connections between two nodes in two adjacent layers. Note that the circles having a value of one in the Input, 1st, and 2nd layer are simple tricks in order to process bias connection (latin small letter b) of each layer as weight connection (latin small letter w) for the sake of simple notation and matrix calculation.

Secondly, it will be helpful to check the notation of weight & bias connections for easy understanding of mathematical expression related to the MLP training operation. For the notation of weight & bias connections, the right subscript and left subscript mean the sequential number of source node in the previous layer and target node in the next layer, respectively. Also, number in the bracketed superscript means the layer number. For example, $w^{(1)}_{21}$ means the weight connection between the first node (x1) in the input layer and second node in the 1st layer.

Next, detailed feedforwarding operation of MLP training will be explained with a step by step approach. In step 1, training dataset, which is composed of N pairs of input and labeled answer is generated, and then, one pair of input ({x1, x2}) and labeled answer ({y1, y2}) is randomly selected for MLP training operation. In step 2, all weight & bias values in the MLP are properly initialized (with Xavier or He's initialization method), and feedforwarding matrix multiplication is carried out in order to obtain estimated output value (y1 hat, y2 hat) at the output layer or 3rd layer of MLP as shown in the bottom-right figures, step 2-1, 2-2, 2-3 (the v, φ, and o mean weighted summation of input, non-linear activation function, and output of specific node, respectively). After abovementioned feedforwarding operation, we can define error (y1 – y1 hat, y2 – y2 hat) between the estimated output value and labeled answer in the one pair of dataset, which is selected in step 1. Especially, cost function, E is defined as a summation of squared errors at the output layer's all nodes with a multiplication of 0.5 in order to express total error of deep neural network. Training operation, which is a process to obtain weight & bias minimizing deep neural network's error, is conducted in order to reduce abovementioned cost function, E.

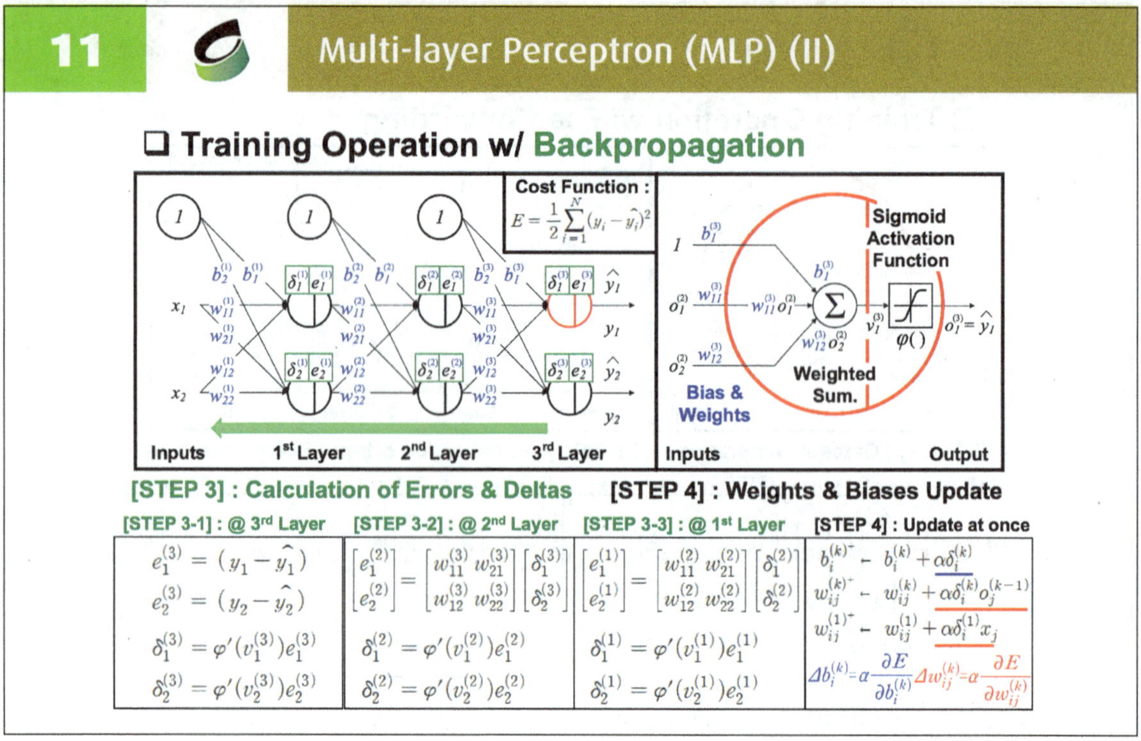

Next, detailed backpropagation operation of MLP training will be explained with a step by step approach. In order to obtain all weights & biases minimizing cost function of deep neural network, we need to repeatedly calculate updating values of all weights & biases and update all weights & biases according to the 'training rule' such as delta rule, which requires all errors of nodes in the hidden layers and output layer. Abovementioned feedforwarding operation can be regarded as a specific procedure to calculate all errors of nodes in the output layer. On the other hand, backpropagation operation can be regarded as a specific procedure to define and calculate all errors of nodes in hidden layers. In step 3, we will define and calculate 'error's and 'delta's of each hidden layer and output layer. In step 3-1, errors (latin small letter e) of nodes in the output layer (3rd layer) are defined, and then deltas (greek small letter delta, δ) of nodes in the output layer are defined as errors multiplied by first derivative of non-linear activation function (greek small letter phi, φ) with respect to the weighted summation (latin small letter v) before activation function. In step 3-2, errors of nodes in the hidden layer 2 (2nd layer) are defined as a matrix multiplication of output layer's delta and transposed weight matrix of the output layer, after that, deltas of nodes in the hidden layer 2 are identically defined as the previous case of step 3-1. In step 3-3, errors and deltas of nodes in the hidden layer 1 (1st layer) are identically defined as the previous case of step 3-2. Finally, in step 4, all weights & biases in deep neural network can be updated according to the delta rule, which means that if a specific node j (in the K-1th layer) contributes to the error of a specific node i (in the Kth layer), updating value of the weight (or bias) between the two nodes should be defined in proportion to learning rate (α, 0<α<1), delta of the node i (delta includes error term), and output of the node j as shown in the highlighted mathematical expression in the step 4. Note that partial derivative of cost function with respect to the specific weight (or bias) mathematically equal to the delta (connected to the specific weight) multiplied by the output of the previous layer's node (also connected to the specific weight). Note that abovementioned detailed explanation about backpropagation operation assumes that activation function in the hidden and output layers is sigmoid function.

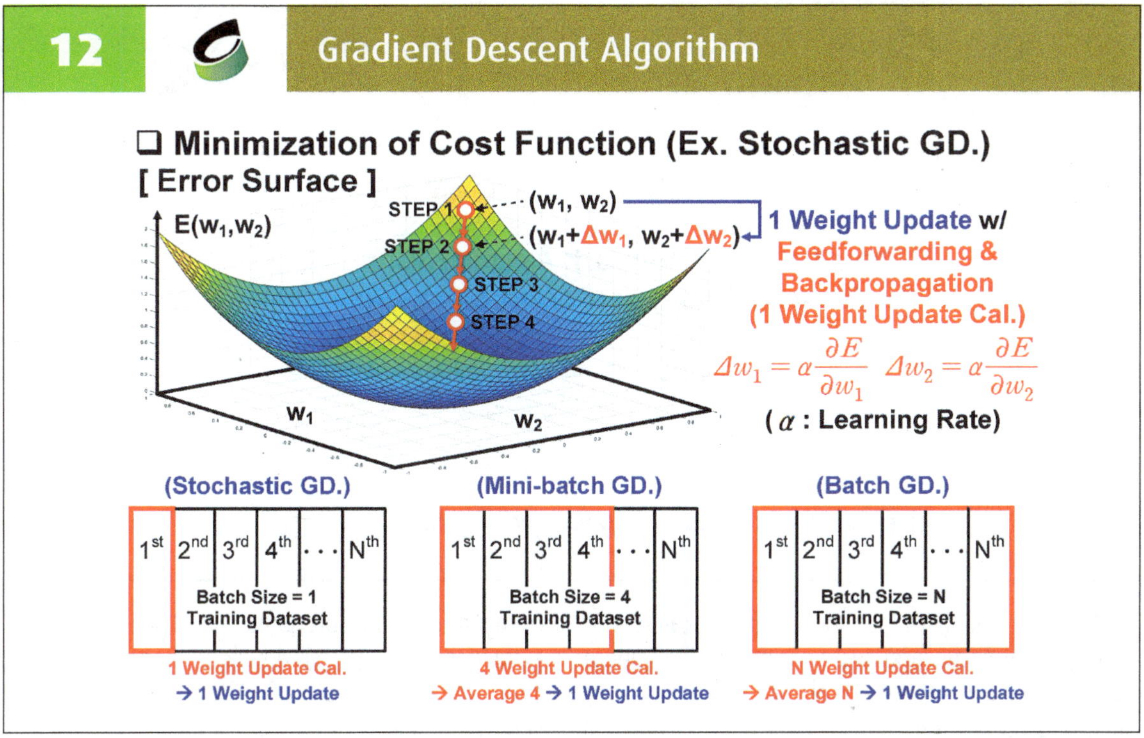

Eventually, training operation of deep neural network can be regarded as a specific procedure, which repetitively updates all weights & biases in order to obtain all weights & biases minimizing the cost function (in this example, sum of squared error multiplied by 0.5) of deep neural network. Abovementioned procedure can be described as specific situation that a basketball goes down the slope toward bottom surface with a step by step approach. In this situation, the slope and the bottom surface mean error surface (which is composed of weights & biases, and cost function) and global minimum point (which is the weights & biases of deep neural network after successful training operation), respectively. Especially, abovementioned specific procedure is called by gradient descent algorithm because the gradient (which is mathematically partial derivative of the cost function with respect to weights & biases) is used to determine the direction and step size of descent.

In this slide, gradient descent algorithm will be explained with simple error surface (which is not error surface of deep neural network) in order to get an intuition about gradient descent algorithm. In the stochastic gradient descent (GD) algorithm, one-time feedforwarding & backpropagation (with randomly selected one pair of input and labeled answer) generates one-time updating of weights & biases as shown in the upper figure. On the other hand, in the batch gradient descent (GD) algorithm, N-times feedforwarding & backpropagation (with randomly selected N pairs of input and labeled answer) followed by N-average of updating values of weights & biases generates one-time updating of weights & biases. In the case of the batch GD, variance of the weights & biases updating is small, thus enabling stable convergence toward the lowest point of the error surface. However, the batch GD also shows disadvantage of slow convergence speed due to large amount of calculation. On the other hand, the stochastic GD shows fast convergence speed, but convergence stability is poor due to high variance of weights & biases updating. Therefore, in order to compensate defects of stochastic GD and batch GD, mini-batch gradient descent (GD) is widely used in gradient descent algorithm due to its appropriate stability and convergence speed.

13 Outline

❑ **Introduction :**
Artificial Intelligence, Machine Learning, and Deep Learning

❑ **(Deep) Neural Network Applications for Bio-medical Circuit and Systems**

❑ **H/W Implementation of Deep Learning Inference Processors :** 1) DNPU, 2) UNPU

❑ **Future Direction :**
Neuromorphic Computing with Nonvolatile Memory

Next, we will look into some representative bio-medical circuit and systems introducing (deep) neural networks. Bio-medical circuit and systems including (deep) neural network will be covered in terms of 1) specific application and overall system configuration, 2) pre-processing and/or feature extraction strategy, 3) detailed information about deep neural network, and 4) measurement or verification results.

14 Non-invasive Blood Glucose (BG) Estimation (I)

❑ **Overall System**

– [HW] Bio-Z+NIRS Sensing w/ Customized Sensor Board

– [SW] BG Estimation w/ Shallow NN Training & Inference

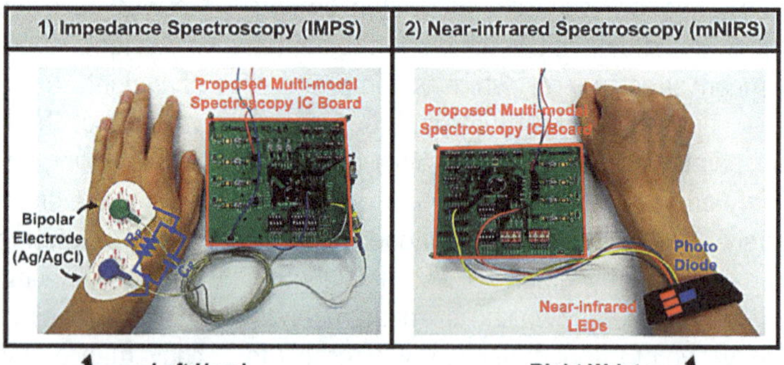

[1] K. Song et al., JSSC 2015

Firstly, a non-invasive blood glucose estimation system (Kiseok et al., 2015) will be introduced. This pioneer work was proposed in order to non-invasively estimate blood glucose level with customized sensors and neural network techniques. The system is composed of two main parts; 1) hardware part and 2) software part. The hardware part includes multi-modal bio-medical sensors, which are bio-impedance (Bio-Z) sensor and multi-wavelength near-infrared spectroscopy (mNIRS) sensor, to measure electrical characteristic variation of surrounding tissues of blood vessel by glucose variation and optical characteristic variation of glucose itself, respectively. The software part includes blood glucose level estimation algorithms with neural network training and inference operation.

15

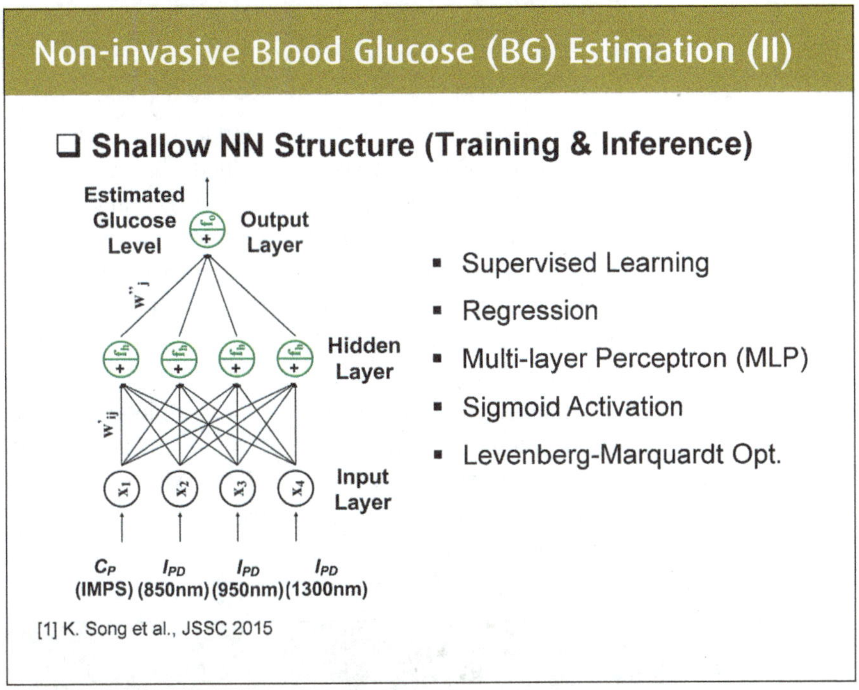

This slide shows a detailed information about the neural network, which is used for non-invasive blood glucose level estimation. The neural network has a structure of MLP, which is composed of three layers (input layer, one hidden layer, output layer). Especially, the neural network is called shallow (or vanilla) neural network because it has only one hidden layer. Capacitance value from Bio-Z sensor and DC current values from mNIRS sensors are inserted into the input layer of the shallow neural network, which outputs continuous value of estimated glucose level with sigmoid activation. In a training mode, the neural network for regression is trained by training dataset and supervised learning with Levenberg-Marquardt optimization algorithm. In an inference mode, input data, which is not included in the training dataset is inserted into the trained shallow neural network, thus enabling non-invasive estimation of blood glucose level.

16

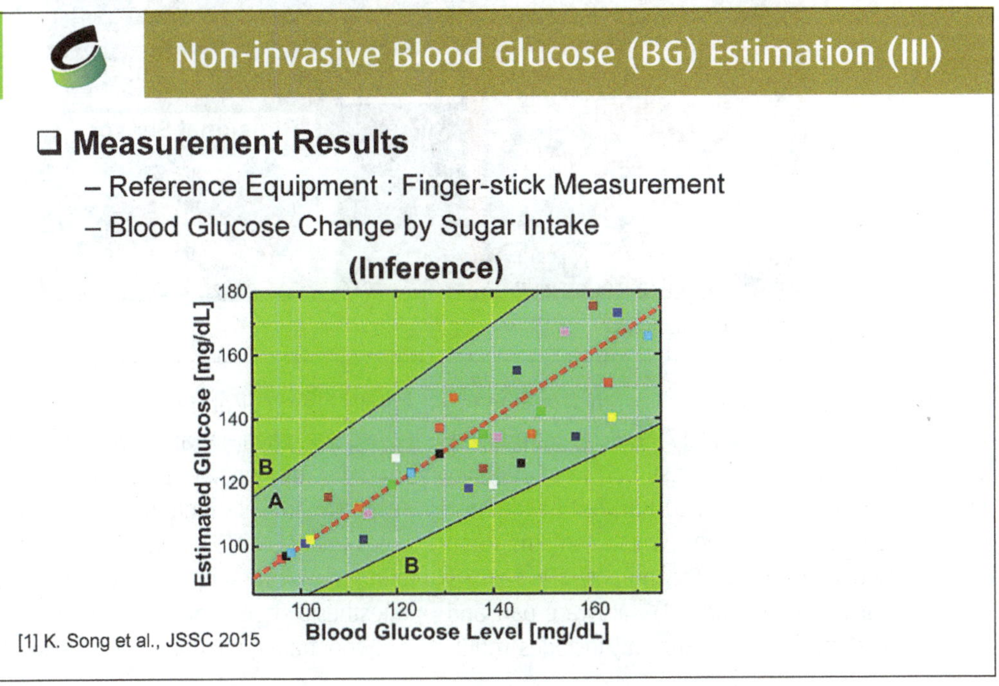

16 Non-invasive Blood Glucose (BG) Estimation (III)

This slide shows measurement results of the proposed non-invasive blood glucose system. In order to quantify the accuracy of the glucose estimation, two accuracy performance metric, mean absolute relative difference (mARD) and Clarke grid error (CGE) are used. Initial (training) dataset consists of 50 pairs of input (capacitance value & three kinds of DC currents values from 850nm, 950nm, and 1300nm photo diodes) & labeled output (blood glucose level changed by sugar intake and measured by finger-stick reference equipment) from one volunteer. The number of training, validation, and test set are 15, 17, and 17 pairs of input & labeled output, respectively. After training operation, (pre) trained shallow neural network is applied to 10 additional volunteers in order to measure inference performance. Since each volunteer's skin shows difference electrical and optical characteristics, the pre-trained shallow neural network is further trained with 5 pairs of input & labeled output from each additional volunteers, and finally tested with 5 pairs of input & labeled output. As a result, all test points from 10 additional volunteers are included in A area of CGE plot, thus showing successful non-invasive blood glucose estimation.

17 Anesthesia Depth (AD) Estimation (I)

❑ **Overall System**

– [HW] EEG+NIRS Sensing w/ Customized Patch Sensor

– [SW] AD Estimation w/ DNN Training & Inference

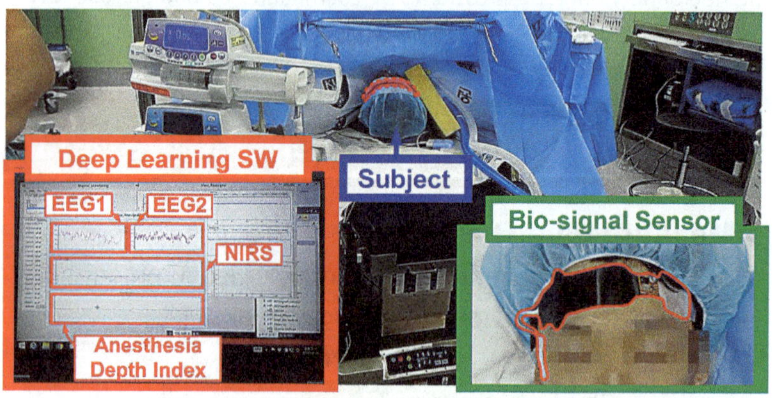

[2] U. Ha et al., ISSCC 2017

Secondly, an anesthesia depth monitoring system (Unsoo et al., 2017) will be introduced. This work was proposed to estimate and monitor anesthesia depth index with customized head patch sensor and software package which contains deep neural network technique. The proposed system is composed of two main parts; 1) hardware part and 2) software part. The hardware part includes multi-modal bio-medical sensors, which are 2-channel EEG sensors and near-infrared spectroscopy (NIRS) sensor, in order to improve bispectral index (BIS)'s false response to ketamine. The software part contains real-time EEG/NIRS signal acquisition/monitoring part, pre-processing & feature extraction part, and anesthesia depth calculation part with deep neural network training and inference, respectively.

18 Anesthesia Depth (AD) Estimation (II)

❑ **Feature Extraction for DNN**

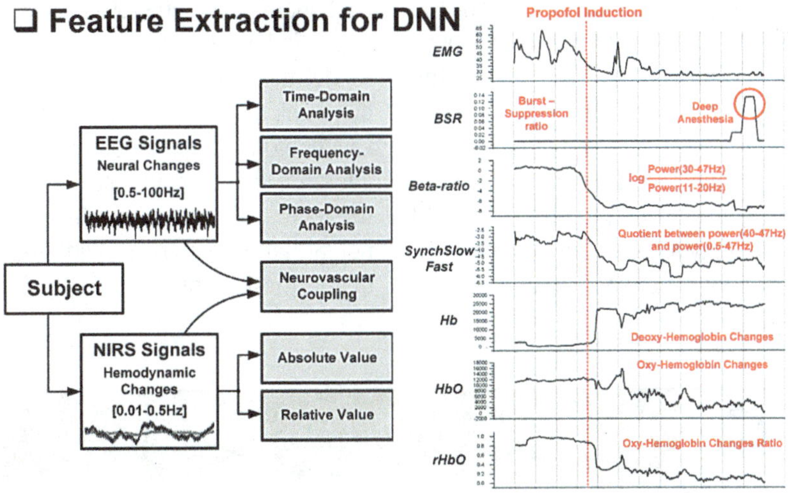

[2] U. Ha et al., ISSCC 2017

This slide shows pre-processing & feature extraction strategy for deep neural network in the proposed system. For the EEG signal, 3 kinds of domains (time, frequency, and phase) are used to extract 5 kinds of features such as burst suppression ratio (BSR), quazi, beta ratio, electromyogram (EMG), and SynchFastSlow. For the NIRS signal, 4 kinds of features such as absolute value of oxygenated & deoxygenated hemoglobin concentration (HbO2 & Hb) and relative value of oxygenated & deoxygenated hemoglobin concentration (rHbO2 & rHb) are extracted and used to reflect hemodynamic changes in the brain. Lastly, neurovascular coupling coefficient is calculated by temporal kernel canonical correlation analysis (tkCCA) as a last features. Abovementioned features are calculated by real-time signal processing algorithms and inserted into the purpose-built deep neural network.

19 Anesthesia Depth (AD) Estimation (III)

❑ **DNN Structure (Training & Inference)**

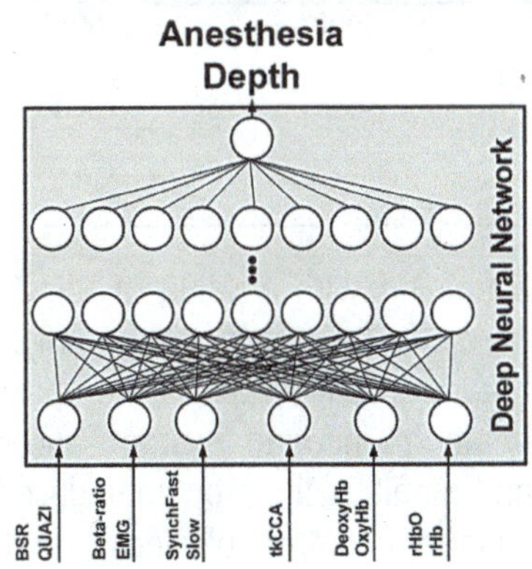

- Supervised Learning
- Regression
- Multi-layer Perceptron (MLP)
- Input Layer's # of Nodes : 10
- Hidden Layer's # of Nodes : 5
- Sigmoid Activation
- Levenberg-Marquardt Opt.

[2] U. Ha et al., ISSCC 2017

19 Anesthesia Depth (AD) Estimation (III)

This slide shows a detailed information about the purpose-built deep neural network, which is used for anesthesia depth estimation and monitoring. The deep neural network as shown in the left figure has a structure of MLP, which is composed of input layer, output layer, and hidden layers. Total 10 kinds of features which are extracted by pre-processing and feature extraction strategy are inserted into the input layer of the deep neural network, which outputs continuous value (0~100) of estimated anesthesia depth index with sigmoid activation. In a training mode, the neural network for regression is trained by training dataset and supervised learning with Levenberg-Marquardt optimization algorithm. In an inference mode, input data, which is not included in the training dataset is inserted into the trained deep neural network, thus enabling real time monitoring of anesthesia depth.

20 Anesthesia Depth (AD) Estimation (IV)

❑ Measurement Results 1

– Reference Equipment : Bispectral Index (BIS)
– Propofol-induced Anesthetic Depth Change

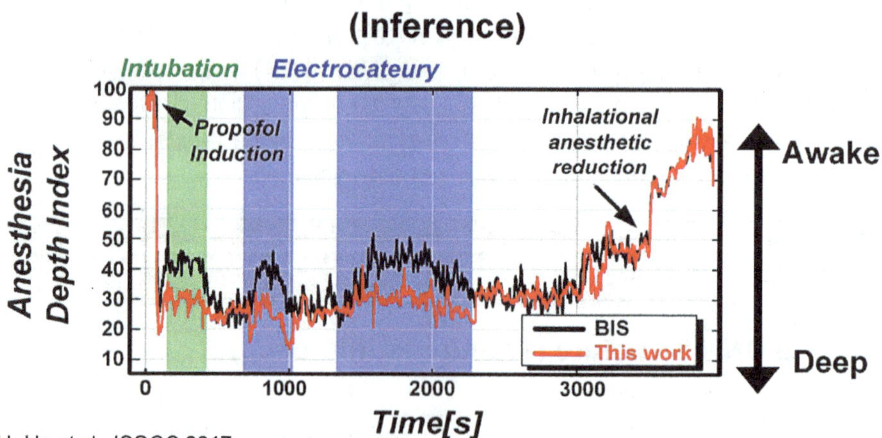

[2] U. Ha et al., ISSCC 2017

This slide shows measurement results of the proposed anesthesia depth monitoring system. Total 1210 hours of recording data from 380 patients was successfully collected for a training and test of the deep neural network. Skilled anesthetist's comments including anesthetic agent (propofol, ketamine) concentration and patient's response to external stimulation are used as a ground truth for a training with a value of anesthesia depth index from BIS (reference equipment). As shown in the measurement results, the proposed anesthesia depth monitoring system successfully estimates and monitors the propofol-induced anesthetic depth change regardless of electrical interference such as electrocateury.

21

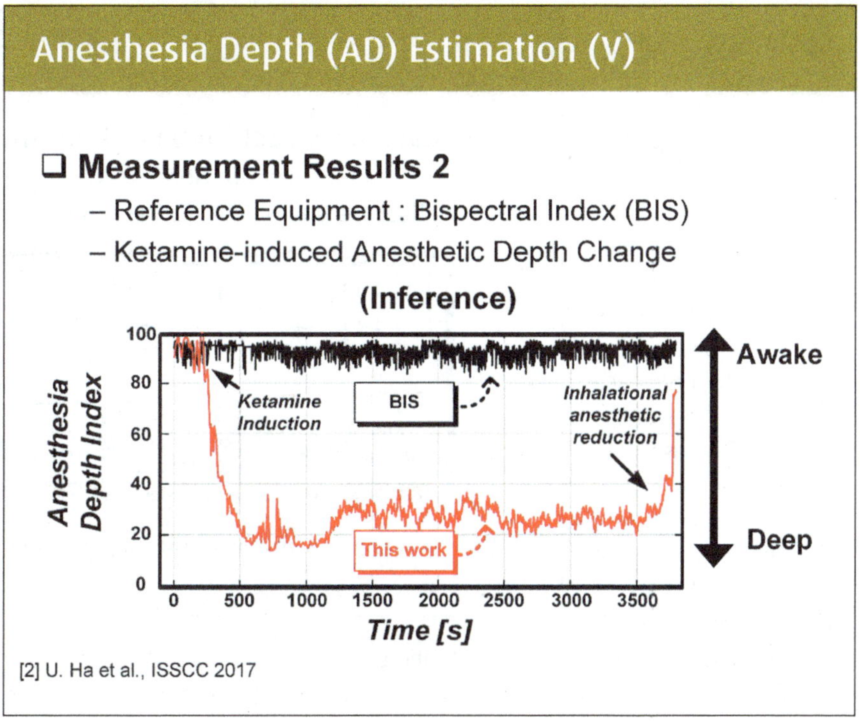

This slide shows another measurement results about ketamine-induced anesthetic depth change. As shown in the measurement results, the proposed system successfully estimates and monitors the ketamine-induced anesthetic depth change, while conventional BIS shows fault estimation in the condition of ketamine-induced anesthetic depth change, thus illustrating superiority of the proposed anesthesia depth monitoring system.

22

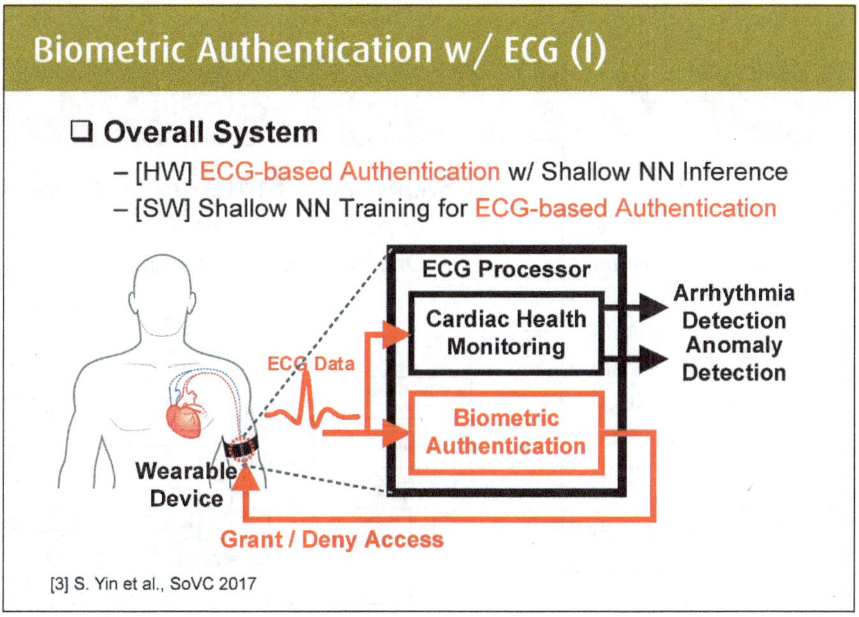

Thirdly, ECG based biometric authentication system (Shihui et al., 2017) will be introduced. This work was proposed for functions of 1) a few kinds of arrhythmia detections and 2) biometric authentication with ECG pre-processing algorithms and (shallow) neural network technique. In this chapter of tutorial, we will focus on a function of biometric authentication. The proposed system is composed of two parts; 1) hardware part and 2) software part. The hardware part includes pre-processing of ECG, inference operation of (shallow) neural network, and decision (granting or denying) making operation with cosine similarity calculation for ECG based biometric authentication. The software part contains training operation of (shallow) neural network for ECG based biometric authentication.

23

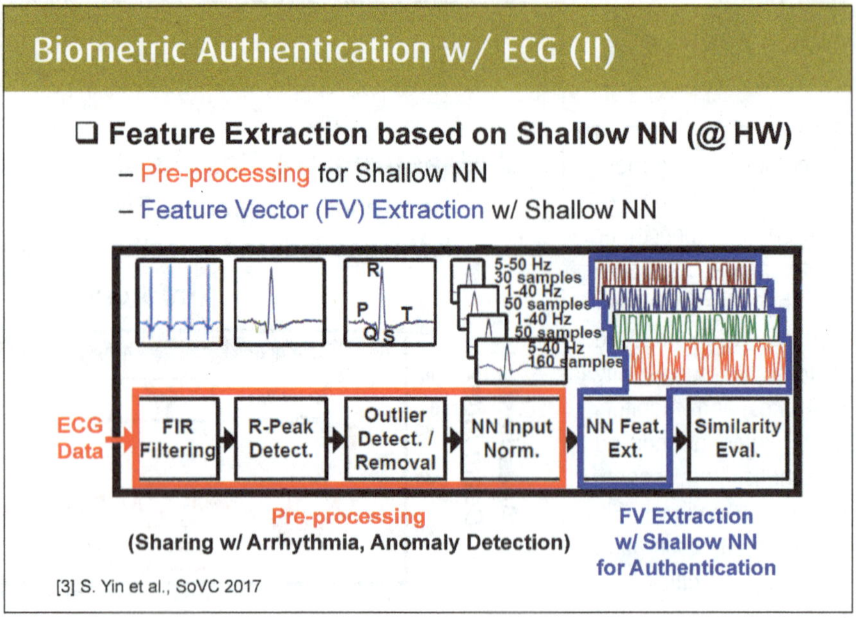

This slide shows pre-processing & feature extraction strategy for ECG based biometric authentication. In this work, (shallow) neural network extracts a feature vector from each subject's ECG waveform in order to use the extracted feature vector as authentication key, thus showing different utilization of (shallow) neural network which is often applied to regression or classification. Before the extraction of feature vector with (shallow) neural network, pre-processing which contains FIR filtering, R-peak detection, outlier detection & removing, and normalization for the (shallow) neural network's input layer is conducted as shown in the figure. As a result, 4 kinds of pre-processed ECG waveforms (30 samples @ 5-50 Hz, 50 samples @ 1-40 Hz, 50 samples @ 1-40 Hz, 160 samples @ 5-40 Hz) are inserted into 4 kinds of different (shallow) neural networks, respectively.

24

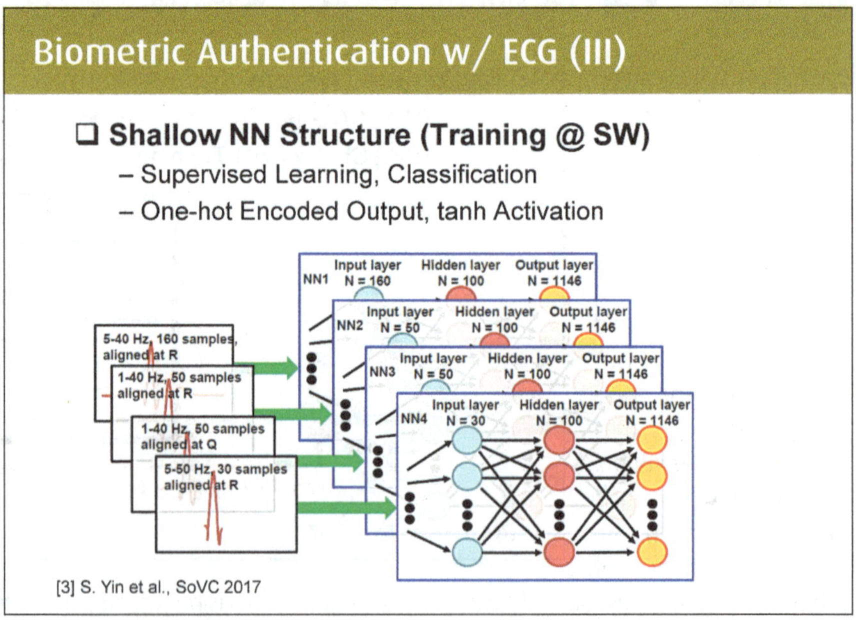

This slide shows a detailed structure of (shallow) neural network which is used in training operation. Training dataset is composed of pre-processed ECG waveforms from each subject and one-hot encoded ID from each subject, thus representing supervised learning and classification application. After training operation, however, weights & nodes of output layer are eliminated and only hidden layer outputs are used as feature vector of registered subject, thus using (shallow) neural network as a measure of feature vector extraction for ECG based biometric authentication. Note that the (shallow) neural network uses tangent hyperbolic (tanh) activation.

25

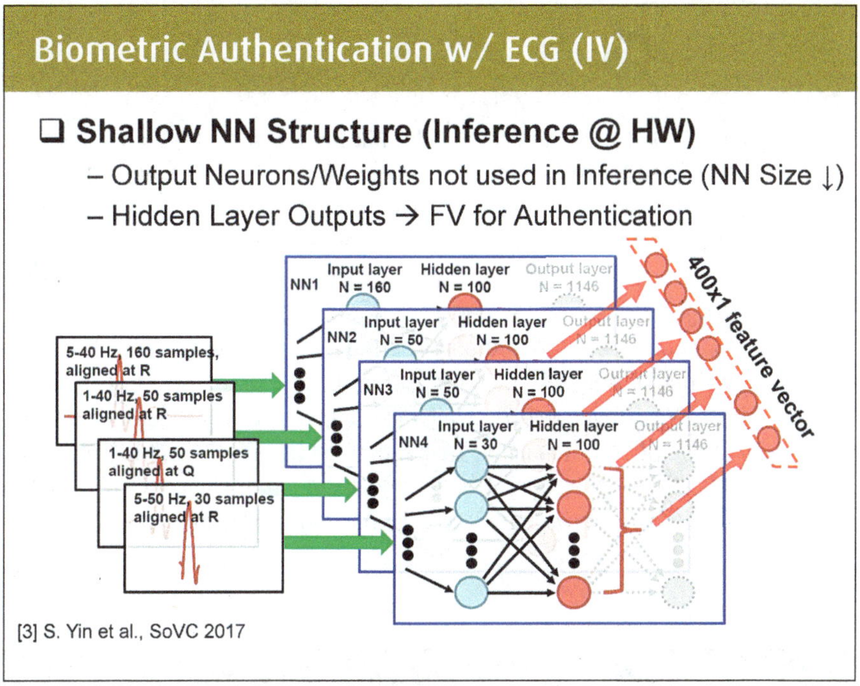

This slide shows a detailed structure of (shallow) neural network which is used in inference operation. As mentioned in the previous slides, the (shallow) neural network from which weights & nodes of the output layer are eliminated is used for feature vector extraction according to the pre-processed ECG waveforms. In the inference mode, pre-processed ECG waveforms from subject, who are not included in the training dataset, are inserted into the trained (shallow) neural network, thus generating new feature vector to be compared to existing feature vectors generated from registered subjects.

26

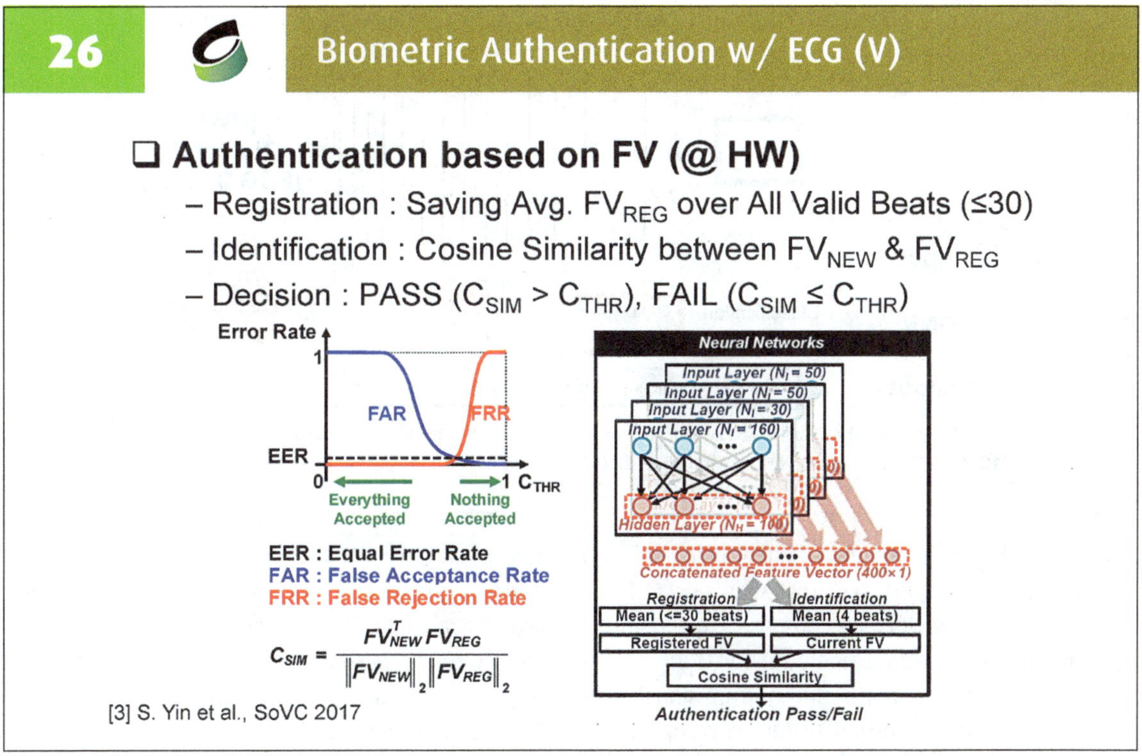

26 Biometric Authentication w/ ECG (V)

This slide shows a detailed authentication procedure based on feature vector. The authentication procedure can be divided into 3 modes; 1) registration mode, 2) identification mode, and 3) decision making mode to grant or deny access permission to user. In the registration mode, 30 beats of ECG are pre-processed and inserted into the (shallow) neural network, thus generating final feature vector FV_{REG} (averaged over all valid beats) with training operation of (shallow) neural network. In the identification mode, only 4 beats of ECG are pre-processed and inserted into the trained (shallow) neural network, thus generating final feature vector FV_{NEW} (averaged over all valid beats) with inference operation of (shallow) neural network. Lastly, in the decision making mode, cosine similarity (C_{SIM}) between the registered feature vector (FV_{REG}) and the new feature vector (FV_{NEW}) are calculated as shown in the bottom-left equation, thus granting access when the calculated value of cosine similarity (C_{SIM}) is larger than specific threshold value (C_{THR}). On the other hand, if the calculated value of cosine similarity (C_{SIM}) is not larger than specific threshold value (C_{THR}), access is denied.

27 Cardiologist-level Arrhythmia Detection w/ CNN (I)

❑ Overall System

– [HW] ECG Sensing w/ Zio Patch Monitor [5] (Single-lead)

– [SW] Arrhythmia Detection w/ CNN Training & Inference

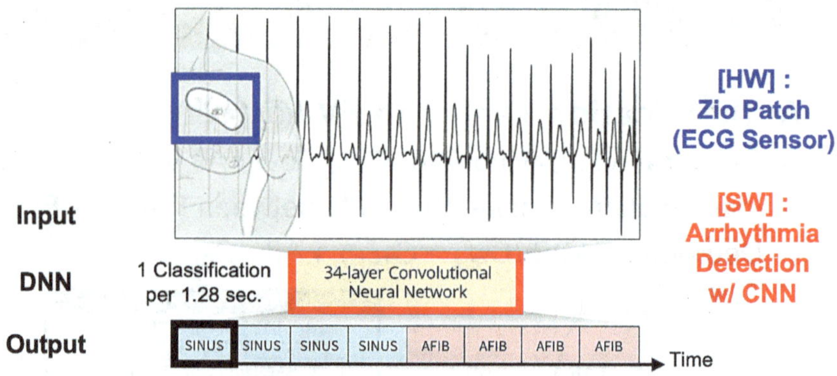

[4] P. Rajpurkar et al., arXiv 2017, with Author's Permission [5] M. Turakhia et al., AJC 2013
[10] A. Y. Hannun et al., Nature Medicine 2019

Lastly, highly accurate (cardiologist-level) arrhythmia detection system based on CNN will be introduced. This work was proposed to classify 12 rhythm classes from raw single-lead ECG inputs (10 kinds of arrhythmias, noise, and normal sinus) with high accuracy. The proposed system is composed of two main parts; 1) hardware part and 2) software part. The hardware part is the Zio Patch monitor which is a single-lead, chest patch-type, ECG sensor. The software part contains training and inference of 34-layer CNN to detect arrhythmia every 256 samples (~1.28 second).

28

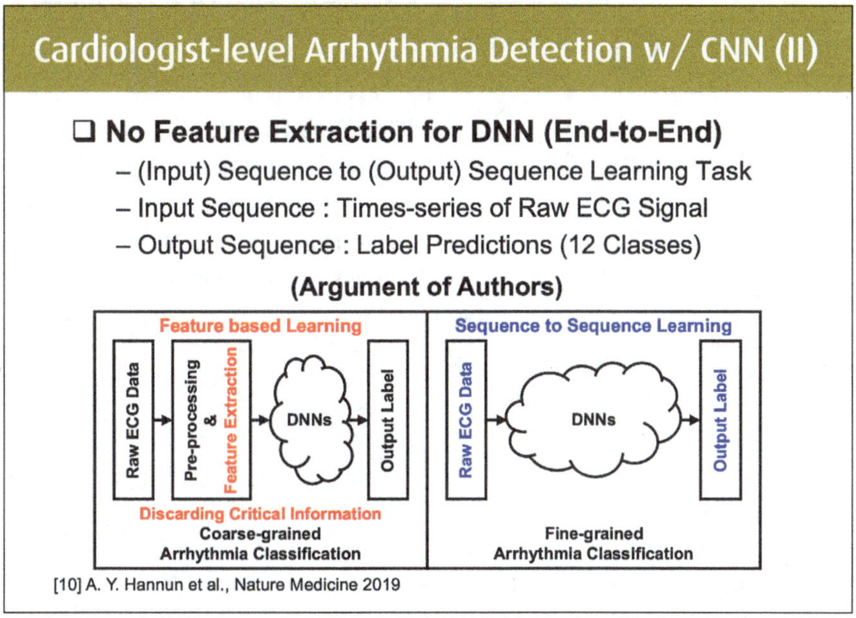

This slide shows the feature extraction strategy for deep neural network in the proposed system. This work is distinguished from other previously introduced works in that there is no feature extraction stage before input of deep neural network. According to the argument of the authors, pre-processing and feature extraction can discard critical information in the ECG waveforms, thus leading to coarse-grained arrhythmia classification. On the other hand, (input) sequence to (output) sequence learning, which does not apply pre-processing and feature extraction, uses raw ECG data and 12 kinds of label as input sequence and output (label) sequence, respectively, thus enabling fine-grained arrhythmia classification with very large-scale training dataset and large-scale deep neural network as a cost.

29

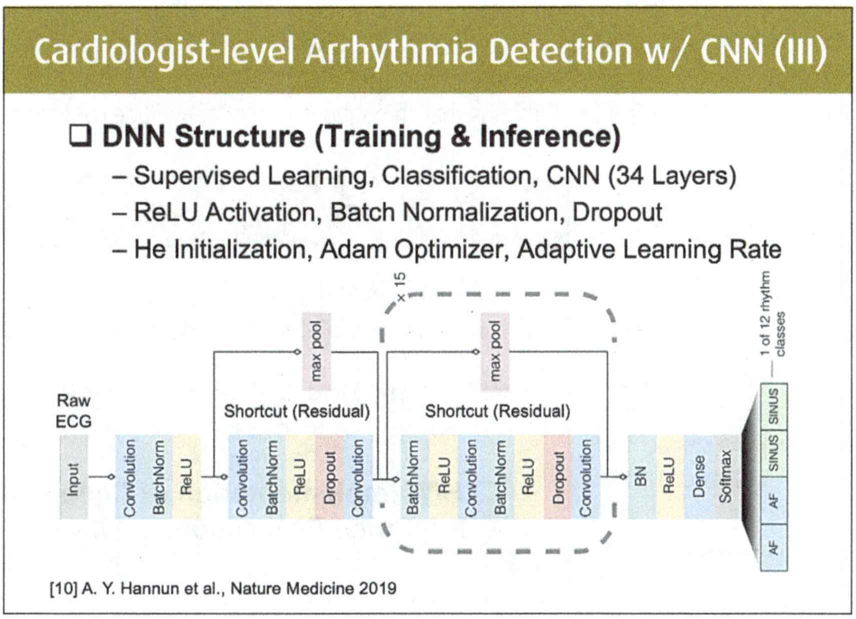

This slide shows a detailed information about a purpose-built deep neural network (CNN), which is used for arrhythmia classification based on sequence to sequence (or end to end) supervised learning. The CNN based deep neural network is composed of 34 layers including input and output layer. In order to facilitate deep neural network training (especially, in the case of large numbers of hidden layers), shortcut connections, which are called residual connections, are applied with ReLU activation, He initialization, and batch normalization. Also, dropout technique is used to suppress overfitting problem. Finally, Adam optimizer and adaptive learning rate technique are used for fast & better optimization.

30

Cardiologist-level Arrhythmia Detection w/ CNN (IV)

❑ Verification Results

- Training Dataset (91,232 records from 53,549 patients)
- Test Dataset (328 records from 328 unique patients)
- Golden Standard* vs. DNN vs. Average Cardiologist
- Overall Performance : DNN ≥ Average Cardiologist

Table 1 | Diagnostic performance of the DNN and averaged individual cardiologists compared to the cardiologist committee consensus ($n = 328$)

	Algorithm AUC (95% CI)[a]		Algorithm F_1[b]		Average cardiologist F_1	
	Sequence[a]	Set[b]	Sequence	Set	Sequence	Set
Atrial fibrillation and flutter	0.973 (0.966-0.980)	0.965 (0.932-0.998)	0.801	0.831	0.677	0.686
AVB	0.988 (0.983-0.993)	0.981 (0.953-1.000)	0.828	0.808	0.772	0.761
Bigeminy	0.997 (0.991-1.000)	0.996 (0.976-1.000)	0.847	0.870	0.842	0.853
EAR	0.913 (0.889-0.937)	0.940 (0.870-1.000)	0.541	0.596	0.482	0.536
IVR	0.995 (0.989-1.000)	0.987 (0.959-1.000)	0.761	0.818	0.632	0.720
Junctional rhythm	0.987 (0.980-0.993)	0.979 (0.946-1.000)	0.664	0.789	0.692	0.679
Noise	0.981 (0.973-0.989)	0.947 (0.898-0.996)	0.844	0.761	0.768	0.685
Sinus rhythm	0.975 (0.971-0.979)	0.987 (0.976-0.998)	0.887	0.933	0.852	0.910
SVT	0.973 (0.960-0.985)	0.953 (0.903-1.000)	0.488	0.693	0.451	0.564
Trigeminy	0.998 (0.995-1.000)	0.997 (0.979-1.000)	0.907	0.864	0.842	0.812
Ventricular tachycardia	0.995 (0.980-1.000)	0.980 (0.934-1.000)	0.541	0.681	0.566	0.769
Wenckebach	0.978 (0.967-0.989)	0.977 (0.938-1.000)	0.702	0.780	0.591	0.738
Frequency-weighted average	0.978	0.977	0.807	0.837	0.753	0.780

[a]DNN algorithm area under the ROC compared to the cardiologist committee consensus. [b]DNN algorithm and averaged individual cardiologist F_1 scores compared to the cardiologist committee consensus. Sequence-level describes the algorithm predictions that are made once every 256 input samples (approximately every 1.3 s) and are compared against the gold-standard committee consensus at the same intervals. Set-level describes the unique set of algorithm predictions that are present in the 30-s record. Sequence AUC prediction, n = 7,544; set AUC prediction, n = 328.

Table 2 | DNN algorithm and cardiologist sensitivity compared to the cardiologist committee consensus, with specificity fixed at the average specificity level achieved by cardiologists

	Specificity	Average cardiologist sensitivity	DNN algorithm sensitivity
Atrial fibrillation and flutter	0.941	0.710	0.861
AVB	0.981	0.731	0.858
Bigeminy	0.996	0.829	0.921
EAR	0.993	0.380	0.445
IVR	0.991	0.611	0.867
Junctional rhythm	0.984	0.634	0.729
Noise	0.983	0.749	0.803
Sinus rhythm	0.859	0.901	0.950
SVT	0.983	0.408	0.487
Ventricular tachycardia	0.996	0.652	0.702
Wenckebach	0.986	0.541	0.651

[10] A. Y. Hannun et al., Nature Medicine 2019

*Cardiologist Consensus Committee

This slide shows verification results of the proposed cardiologist-level arrhythmia detection system. For training, 760 hours of recording data (91,232 records multiplied by 30 seconds) from 53,549 patients was successfully collected. For the validation of the proposed DNN, test dataset of 328 records from 328 unique patients was used. As shown in the Table 1, the AUC (sequence-level, set-level) performance of the DNN was firstly compared against the gold standard (cardiologist consensus committee), and the F1 score (sequence-level, set-level) of the DNN was compared against that of the average cardiologists. The 'sequence-level' means that predictions of the DNN at each output interval (every 256 samples, ~1.28 seconds) were compared with the corresponding labels of committee consensus. On the other hand, the 'set-level' means that throughout a given ECG record, the set of unique rhythm classes predicted by the DNN was compared with the set of rhythm classes annotated by the committee consensus. The Table 2 shows performance of the DNN and average cardiologists sensitivity compared to the cardiologist committee consensus, with specificity fixed at the average specificity level achieved by average cardiologists. The verification results show that performance of the DNN model generally outperforms that of average cardiologists with reasonable mistakes.

31

Outline

- ❑ **Introduction :**
  Artificial Intelligence, Machine Learning, and Deep Learning
- ❑ **(Deep) Neural Network Applications for Bio-medical Circuit and Systems**
- ❑ **H/W Implementation of Deep Learning Inference Processors :** 1) DNPU, 2) UNPU
- ❑ **Future Direction :**
  Neuromorphic Computing with Nonvolatile Memory

Next, we will look into some representative deep learning inference processors such as the DNPU and the UNPU, accelerating inference operation of deep neural network. The proposed deep learning inference processors will be covered in terms of 1) motivations and design challenges/improvements, 2) overall architecture, and 3) chip photograph & measurement results. Firstly, we will introduce the DNPU, which is a pioneer work accelerating inference operation of both convolution layers and recurrent layers/fully connected layers in mobile application.

32

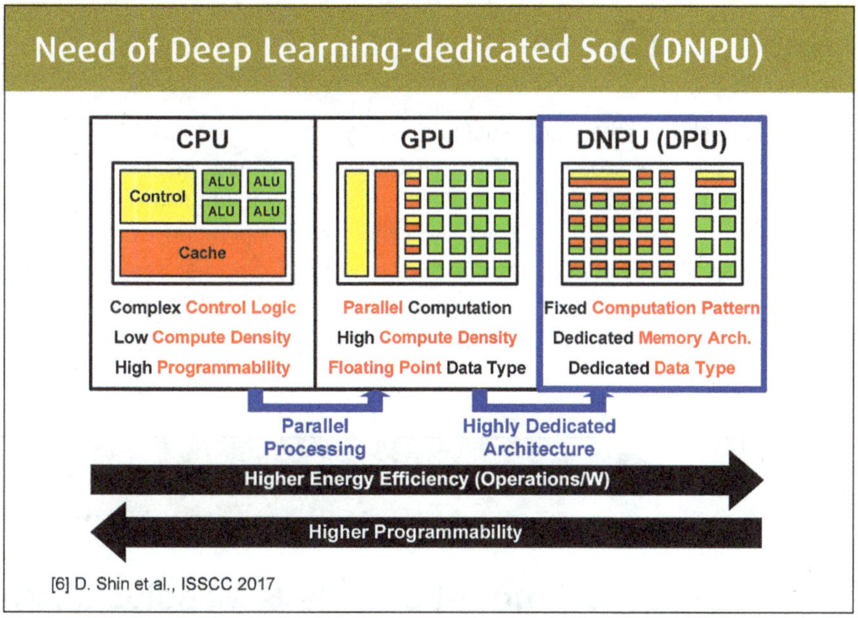

This slide shows a necessity of deep learning-dedicated inference processors. As mentioned before, the deep learning algorithm consists of 1) training operation and 2) inference operation. Especially, inference operation of deep learning algorithm requires massive amount of computation according to the increase of deep neural network complexity. To cope with the massive amount of computation, the general purposed processors such as CPU and GPU have been used. Generally, GPU is more preferred for inference operation of deep learning algorithms due to GPU's capability of parallel computation and high compute density. Although GPU can be used for an inference operation of the deep learning algorithms, the inference operation does not require a 32 bit floating point operation, thus generating inefficient computation of GPU which uses data-type of the 32 bit floating point. Therefore, with sacrificing high programmability, deep learning-dedicated processing unit (DPU), which has unique characteristics of fixed computation pattern, dedicated memory architecture, and dedicated data-type, is highly required to achieve higher energy efficiency for the inference operation of deep learning algorithms.

33

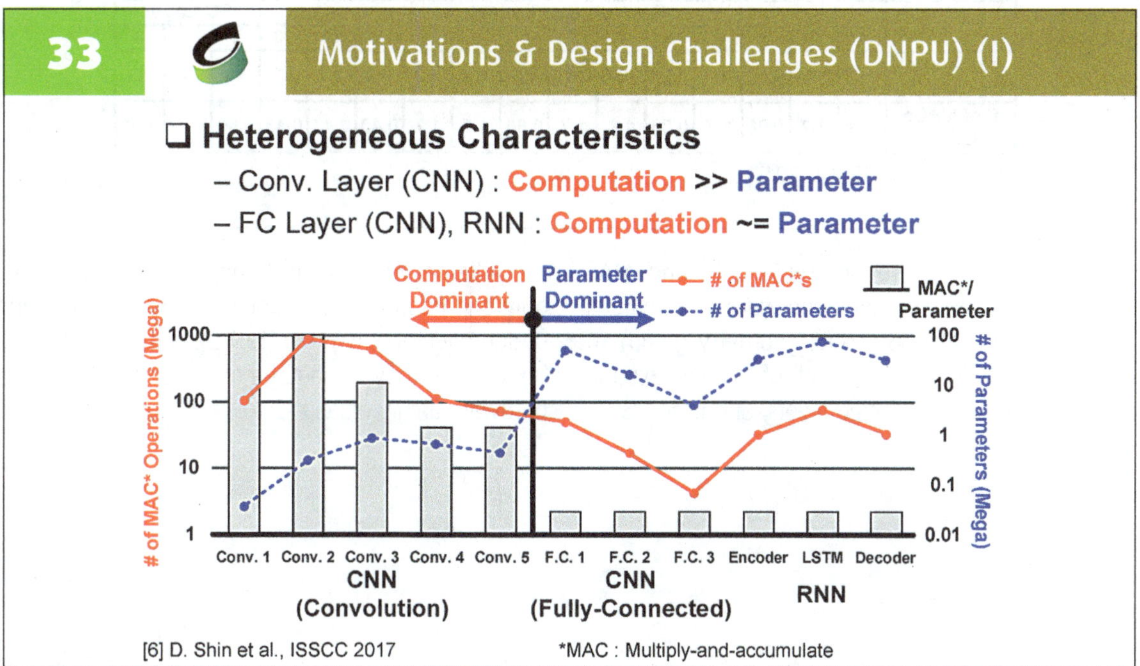

33 Motivations & Design Challenges (DNPU) (I)

This slide shows the motivations and design challenges of the proposed DNPU. Firstly, layers of deep neural networks such as convolution layers (CLs) and recurrent layers (RLs)/fully connected layers (FCLs) have heterogeneous characteristics. The CLs have relatively small numbers of parameters than RLs/FCLs, but its computation workload is more heavy. On the other hands, RLs/FCLs relatively have much more parameter and less multiply-and-accumulate (MAC) operation. Generally, more MAC operations need more process units, and more parameters need more external memory accesses. Therefore, optimizing all kinds of layers with single hardware architecture is difficult.

34 Motivations & Design Challenges (DNPU) (II)

❑ Large Amount of Data & Massive MACs

- Total # of Kernel Weights : **70 M** (1 Inference @ VGG-16)
- Total # of MACs : **14 G** (1 Inference @ VGG-16)

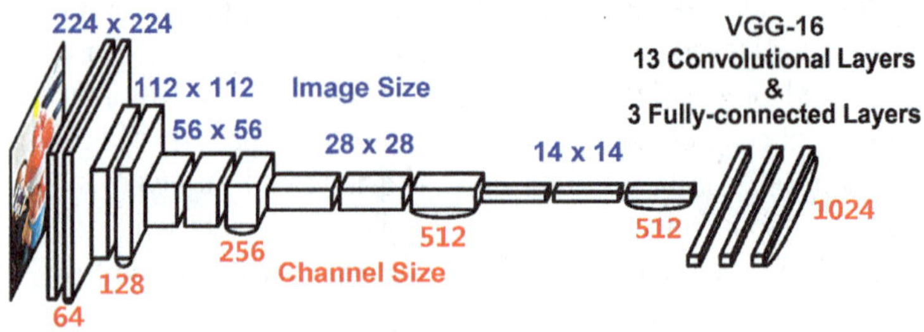

	Convolution Layer													FC Layer		
	1	2	3	4	5	6	7	8	9	10	11	12	13	1	2	3
# of Kernel Weights (M)	0.01	0.04	0.07	0.14	0.28	0.56	0.56	1.1	2.2	2.2	2.2	2.2	2.2	49	8	2
# of MACs (GOPS)	0.08	1.7	0.86	1.7	0.86	1.7	1.7	0.86	1.7	1.7	0.43	0.43	0.43	0.09	0.02	0.01

[6] D. Shin et al., ISSCC 2017

Not only that, there are huge data and MAC operations in deep neural network. The figure is simple diagram of VGG (visual geometry group)-16 network. VGG-16 is composed of 13 convolution layers and 3 fully-connected layers, and it needs 70 mega kernel weights and 14 giga MAC operations for only one inference operation. To sum up, workloads of deep neural networks (CLs, RLs, and FCLs) are very heavy, but they cannot be optimized with single hardware architecture.

35 Overall Architecture (DNPU)

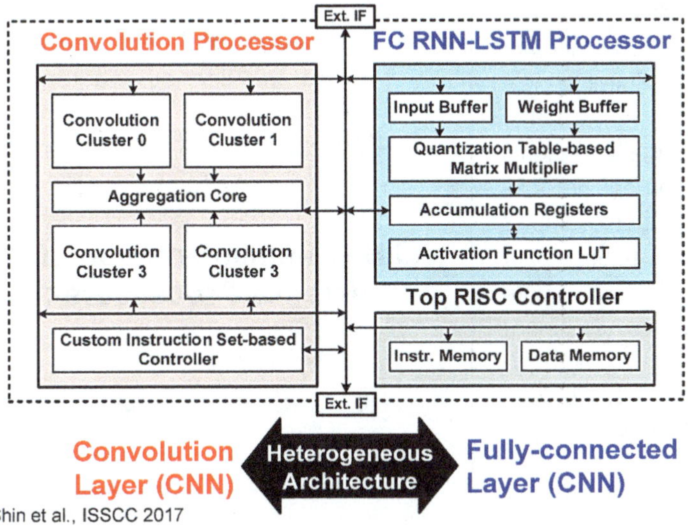

[6] D. Shin et al., ISSCC 2017

To solve abovementioned design challenges, the proposed architecture is heterogeneous one. The proposed design has convolution processor, fully connected & LSTM processor, and a top RISC controller. The convolution processor is composed of 4 convolution clusters, aggregation core, and custom instruction set-based controller, while the FC RNN-LSTM processor has quantization-table-based matrix multiplier, accumulation register, and activation look-up table as shown in the figure.

36 Chip Photograph & Measurement Results (DNPU)

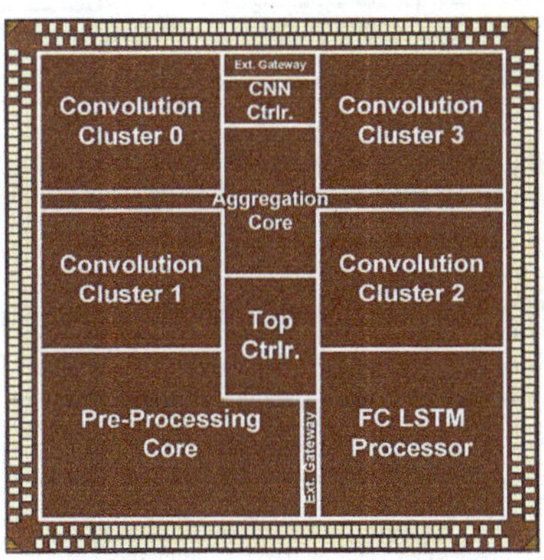

	Specifications		
Process	65nm 1P8M CMOS		
Die Area	4mm x 4mm (16mm^2)		
SRAM	CLP	FCRLP	
	280 KB	10 KB	
Supply	0.77V ~ 1.1V		
Frequency	50 ~ 200MHz		
Power	50MHz @ 0.77V	34.6mW	
	200MHz @ 1.1V	279mW	
Energy Efficiency		CLP Word Length: 16b	CLP Word Length: 4b
	50MHz @ 0.77V	2.1 TOPS/W	8.1 TOPS/W
	200MHz @ 1.1V	1.0 TOPS/W	3.9 TOPS/W

[6] D. Shin et al., ISSCC 2017

This slide shows chip microphotograph and measurement results of the proposed deep learning inference processor. The DNPU shown in the photo is fabricated with 65nm CMOS technology and it occupies 16mm^2 die area. The lowest operation voltage is 0.77V. The DNPU consumes 34.6mW at 50MHz with 0.77V supply voltage and shows 8.1TOPS/W energy efficiency with 4 bit word length at 0.77V supply voltage.

37

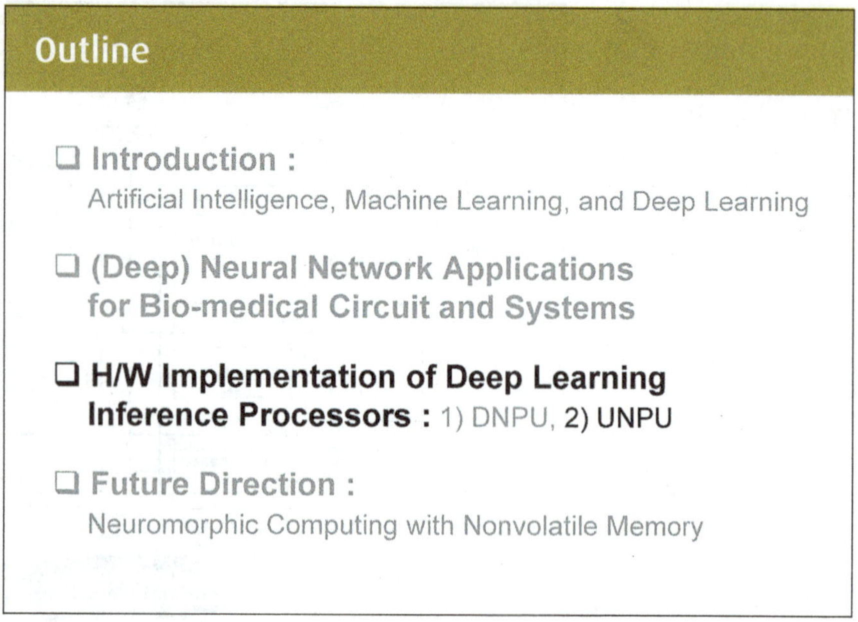

Secondly, we will introduce another deep learning inference processor, UNPU that partially inherits motivations of previous DNPU but shows different solution for acceleration of both computation-dominant CLs and memory-dominant RLs/FCLs with single accelerator chip.

38

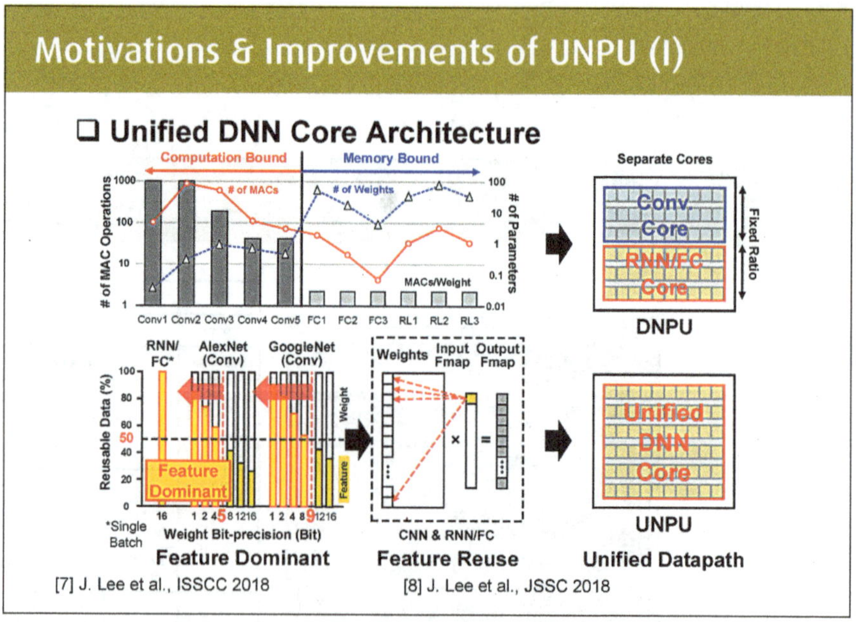

In this slide, motivations and improvements of UNPU will be introduced. The DNPU, which is the previously mentioned deep learning inference processors, has heterogeneous core architecture in order to cope with acceleration of both computation-dominant CLs and memory-dominant RLs/FCLs. In real-world applications, however, there are so many kinds of workload combinations between CLs and RLs/FCLs. In the case of DNPU, which has a heterogeneous core architecture, if any application requires only CLs or RLs/FCLs, it suffers from low utilization problem because the number of processing engines (PEs) are limited by their area. Therefore, the UNPU proposes unified DNN core architecture based on feature reuse as shown in the lower-left figure. There are two kinds of data for external memory access; 1) weights and 2) features. When the weight bit precision is reduced to operate DNN on the optimal weight bit precision, the feature-map proportion increases. On the other hand, RLs/FCLs are always feature dominant because their weights are not reused in a single batch operation. In this work, the proposed processor, UNPU, reuses input features as many as possible to minimize the external memory access. By doing so, the DNN cores can be unified.

39

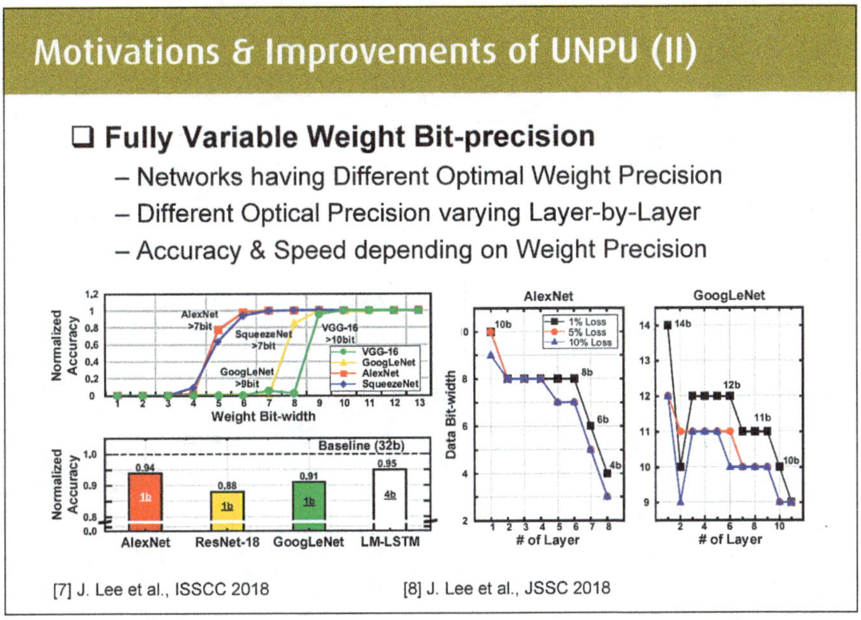

In this slide, another key motivations and improvements of the UNPU will be introduced. Firstly, as shown in the upper-left figure, different networks require different weight precisions. For example, AlexNet only requires more than 7 bit, while VGG-16 requires more than 10 bit weight precisions to guarantee the accuracy. Secondly, different layers in DNN has different optimal precisions. The minimum precision required for minimum accuracy loss is very different among the layers as show in the right figure. Lastly, weight precision is varied according to the accuracy requirement for the given DNN. For example, AlexNet can be operated using 1 bit weight precision, but it sacrifices some degrees of its network accuracy as show in the lower-left figure. For these reasons, the UNPU support fully variable weight bit-precision.

40

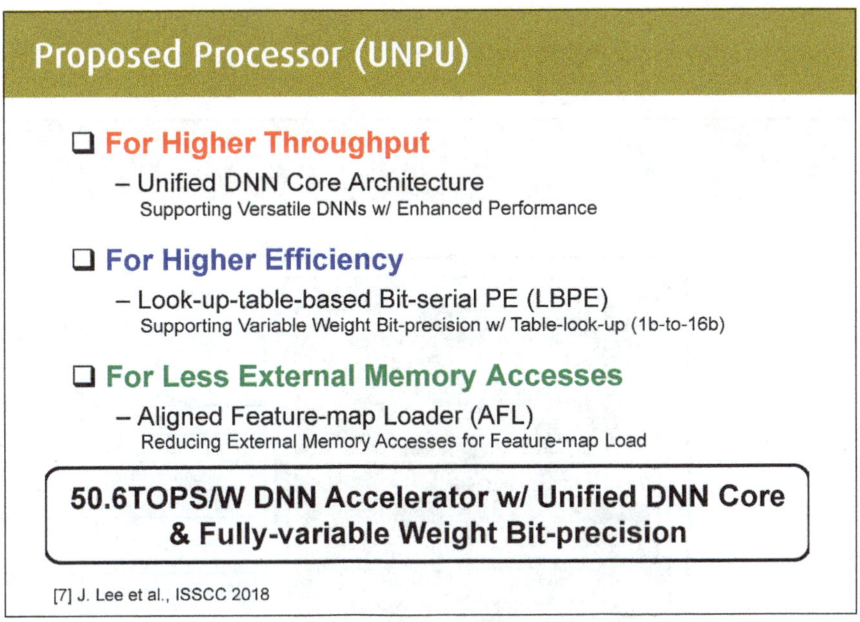

The UNPU proposes 3 key features. For high throughput, unified DNN core architecture is proposed and implemented in order to support various workload combination between CLs and RLs/FCLs with improved performance. For high efficiency, loop-up-table-based bit-serial processing engine is implemented to support variable weight bit precision from 1 bit to 16 bit with improved energy-efficiency. Lastly, aligned feature-map loader is implemented to reduce external memory access, which is required for feature-map load. Consequently, the proposed UNPU achieves 50.6TOPS/W with unified DNN core and fully-variable weight bit-precision.

41

Overall Architecture of UNPU

❑ Unified Architecture

– LUT-based Bit-serial PE (LBPE) — Reusing Feature-map (Fmap) → Enhancing Energy Efficiency

– Aligned Feature Map Loader (AFL) — Fetching Input Fmaps into LBPEs → Reducing External Memory Access

– Weight Memory (48KB SRAM)

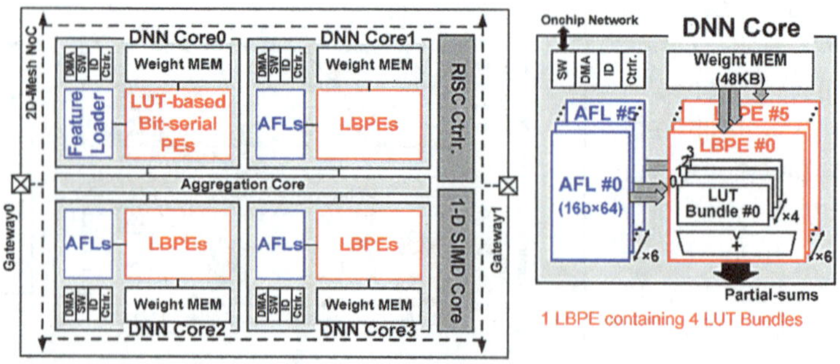

[7] J. Lee et al., ISSCC 2018 [8] J. Lee et al., JSSC 2018

In this slide, overall architecture of the UNPU will be introduced. The UNPU consists of 4 unified DNN cores, and each DNN core includes look-up-table-based bit-serial processing engine (LBPE), aligned feature loader (AFL), and weight memory (48KB SRAM). They are connected with 2D mesh NoC, and 4 DNN cores transfer partial-sums to aggregation core. Through the dual external gateway, the UNPU is connected to external memory interfaces such as DDR3 memory controller in FPGA. A RISC controller and 1-D SIMD core perform remaining operations such as activation and normalization. Also, detailed information of the unified DNN core architecture is described in this slide. In the DNN core, 6 pairs of AFL and LBPE are included. The AFL is a buffer to give input activations to corresponding PE, while supporting in-buffer data movement to reduce redundant external memory accesses. In the LBPE, 4 look-up-table (LUT) bundles are included. They energy-efficiently make partial sums by table-lookup, and weight values are broadcasted from the local weight memory.

42

Chip Photograph & Measurement Results

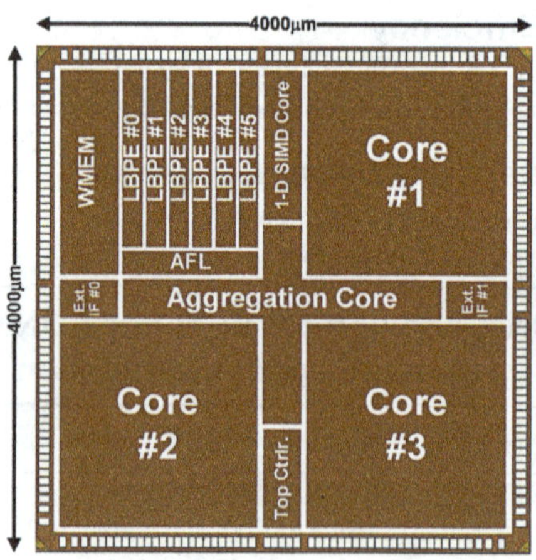

	Specifications
Technology	65nm Logic CMOS
Area [mm²]	16
SRAM [KB]	256
Supply voltage [V]	1.1
Frequency [MHz]	200
Weight Precision [Bit]	**1 – 16**
Power [mW]	**297 @ 200MHz, 1.1V**
	3.2 @ 5MHz, 0.63V
Power Efficiency [TOPS/W]	**50.6 (1b Weight)**
	3.08 (16b Weight)

[7] J. Lee et al., ISSCC 2018

42 Chip Photograph & Measurement Results

This slide shows chip microphotograph and measurement results of the UNPU. The UNPU is fabricated with 65nm logic CMOS with $16mm^2$ die area. The UNPU includes totally 256KB on-chip scratch pad SRAM. Maximum and minimum operating frequency of the UNPU is 200MHz and 5MHz, respectively. As mentioned before, the UNPU supports fully variable weight precision from 1 bit to 16 bit by adopting bit-serial manner operation. The power consumption at the nominal supply voltage is 297mW, and it can be lowered to 3.2mW at the 0.63V supply voltage. The maximum power efficiency of the UNPU is 50.6 TOPS/W with 1 bit weight precision and 3.08 TOPS/W with 16 bit weight precision.

43 Outline

- **Introduction :**
  Artificial Intelligence, Machine Learning, and Deep Learning
- **(Deep) Neural Network Applications for Bio-medical Circuit and Systems**
- **H/W Implementation of Deep Learning Inference Processors :** 1) DNPU, 2) UNPU
- **Future Direction :**
  Neuromorphic Computing with Nonvolatile Memory

Finally, we will talk about a neuromorphic computing with emerging nonvolatile memory (eNVM) as a future direction of AI (deep learning) processors in bio-medical application. Recently, wearable healthcare technologies has emerged as a killer application of various healthcare services and care of chronic diseases. Wearable healthcare devices such as patch-type bio-signal monitoring devices and smartwatch are a core of wearable healthcare service, becoming more 'convenient', 'smart', 'active', and 'accurate' with the assistance of AI. Especially, as mentioned in the second subsection which is "(Deep) Neural Network Application for Bio-medical Circuit and Systems", deep neural network technology provides more convenient (non-invasive) and accurate monitoring/detection of patient's biometric information such as non-invasive estimation of blood glucose level and detection of arrhythmia, respectively. For these reasons, combination of wearable healthcare devices and deep neural network technology is inevitable. However, in many cases, direct application of deep learning processor to the wearable healthcare devices is limited, considering key requirements of wearable healthcare devices such as extremely low power consumption and small PCB area. In this situation, the neuromorphic computing with eNVM can be promising alternative due to its inherent efficiency.

44

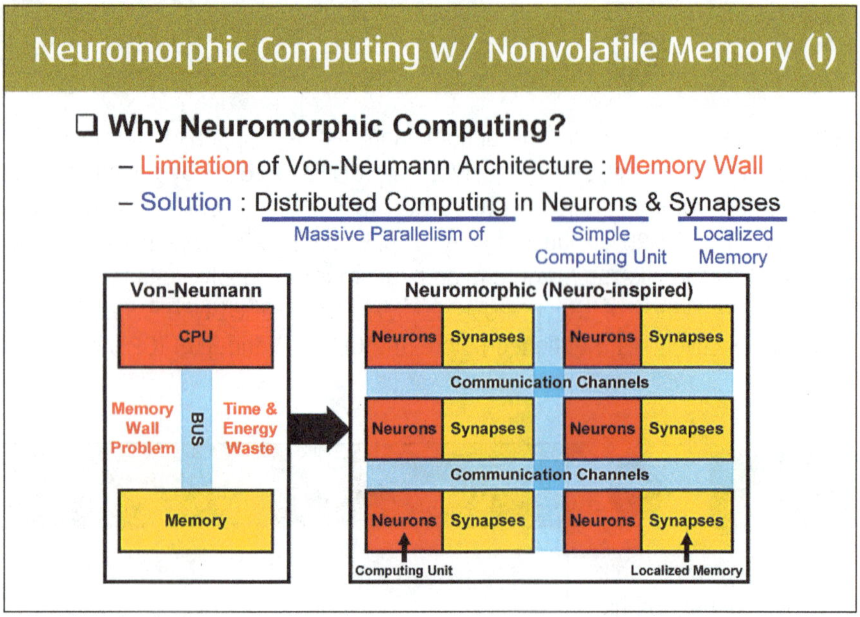

This slide shows an appropriateness of paradigm shift from traditional Von-Neumann architecture-based deep learning operation to the neuromorphic computing with eNVM. As shown in the left figure, the well-known Von-Neumann architecture's memory wall problem arises from communication bottleneck between the processing unit (CPU) and on-chip SRAM which has insufficient storage capacity (typically under a few megabytes) with large external (off-chip) memory accesses, generating a waste of processing time and energy. Although SRAM has been following the CMOS scaling trend, extremely large number of parameters in deep learning algorithms (typically hundreds of megabytes) cannot be loaded into the SRAM. For this reason, as shown in the right figure, the neuromorphic computing with eNVM proposes novel computing paradigm, which is a massive parallelism of simple computing unit ('Neurons') and localized memory ('Synapses') as a solution.

45

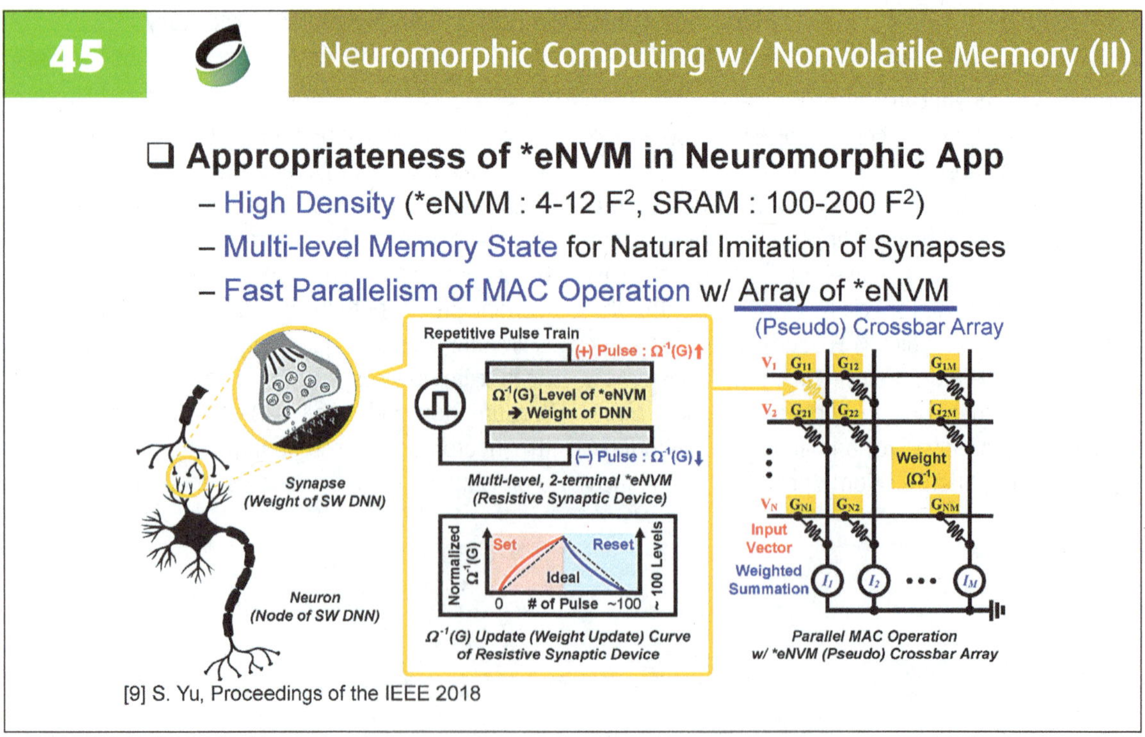

45 Neuromorphic Computing w/ Nonvolatile Memory (II)

The eNVM can maintain a written data with no power consumption, and this kind of eNVM's characteristic have an advantage in deep learning application because eNVM can hold deep neural network's parameters such as weights and biases, which are the deep neural network itself. Although eNVMs used in the neuromorphic computing include many kinds of nonvolatile memory, resistance-based eNVMs such as phase change memory (PCM) and resistive random access memory (RRAM) are preferred because it can represent information as multi-level resistance (or conductance) value. An appropriateness of the eNVM in neuromorphic computing application can be explained with three kinds of perspectives: 1) eNVM's High Density, 2) eNVM(especially in the case of PCM and RRAM)'s multi-level expression capability, and 3) capability of fast parallelism of MAC (weighted sum) operation with an array of eNVM. Firstly, the eNVM (typically 4-12 F2 per bit cell) shows more higher cell density than the traditional SRAM (typically 100-200 F2 per bit cell), thus enabling storage of extremely large number of parameters of today's deep learning algorithm without external (off-chip) memory accesses. Secondly, eNVM(PCM and RRAM)'s multi-level memory state which is represented by multiple levels of resistance (or conductance) enables natural implementation of synapses between two adjacent neurons which is equally corresponded to weights of deep neural network. As shown in the left & middle figure, in the neuromorphic computing with eNVM application, deep neural network's weight parameters can be implemented with two-terminal eNVM(PCM and RRAM) which called by resistive synaptic device (RSD), and conductance value of RSD (weight of the deep neural network) can be gradually changed by repetitive pulse train. In order to increase and decrease conductance of RSD, positive and negative pulse train should be applied to the RSD, respectively. The number of conductance levels of the RSD can be up to approximately one hundred levels (order of 6 Bits), and increasing phase & decreasing phase of conductance level of RSD are called by set phase and reset phase, respectively. Note that actual update curve of conductance level shows nonlinear and asymmetric characteristic, thus generating inaccuracy in inference and training phase (for further information, refer to [9]). Thirdly, large scale array implementation of RSD (eNVM) enables fast parallelism of MAC operation. As shown in the right figure, input vector which is represented by voltage information in a the row of the eNVM array can be multiplied by first column of weight parameters which are expressed by conductance information of RSD in the eNVM array, thus generating weighted summation which is represented by current summing, ***I1***, at the first column of the eNVM array. Because the MAC operation can be conducted in a parallel manner, array implementation of RSD enables inherently efficient deep learning inference operation. Note that the abovementioned parallel MAC operation based on RSD array, which is often called by crossbar array, is free from the well-known sneak path problem when all rows and columns of the crossbar array are activated for parallel MAC operation.

46 Neuromorphic Computing w/ Nonvolatile Memory (III)

❑ **Operation of Neuromorphic Computing w/ *eNVM**

– Inference Operation (Parallel MAC Operation & Activation)

– Training Operation (Row-by-row Weight Update)

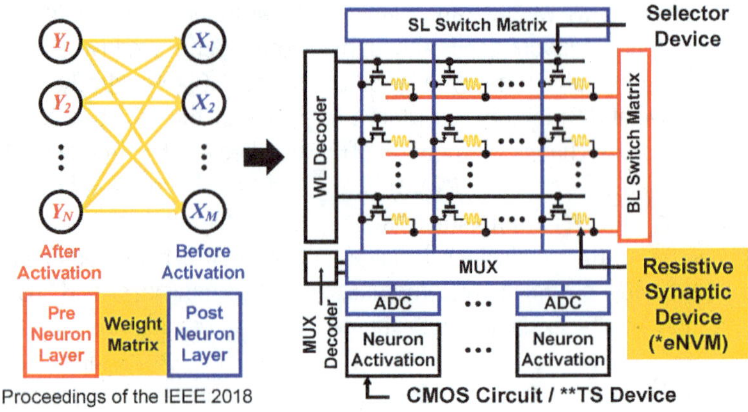

[9] S. Yu, Proceedings of the IEEE 2018

This slide shows two kinds of major operations of neuromorphic computing with eNVM: 1) inference operation that consists of parallel MAC operation & activation, and 2) training operation which is represented by row-by-row weight update. As shown in the figure, weight matrix with pre-neuron layer (after activation) and post-neuron layer (before activation), which are basic elements of deep neural network, are implemented with a pseudo crossbar array architecture based on eNVM (RSD). The pseudo crossbar array architecture is proposed to improve half-selection problem of the conventional crossbar array architecture. In the conventional crossbar array architecture as shown in the right figure of the previous slide, many half-selected RSDs can be generated. For example, if VDD & VDD/2 are applied to the first row & the other rows, respectively, also if GND & VDD/2 are applied to the first column & the other columns, respectively, (G_{12} ~ G_{1M}) and (G_{21} ~ G_{2N}) will be half-selected devices, thus generating crosstalk, IR drop problem, and large additional power consumption in large scale array implementation. As an alternative, the pseudo crossbar array architecture with 1 transistor and 1 RSD (1T1R) is proposed as a near term solution because IR drop and additional power consumption by half-selected devices can be minimized with a selection function of the transistor (also, threshold switching I-V selector can be used as a selector device). The pseudo crossbar array mainly consists of 3 parts: 1) array core which consists of unit cells (1 unit cell : 1 transistor as a selector device and 1 RSD), 2) peripheral switches which consist of selection line (SL) switch matrix, word line (WL) decoder, bit line (BL) switch matrix, and MUX, 3) neuron circuit which consists of ADCs & neuron activation circuits. For inference operation, the WL decoder turns on the all selection transistors, thereby operating all selection transistors in the deep triode region and connecting the bit lines with selection lines for parallel MAC operation. As mentioned in the previous slide, the parallel MAC operation is implemented with a multiplication of bit line voltages (actually, voltage pulse is preferred as input vector to prevent distortion of current summing by the I-V nonlinearity of RSD) and RSD's conductances, thus generating current summing in the selection lines. The current summing which represents the results of weighted summation passes through the MUX, and then the ADC convert the current summing into the application specific information such as the number of pulses, pulse width, digital bit, and spike signal with activation in the neuron activation circuit.

46 Neuromorphic Computing w/ Nonvolatile Memory (III)

Note that if the ADC & the neuron activation circuit (neuron circuit) occupy much larger area than the column pitch of the pseudo crossbar array, column pitch matching problem which means that multiple columns have to share one neuron circuit occur, thus obstructing parallelism of MAC operation. To avoid column pitch matching problem, it is important to implement neuron circuits with reduced area, and various approaches such as implementation of neuron circuit with single threshold switching (TS) device are being tried. For the training operation, although the update operation of the RSD's conductance can be performed in a fully parallel manner, fully parallel training operation requires huge instantaneous power from the peripheral circuits. Therefore, the update operation of the RSD's conductance is performed row-by-row (column-by-column) with the voltage pulse between the bit line and the selection line.

47 Neuromorphic Computing w/ Nonvolatile Memory (IV)

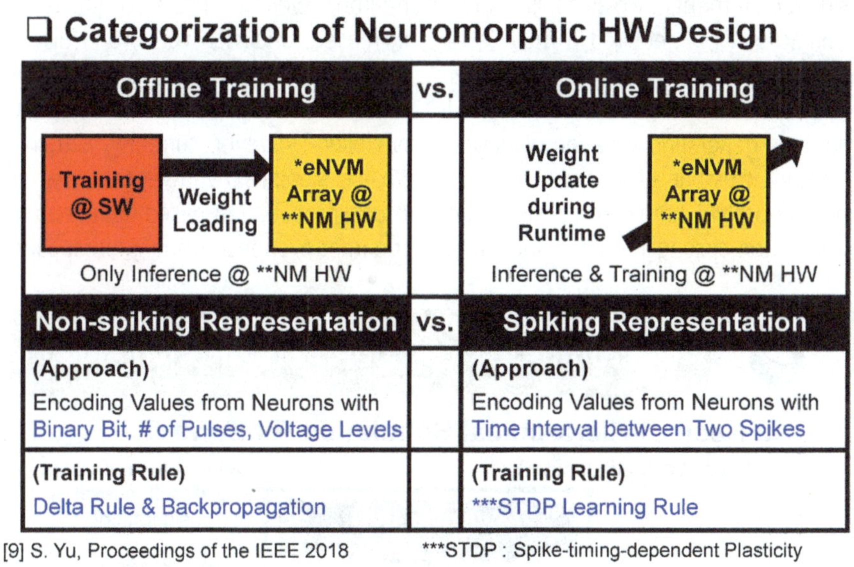

A categorization of neuromorphic hardware is meaningful because hardware design requirement for the neuromorphic computing significantly depends on 1) application-dependent operation scenario & algorithm as well as 2) how to encode information which is generated from the neuron circuit. Firstly, there are two kinds of training scenario: offline training and online training. In the offline training operation scenario, the training is performed with software, and the trained deep neural network's weights which are the deep neural network itself are mounted on the pseudo crossbar array of the neuromorphic hardware by one-off writing which is correspond to the conductance update of the RSDs. Therefore, in the offline training scenario, the neuromorphic hardware performs only inference operation after one-off writing of the deep neural network's weight parameters (no training operation). On the other hand, in the online training scenario, according to the application-dependent operation scenario & algorithm, training can be performed many times on the neuromorphic hardware (on-chip training) during a runtime or a configured time period. Therefore, the neuromorphic hardware should provide

47 Neuromorphic Computing w/ Nonvolatile Memory (IV)

a function of both inference and training operation in the online training scenario. Abovementioned different characteristic between offline training and online training generates different requirements about RSD's conductance (deep neural network's weight) update. In the case of offline training, accuracy of conductance update relatively more important than speed because only inference operation is performed without training operation which provide change of RSD's conductance value. On the other hand, in the online training, speed of conductance update relatively more important than accuracy because RSD's conductance can be revised by subsequent training operations. Note that the RSD's other design requirements such as nonlinearity/asymmetry characteristic, required multi-level state, retention time & endurance, and uniformity & variability are also dependent on abovementioned training scenario, thus generating variation of design requirement of pseudo crossbar array as well.

Secondly, neuromorphic hardware design can be categorized into non-spiking representation and spiking representation according to how to encode information from the neuron. In the non-spiking representation, information from the neuron is encoded by digital bit or the number of voltage pulses or voltage levels, and the encoded information is used by the delta rule & backpropagation algorithm which are used as a training rule. On the other hand, in the spiking representation, information from the neuron is encoded by time interval between spikes from neighboring neurons, and the encoded information is used by a training rule such as the spike-timing-dependent plasticity (STDP) whose rule is that if the postsynaptic neuron fires earlier than the presynaptic neuron, weight (RSD's conductance) should be increased, and vice versa, and also, the change of weight (RSD's conductance) is larger in the case that the timing between the two adjacent neurons firing is closer. However, currently, a training accuracy of STDP is significantly lower than that of the delta rule & backpropagation-based deep learning algorithm in the image/speech recognition application.

48 Neuromorphic Computing w/ Nonvolatile Memory (V)

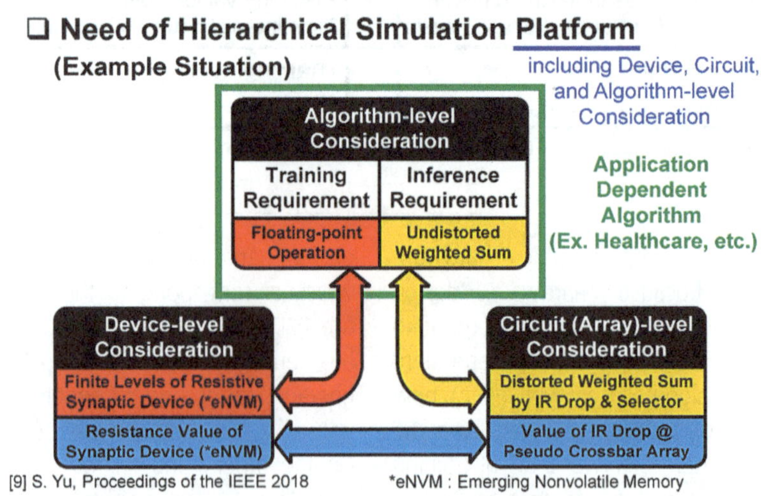

In order to successfully design the neuromorphic hardware, comprehensive analysis including algorithm-level, device-level, and circuit(array)-level considerations is essential because the design requirements of each level are closely related. For example, application-dependent algorithm-level

48 Neuromorphic Computing w/ Nonvolatile Memory (V)

consideration can include two kinds of requirements: training requirement and inference requirement. For training requirement, state-of-the-art software deep learning algorithms require 32bit or 16bit floating point operation in the training operation, however, this algorithm-level requirement is limited by the problem of RSD's finite levels, which is included in the device-level consideration. On the other hand, in the case of abovementioned inference requirement, software deep learning algorithms require undistorted weighted sum (MAC) operation, however, in the pseudo crossbar array architecture of the neuromorphic hardware, distortion of weighted sum operation is generated by IR drop problem and selector device's nonlinear I-V characteristic, which are included in the circuit(array)-level consideration. Also, a resistance value of the RSD which is included in the device-level consideration should be determined considering the degree of IR drop in the pseudo crossbar array architecture to prevent accuracy degradation of the inference and training operation, which shows close relationship between the device-level consideration and the circuit(array)-level consideration. To reflect and support abovementioned comprehensive analysis on the hierarchical neuromorphic hardware design, the NeuroSim, which is a simulation platform for the neuromorphic hardware design, is being used, and interdisciplinary collaborations are actively being performed.

49 Neuromorphic Computing w/ Nonvolatile Memory (VI)

❑ **Current Status & Limitations (Online Training)**

	Algorithm-level	Device-level	Circuit (Array)-level
Current Status	1. Pruning of DNN and Weight Precision Reduction @ Inference 2. Floating Point Operation @ Training	1. Development of Resistance-based *eNVM focusing on Digital Memory Application (< 8 Levels)	1. Single Device or Small Scale Array Implementation with Software Neuron (Activation, Output)
Major Limitations	1. Requirement of High Precision Weight (≥ 6 Bit) @ Training	1. Finite Levels of **RSD (*eNVM) 2. Too Slow Training by Long Pulse Width for **RSD Ω^{-1} Update	1. No Implementation of Large Scale Pseudo Crossbar Array with Peripheral Neuron Circuits and ***TS I-V Selector

Future Direction : Fundamental Reconsideration about Algorithm and Hardware Co-optimization

[9] S. Yu, Proceedings of the IEEE 2018 ***TS : Threshold Switching

Finally, we will check current status and limitations of the neuromorphic computing with the eNVM. Firstly, in the algorithm level, pruning of deep neural network and reduction of weight precision for the efficient inference operation are widely attempted, however, training operation is still and generally performed with floating point operation, thus generating limitation of implementing software deep learning algorithm with neuromorphic hardware which only supports approximately 100 levels (6 bit) of the RSD's conductance. Secondly, in the device-level, it is the fact that development of resistance-based eNVM is mainly focusing on digital memory application (only supports 1~8 levels of the RSD's

49 Neuromorphic Computing w/ Nonvolatile Memory (VI)

conductance), thus generating the limitation of finite levels of the RSD's conductance and the problem of too slow training time caused by requirement of long pulse width for the RSD's conductance update. Thirdly, in the circuit(array)-level, only single device or small scale array implementation with software neuron is being developed without implementation of large scale pseudo crossbar array including peripheral neuron circuits and threshold switching I-V selector. Given abovementioned status and limitations, main bottleneck of implementing the software deep learning algorithm with neuromorphic hardware is related to the training operation, and therefore, fundamental reconsideration about co-optimization of software algorithm and hardware should be performed in the near future.

50 References

1. K. Song, U. Ha, S. Park, J. Bae, and H.-J. Yoo, "An Impedance and Multi-Wavelength Near-Infrared Spectroscopy IC for Non-Invasive Blood Glucose Estimation," in *IEEE J. Solid-State Circuits*, vol. 50, no. 4, pp. 1025-1037, Apr. 2015.
2. U. Ha, J. Lee, J. Lee, K. Kim, M. Kim, T. Roh, S. Choi, and H.-J. Yoo, "A 25.2mW EEG-NIRS Multimodal SoCfor Accurate Anesthesia Depth Monitoring," in *IEEE Int. Solid-State Circuits Conf. (ISSCC) Dig. Tech. Papers*, Feb. 2017, pp. 450–451.
3. S. Yin, M. Kim, D. Kadetotad, Y. Liu, C. Bae, S. Kim, Y. Cao, and J.-S. Seo, "A 1.06 μWSmart ECG Processor in 65 nm CMOS for Real-Time Biometric Authentication and Personal Cardiac Monitoring," in *Proc. IEEE Symp. VLSI Circuits (VLSI-Circuits)*, Jun. 2017, pp. C102-C103.
4. P. Rajpurkar, A. Hannun, M. Haghpanahi, C. Bourn, and A. Ng. (Jul. 2017). "Cardiologist-Level Arrhythmia Detection with Convolutional Neural Networks," [Online]. Available : https://arxiv.org/abs/1707.01836
5. M. P. Turakhia, D. D. Hoang, P. Zimetbaum, J. D. Miller, V. F. Froelicher, U. N. Kumar, X. Xu, F. Yang, P. A. Heidenreich, "Diagnostic Utility of a Novel Leadless Arrhythmia Monitoring Device," in *The American Journal of Cardiology*, 112(4):520-524, 2013.
6. D. Shin, J. Lee, J. Lee, and H.-J. Yoo, "DNPU: An 8.1 TOPS/W Reconfigurable CNN-RNN Processor for General-purpose Deep Neural Networks," in *IEEE Int. Solid-State Circuits Conf. (ISSCC) Dig. Tech. Papers*, Feb. 2017, pp. 240–241.
7. J. Lee, C. Kim, S. Kang, D. Shin, S. Kim, and H.-J. Yoo, "UNPU: An Energy-Efficient Deep Neural Network Accelerator With Fully Variable Weight Bit Precision," in *IEEE Int. Solid-State Circuits Conf. (ISSCC) Dig. Tech. Papers*, Feb. 2018, pp. 218–219.
8. J. Lee, C. Kim, S. Kang, D. Shin, S. Kim, and H.-J. Yoo, "UNPU: An Energy-Efficient Deep Neural Network Accelerator With Fully Variable Weight Bit Precision," in *IEEE J. Solid-State Circuits*, vol. 54, no. 1, pp. 173-185, Jan. 2019.
9. S. Yu, "Neuro-inspired Computing with Emerging Nonvolatile Memory," in *Proceedings of the IEEE*, vol. 106, no. 2, pp. 260-285, Feb. 2018.
10. A. Y. Hannun, P. Rajpurkar, M. Haghpanahi, G. H. Tison, C. Bourn, M. P. Turakhia, and A. Y. Ng, "Cardiologist-level Arrhythmia Detection and Classification in Ambulatory Electrocardiograms using a Deep Neural Network," in *Nature Medicine*, 25, 65-69, 2019.

CHAPTER 06

Neuro-Inspired Computing and Neuromorphic Processors for Biomedical Circuits and Systems

Jae-sun Seo and Shimeng Yu

Arizona State University, USA

In this chapter, we present neuro-inspired computing and neuromorphic processors in the literature, which can be suitable for biomedical circuits and systems that necessitate real-time performance, low power, and small footprint.

First, we will present the TrueNorth neuromorphic processor designed by IBM. TrueNorth processor includes 1 million neurons and 256 million synapses in the 28nm CMOS chip, which is implemented in all-digital fashion without on-chip learning support. Second, we will present the ROLLS processor designed by ETH Zurich. ROLLS processor implemented 256 neurons and 128K synapses in the 180nm CMOS chip with analog/mixed-signal circuits and support long/short-term on-chip learning. Third, we will present the recent Loihi processor designed by Intel. Loihi processor encompasses 128 thousand neurons and 128 million synapses in the 14nm CMOS chip with all-digital implementation and supports on-chip learning. Since neuromorphic processors are designed based on our understanding of the brain operation/activity, they have potentials to make machines resemble the intelligence of brains and help us understand brain injury/disorders. Furthermore, these neuromorphic processors that exhibit spike-based computing are advantageous for biomedical circuits and systems that interface with brains, since they can directly communicate via spikes and thus can be integrated in the brain-machine-interface or brain stimulator systems.

Finally, we will present RRAM-based computing research literature, which includes in-memory mixed-signal computing with small or moderate array demonstrations. We will discuss the ideal targets of synaptic devices and neuronal devices, and the principle of using crossbar array for implementing the weighted sum and weight update operations in the neural networks. We will survey the representative experimental data of the weight update characteristics in the literature and then we will use a small multilayer perceptron model to illustrate the challenges of using RRAM based synaptic devices for online training. These RRAM based computing could be utilized in biomedical systems that exhibit severe area and power constraints.

1 A Shift in Computing Paradigm towards Neuro-inspired Computing

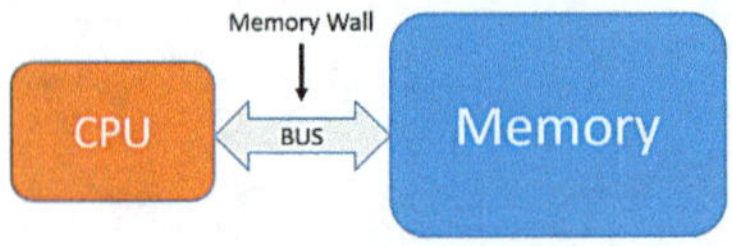

(a) von-Neumann architecture

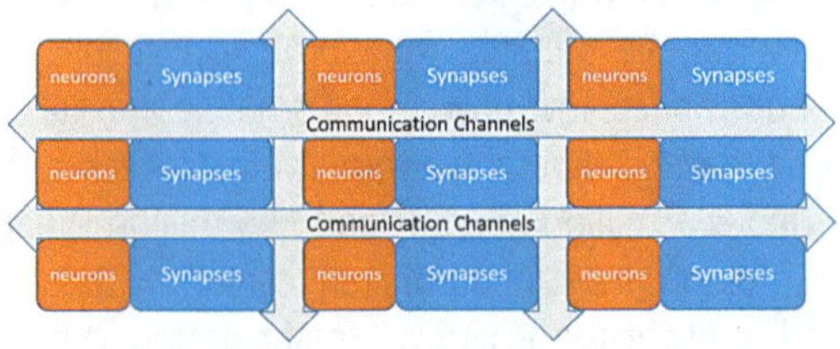

(b) Neuro-inspired Architecture

S. Yu, "Neuro-inspired computing with emerging non-volatile memory," **Proc. IEEE**, vol. 106, no. 2, pp. 260-285, 2018

In the recent years, there is a shift in computing paradigm towards neuro-inspired computing. Traditionally, the von-Neumann architecture has CPU and memory separated, the back and forth data transfer in the data bus has become a memory wall problem when dealing with large dataset in the deep learning algorithms.

Therefore the neuro-inspired architecture distributes the computation into a massively connected neural network, each neurosynaptic core has local neurons (the simple computing units) and synapses between neurons (the local memory). This parallel computing architecture could solve the traffic jam in the data bus in the convectional architecture.

Then how would the resistive device play a role in the neuro-inspired computing?

If we compare a biological synapse with an artificial synapse based on RRAM, there are many similarities, for example, the bio-synapse changes its conductance by releasing the Ca2+ or Na+ ions into the junction, while the resistive synapse changes its conductance by moving the oxygen ions and vacancies in the oxide materials. The learning happens in a neural network by changing the conductance in synapses essentially.

The long-term vision is to build a brain-like computer. There are many large programs in the recent years targeting at this field, e.g. DARPA SyNAPSE, DARPA UPSIDE, NSF Expeditions on Computing, and recently SRC's E2CDA and JUMP program.

2 Learning Algorithms for Cognitive Computing

There are a variety of machine learning and neuro-inspired algorithms as well as hardware implementations that have been presented in recent years.

On the left side, various learning algorithms and neural networks are illustrated, which include feed forward (fully-connected) neural networks, deep convolutional networks, recurrent neural networks, and generative adversarial networks.

On the right side, various neural network and neuromorphic computing hardware implementations have been surveyed for the past 30 years. It can be seen that feed forward (fully-connected) neural networks and spiking neural networks have been the popular choice of neural networks for hardware implementation.

2 Learning Algorithms for Cognitive Computing

- Various learning algorithms & neural networks

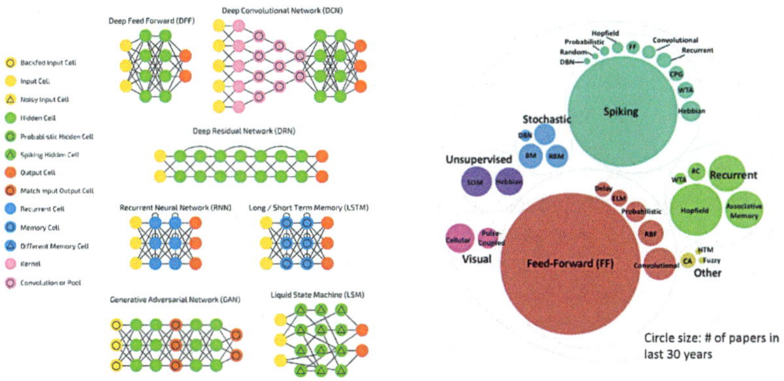

http://www.asimovinstitute.org/neural-network-zoo/

Schuman, et al., "A Survey of Neuromorphic Computing and Neural Networks in Hardware," arXiv 1705.06963, 2017.

3 Summary of Neuromorphic Hardware Approaches

	Off-the-shelf technologies	CMOS ASIC	Emerging resistive synaptic devices
Digital representation (machine/deep learning driven)	GPUs FPGAs	Google's TPU CNN accelerators: MIT's Eyeriss, KU Leuven's Envision, etc.	Analog synapses: IBM's 500×661 PCM array with off-chip neuron control [IEDM 2014] Binary synapses: ASU/Tsinghua's 16 Mb RRAM macro [IEDM 2016]
Spike representation (neuroscience driven)	SpiNNaker	Analog neuron: BrainScales, ROLLS Digital neuron: IBM's TrueNorth, Intel's Loihi Analog synapse: Floating gate memory arrays (Gatech's FPAA)	IBM's 256×256 PCM array with STDP neuron circuits [IEDM 2015]

S. Yu (Editor), Neuro-inspired Computing Using Resistive Synaptic Devices, Publisher: Springer, 2017.
A. Basu, J. Seo et al., "Low-Power, Adaptive Neuromorphic Systems: Recent Progress and Future Directions," IEEE JETCAS, vol. 8, 2018.

This is the taxonomy for neuromorphic hardware approaches. Depending on how we encode the information, we could classify into two categories in rows, one is digital representation for implementing machine/deep learning algorithms; the other one is spike representation that is more computational neuroscience driven.

Depending on the technology choice, we could classify into three categories in columns. The first one is off-the-shelf technologies, such as GPUs and FPGAs as machine/deep learning accelerators, or the European's SpiNNaker project based on clusters of ARM cores. The second one is based on silicon CMOS ASIC solution. Examples are Google's TPU and the reported CNN accelerators such as Eyeriss and Envision reported in ISSCC 2016-2017. Various approaches are also explored for implementing spiking neural networks into CMOS ASIC designs such as European's BrainScales and ROLLS, and IBM's TrueNorth and Intel's Loihi, as well as Gatech's FPAA based on floating gate memory arrays. The third one is based on emerging resistive synaptic devices, including the phase change memory (PCM) and resistive random access memory (RRAM).

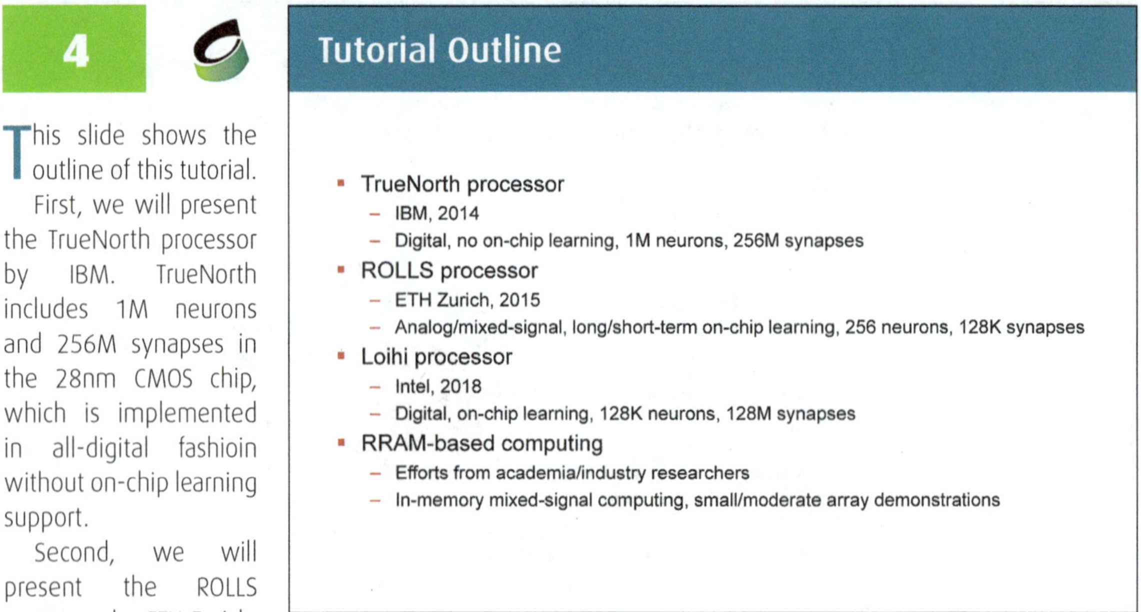

This slide shows the outline of this tutorial.

First, we will present the TrueNorth processor by IBM. TrueNorth includes 1M neurons and 256M synapses in the 28nm CMOS chip, which is implemented in all-digital fashioin without on-chip learning support.

Second, we will present the ROLLS processor by ETH Zurich. ROLLS implemented 256 neurons and 128K synapses in the 180nm CMOS chip with analog/mixed-signal circuits and support long/short-term on-chip learning.

Third, we will present the recent Loihi processor by Intel. Loihi encompasses 128K neurons and 128M synapses in the 14nm CMOS chip with all-digital implementation and supports on-chip learning.

And finally, we will present RRAM-based computing research literature, which includes in-memory mixed-signal computing with small or moderate array demonstrations.

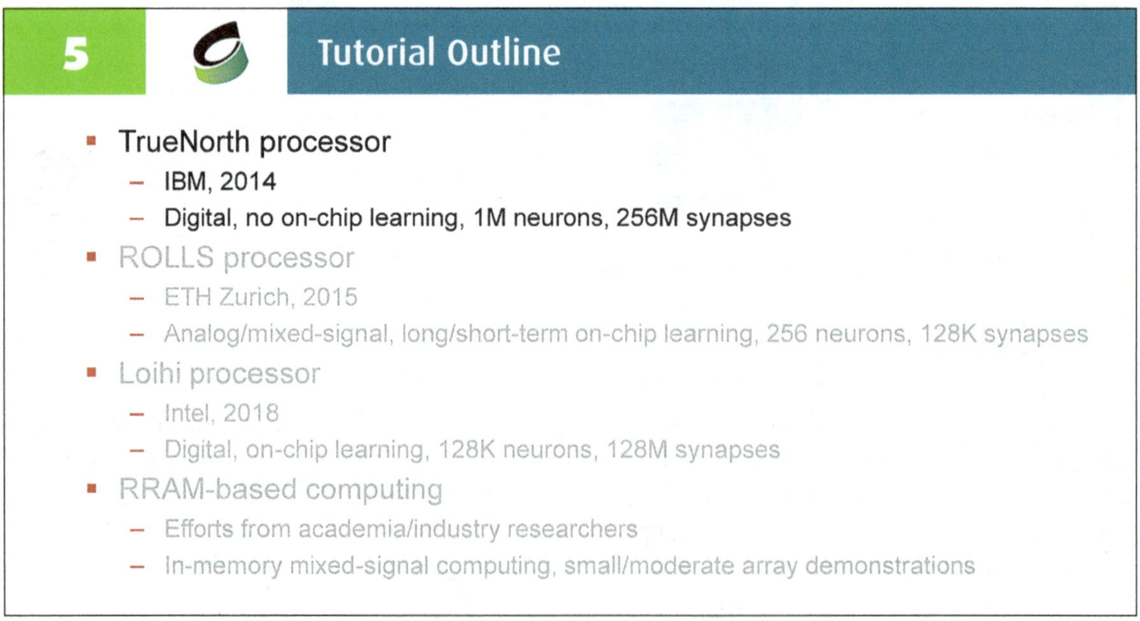

First, we will start with the TrueNorth processor by IBM.

6 TrueNorth: Computation, Communication, Memory

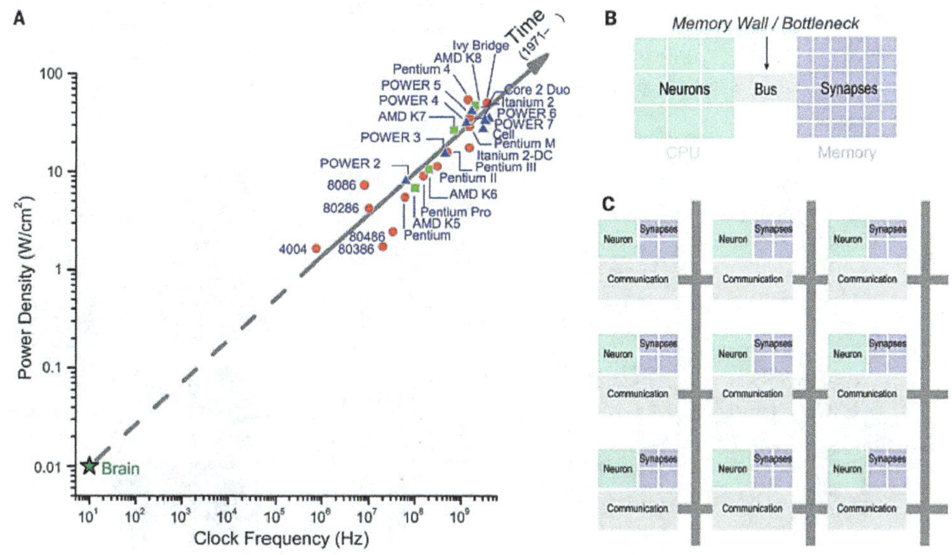

- Non-von-Neumann architecture, mimicking biological nervous systems

The parallel, distributed architecture of the brain is different from the sequential, centralized von Neumann architecture of today's computers. The trend of increasing power densities and clock frequencies of Intel/IBM/AMD processors is headed away from the brain's operating point.

(B) In terms of computation, a single processor has to simulate both a large number of neurons as well as the inter-neuron communication infrastructure. In terms of memory, the von Neumann bottleneck, which is caused by separation between the external memory and processor, leads to energy-hungry data movement when updating neuron states and when retrieving synapse states. In terms of communication, inter-processor messaging explodes when simulating highly interconnected networks that do not fit on a single processor. Overall, the conventional von Neumann architecture is fundamentally inefficient and non-scalable for representing massively interconnected neural networks with respect to computation, memory, and communication.

(C) TrueNorth design combines event-driven communication with co-located memory and computation to mitigate the von Neumann bottleneck. Conceptual blueprint of the TrueNorth architecture is, like the brain, to tightly integrate memory, computation, and communication in distributed modules ("neurosynaptic cores") that operate in parallel and communicate via an event-driven network. This can implement large-scale spiking neural networks that are efficient, scalable, and programmable within today's technology.

7 TrueNorth: Overall Architecture and Design

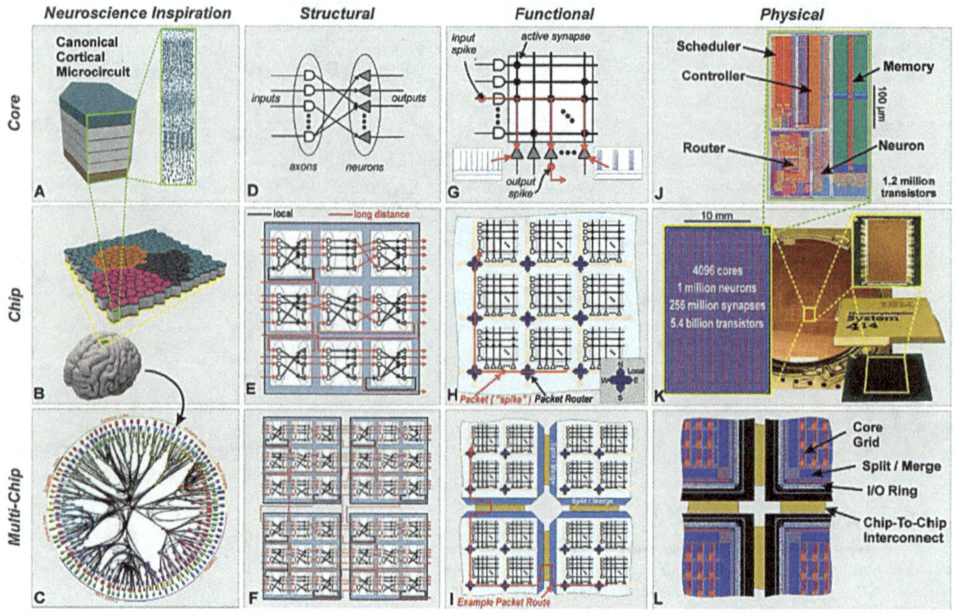

Panels are organized into rows at three different scales (core, chip, and multichip) and into columns at four different views (neuroscience inspiration, structural, functional, and physical).

(A) The neurosynaptic core is loosely inspired by the idea of a canonical cortical microcircuit.

(B) A network of neurosynaptic cores is inspired by the cortex's two-dimensional sheet.

(C) The multichip network is inspired by the long-range connections between cortical regions shown from the macaque brain.

(D) Structure of a neurosynaptic core with axons as inputs, neurons as outputs, and synapses as directed connections from axons to neurons.

Multicore networks at (E) chip scale and (F) multichip scale are both created by connecting a neuron on any core to an axon on any core with point-to-point connections.

(G) Functional view of core as a crossbar where horizontal lines are axons, cross points are individually programmable synapses, vertical lines are neuron inputs, and triangles are neurons. Information flows from axons via active synapses to neurons. Neuron behaviors are individually programmable, with two examples shown.

(H) Functional chip architecture is a two-dimensional array of cores where long-range connections are implemented by sending spike events (packets) over a mesh routing network to activate a target axon. Axonal delay is implemented at the target.

(I) Routing network extends across chip boundaries through peripheral merge and split blocks.

(J) Physical layout of core in 28-nm CMOS fits in a 240-mm-by-390-mm footprint. A memory (static random-access memory) stores all the data for each neuron, a time-multiplexed neuron circuit updates neuron membrane potentials, a scheduler buffers incoming spike events to implement axonal delays, a router relays spike events, and an event-driven controller orchestrates the core's operation.

(K) Chip layout of 64-by-64 core array, wafer, and chip package.

(L) Chip periphery to support multichip networks,

8 TrueNorth: Chip Specification

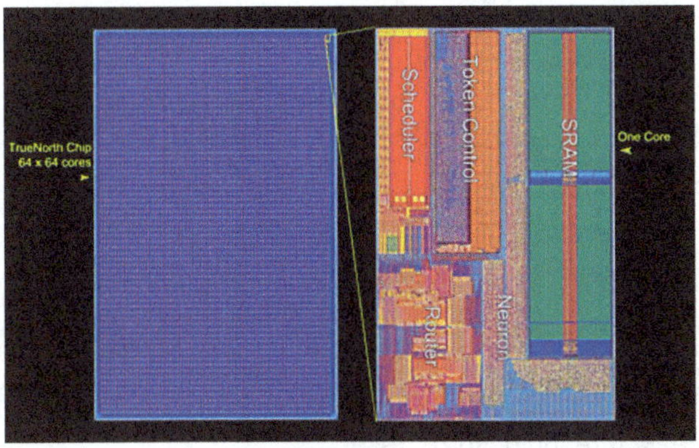

- 4,096 cores (64x64)
- Each core: 256 neurons, 256x256 synapses
- Each neuron fanouts to one axon in one core (256 neurons)
- Each synapse is 1-bit SRAM (binary), but at the axon periphery, there are several configuration strengths (applies to entire row, not individual synapses)
- Built with Samsung 28nm LP CMOS technology

P. Merolla et al., "A million spiking-neuron integrated circuit with a scalable communication network and interface," **Science**, vol. 345, no. 668, 2014.

TrueNorth, a fully functional digital chip with 1 million spiking neurons and 256 million synapses (non-plastic) was developed.

There are 4,096 (64x64) homogeneous cores in TrueNorth, where each core has 256 neurons and 65,536 (256x256) synapses.

Each synapse is implemented with one SRAM bitcell (binary), but at the axon periphery, there are several configurations that can represent different strengths of synapses (applies to entire row, not individual synapses).

With 5.4 billion transistors occupying 4.3cm^2 area in Samsung's 28-nm process technology, TrueNorth has -428 million bits of on-chip memory. Each core has 104,448 bits of local memory to store synapse states (65,536 bits), neuron states and parameters (31,232 bits), destination addresses (6,656 bits), and axonal delays (1,024 bits).

In terms of efficiency, TrueNorth's power density is 20 mW per cm^2, whereas that of a typical central processing unit (CPU) is 50 to 100 W per cm^2. Active power density was low because of the architecture, and passive power density was low because of process technology choice with low-leakage transistors.

To enable an event-driven, hybrid asynchronous-synchronous approach, a custom tool flow was developed, outside the scope of commercial software, for simulation and verification.

9 TrueNorth: Neuron Design

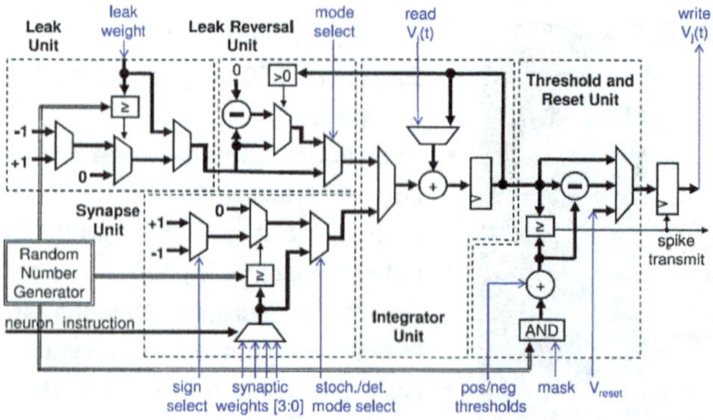

SYNAPTIC INTEGRATION

$$V_j(t) = V_j(t-1) + \sum_{i=0}^{255} A_i(t)\, w_{i,j} \Big[(1 - b_j^{G_i}) s_j^{G_i} + b_j^{G_i} F(s_j^{G_i}, \rho_{i,j}) \operatorname{sgn}(s_j^{G_i})\Big] \quad (10)$$

LEAK INTEGRATION

$$\Omega = (1 - \epsilon_j) + \epsilon_j \operatorname{sgn}(V_j(t)) \quad (11)$$

$$V_j(t) = V_j(t) + \Omega\, [(1 - c_j^{\lambda})\lambda_j + c_j^{\lambda} F(\lambda_j, \rho_j^{\lambda}) \operatorname{sgn}(\lambda_j)] \quad (12)$$

THRESHOLD, FIRE, RESET

$$\eta_j = \rho_j^T \,\&\, M_j \quad (13)$$

if $V_j(t) \geq \alpha_j + \eta_j$ (14)

Spike (15)

$$V_j(t) = \delta(\gamma_j) R_j + \delta(\gamma_j - 1)(V_j(t) - (\alpha_j + \eta_j)) + \delta(\gamma_j - 2) V_j(t) \quad (16)$$

elseif $V_j(t) < -[\beta_j \kappa_j + (\beta_j + \eta_j)(1 - \kappa_j)]$ (17)

$$V_j(t) = -\beta_j \kappa_j + [-\delta(\gamma_j) R_j + \delta(\gamma_j - 1)(V_j(t) + (\beta_j + \eta_j)) + \delta(\gamma_j - 2) V_j(t)](1 - \kappa_j) \quad (18)$$

endif (19)

- Versatile functionality for integration, leak, threshold, spike, reset
- All-digital CMOS, 1272 gates

A. Cassidy et al., "Cognitive computing building block: A versatile and efficient digital neuron model for neurosynaptic cores," **IJCNN,** 2013.
F. Akopyan et al., "TrueNorth: Design and tool flow of a 65 mW 1 million neuron programmable neurosynaptic chip," **IEEE TCAD**, vol. 34, no. 10, 2015.

The neuron block is the TrueNorth core's main computational element. A dual stochastic and deterministic neuron is implemented based on an augmented integrate-and-fire neuron model. Striking a balance, the implementation complexity is less than the Hodgkin–Huxley or Izhikevich neuron models, however, by combining 1–3 simple neurons, all 20 biologically observed spiking neuron behaviors (cataloged by Izhikevich) can be replicated. The physical neuron block uses time-division multiplexing to compute the states of 256 logical neurons for a core with a single computational circuit. In order to simplify implementation of the complex arithmetic and logical operations, the neuron block was implemented in synchronous logic, using a standard ASIC design flow. However, it is event-driven, because the token controller only sends the exact number of clock pulses required for the neuron block to complete its computation.

The block diagram shown in this slide depicts the five major elements of the neuron block, with input/output parameters from/to the core SRAM shown in blue. The synaptic input unit implements stochastic and deterministic inputs for four different weight types with positive or negative values. The leak and leak reversal units provide a constant bias (stochastic or deterministic) on the dynamics of the neural computation. The integrator unit sums the membrane potential from the previous tick with the synaptic inputs and the leak input. The threshold and reset unit compares the membrane potential value with the threshold value. If the membrane potential value is greater than or equal to the threshold value, the neuron block resets the membrane potential and transmits a spike event. The random number generator is used for the stochastic leak, synapse, and threshold functions. The core SRAM, external to this block, stores the synaptic connectivity and weight values, the leak value, the threshold value, the configuration parameters, as well as the membrane potential, *Vj(t)*. At the beginning and end of a neural computation cycle, the membrane potential is loaded from and then written back to the core SRAM.

10 TrueNorth: Synapse/Axon/Neuron (I)

$$V_j(t) = V_j(t-1) + \sum_{i=0}^{255} A_i(t) w_{i,j} s_j^{G_i} - \lambda_j.$$

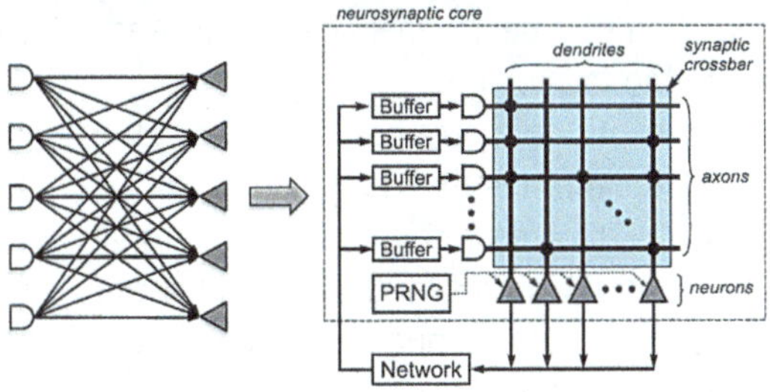

- $w_{i,j}$: ∈ {0, 1} – binary synapse using 1 SRAM cell (per synapse)
  - 256×256 w matrix implemented as a crossbar
- G_i : ∈ {0, 1, 2, 3} – 4 types of axon connections (per axon)
- s_j^{Gi} : each neuron j can individually assign this programmable signed integer, for each axon type G_i

Let's now look into the operations in a neurosynaptic core of 256 neurons and 65,536 (256x256) synapses.

A single neurosynaptic core consists of input buffers that receive spikes from the network, axons which are represented by horizontal lines, dendrites which are represented by vertical lines, and neurons (represented by triangles) that send spikes into the network.

A connection between an axon and a dendrite is a synapse, represented by a black dot. The synapses for each core are organized into a synaptic crossbar. Each synapse $\mathbf{w}_{i,j}$ is a binary synapse (0 or 1 value) that is mapped to one SRAM cell.

At the crossbar periphery, there are 4 types of axons connections Gi (per axon), and for each axon type Gi, each neuron can individually assign a programmable signed integer of $\mathbf{s}_j^{\mathbf{Gi}}$, which is multiplied with $\mathbf{w}_{i,j}$ to represent the effective synaptic weight. The programmability includes synaptic weights being excitatory, inhibitory, or having different relative strengths.

11 TrueNorth: Synapse/Axon/Neuron (II)

$$V_j(t) = V_j(t-1) + \sum_{i=0}^{255} A_i(t) w_{i,j} s_j^{G_i} - \lambda_j.$$

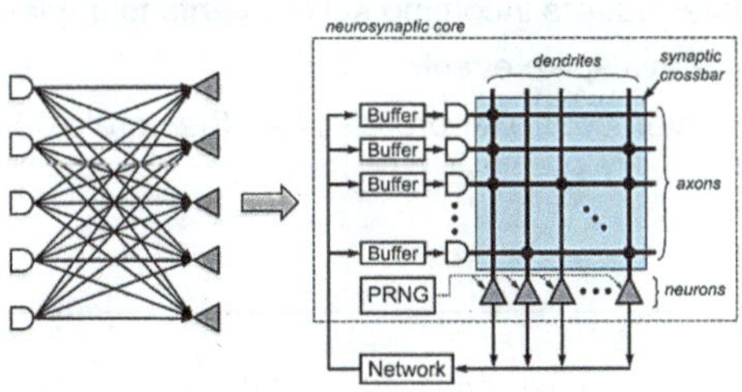

- At each time step t, core processes binary input vector of axon states
- $A_i(t)$: ∈ {0, 1} – each axon is represented by a bit
  - Whether input spike is present or not
- At time t, neuron j receives input: $A_i(t) \times w_{i,j} \times s_j^{Gi}$
- Leak is computed

THRESHOLD, FIRE, RESET

if $V_j(t) \geq \alpha_j$
  Spike
  $V_j(t) = R_j$
endif

11 TrueNorth: Synapse/Axon/Neuron (II)

The computation of a neurosynaptic core proceeds according to the following steps.

1) A neurosynaptic core receives spikes from the network and stores them in the input buffers.

2) When a 1 kHz synchronization trigger signal called a tick arrives, the spikes for the current tick are read from the input buffers, and distributed across the corresponding horizontal axons.

3) Where there is a synaptic connection between a horizontal axon and a vertical dendrite, the spike from the axon is delivered to the neuron through the dendrite.

4) Each neuron integrates its incoming spikes and updates its membrane potential.

5) When all spikes are integrated in a neuron, the leak value is subtracted from the membrane potential.

6) If the updated membrane potential exceeds the threshold, a spike is generated and sent into the network.

The output of each neuron is connected to the input buffer of the axon that it communicates with. This axon may be located in the same core as the communicating neuron or in a different core, in which case the communication occurs over the routing network.

12 TrueNorth: Neurosynaptic Core

- 240µm x 390µm footprint
- SRAM stores all the data for each neuron
  - Synaptic connections, neuron parameters and states (V_M), neuron target address, axonal delays

synaptic connections	$V_j(t)$ & neuron parameters	spike destination	spike delivery tick
256 bits	124 bits	26 bits	4 bits

→ 256x410 bit SRAM (each row for neuron)

- Time-multiplexed neuron circuit updates V_M
  - Single physical neuron
- The scheduler buffers incoming spike events to implement axonal delays
- The router relays spike events
- The event-driven controller orchestrates the core's operation

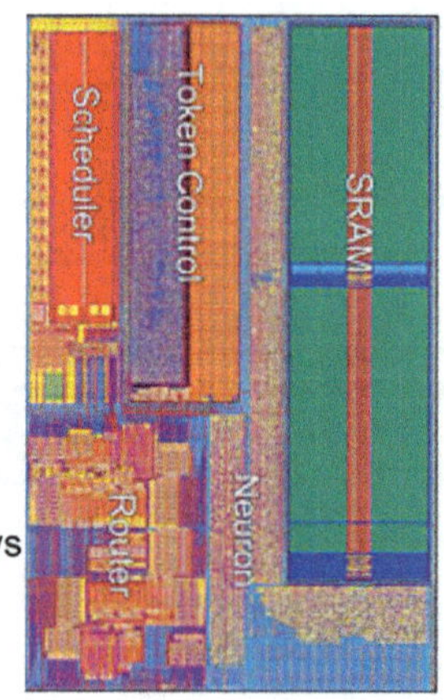

In 28nm CMOS, one neurosynaptic core occupies 240µm x 390µm area.

The primary core memory uses a 0.152 μm2 standard 6-transistor (6T) bitcell.

The SRAM is organized into 256 rows by 410 columns (not including redundant rows and columns). Each row corresponds to the information for a single neuron in the core. The neuron's 410 bits correspond to its synaptic connections, parameters and membrane potential *Vj(t)*, its core and axon targets, and the programmed delivery tick (as detailed in the table in the slide).

Note that there is a single physical neuron in the neurosynaptic core, and this single physical neuron is time-multiplexed to implement the 256 neurons for each neurosynaptic core. The SRAM in the

12 TrueNorth: Neurosynaptic Core

neurosynaptic core stores all the different data and information for each of the 256 neurons.

The scheduler buffers incoming spike events to implement axonal delays, by storing input spikes in a queue until the specified destination tick that is given in the spike packet.

The router communicates spike events with its own core and the four neighboring routers in the east, west, north, and south directions, creating a 2-D mesh network.

The event-driven token controller controls the sequence of computations carried out by a neurosynaptic core. After receiving spikes from the scheduler, it processes 256 neurons one by one.

13 TrueNorth: Overall Timing Diagram

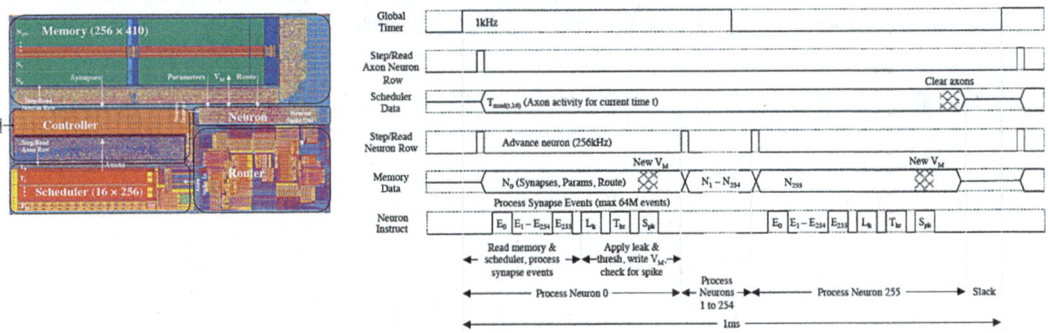

- One physical neuron in each neurosynaptic core, but the parameters of 256 neurons are stored in per-core memory → time-multiplexed operation of neuron circuits
- When a core processes all neurons for the current tick, the controller stops sending enable clock pulses to neurons, halting computation
- Controller instructs scheduler to advance its time pointer to the next tick (global timer)

The overall timing diagram of the neurosynaptic core is illustrated in this slide.

There is one physical neuron in each neurosynaptic core, but the parameters of 256 neurons are stored in per-core SRAM. The controller and scheduler governs the time-multiplexed operation of neuron circuits. Starting from neuron 0, the neurosynaptic core goes through each neuron one by one, till neuron 255, in a sequential fashion.

Each neuron's operation includes read memory, process synaptic events, input spike integration, apply leak, perform thresholding, write membrane potential back to memory, check for spikes in next time-step, etc.

When the neurosynaptic core completes processing of all 256 neurons for the current tick, the controller stops sending enable clock pulses to neurons, thereby halting computation.

The event-driven token controller instructs the scheduler to advance its time pointer to the next tick (global timer).

14 TrueNorth: Power/Energy Results

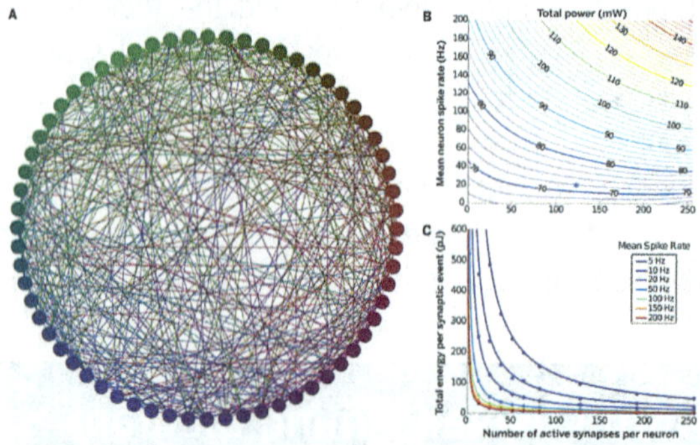

synaptic event: source neuron sending a spike event to a target neuron via a unique (nonzero) synapse.

- (A) nodes: cores, edges: connections, only 64 cores shown
- (B) more power with higher synaptic densities and higher spike rates
- (C) total energy per synaptic decreases with higher synaptic density because leakage and baseline core power amortized over more synapses

(A) Example network topology used for benchmarking power at real-time operation. Nodes represent cores, and edges represent neural connections; only 64 of 4096 cores are shown.

(B) Although power remains low (<150 mW) for all benchmark networks, those with higher synaptic densities and higher spike rates consume more total power, which illustrates that power consumption scales with neuron activity and number of active synapses.

(C) The total energy (passive plus active) per synaptic event decreases with higher synaptic density because leakage power and baseline core power are amortized over additional synapses. For a typical network where neurons fire on average at 20 Hz and have 128 active synapses [marked as * in (B) and (C)], the total energy is 26 pJ per synaptic event.

15 TrueNorth: Object Detection with SVM Classifiers

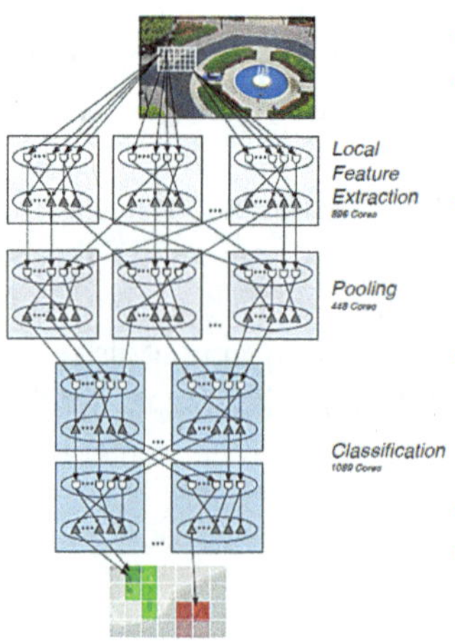

- Performs 64x64 pixel patches at a time
- 1st feature processing layer:
  - Extract local features
  - k-means clustering
- 2nd pooling layer:
  - Weight matrix from features to classes created using support vector machine (SVM)
  - Features are further merged based on covariance between weights
- 3rd classification layer:
  - A total of 1,024 features from 64x64 patch, 256 features from neighboring 8 patches are fed through a discrete weight SVM
  - Results are smoothed with a spatio-temporal filter
- Different cores/neurons perform different functions, programmable
- Classification categorizes detected objects as one of five classes:
  - car, bus, truck, person, cyclist

15 TrueNorth: Object Detection with SVM Classifiers

To demonstrate TrueNorth's suitability for real-world problems, a network is developed for real-time detection and classification of people, bicyclists, cars, trucks, buses, capable of identifying multiple objects per frame.

This slide describes the network that uses a multilayer processing system to provide object classifications at a resolution corresponding to classifying 64x64 pixel patches evenly tiled across the image.

The classification network consists of three layers of integrate-and-fire neurons with their parameters and connections configured to perform local feature extraction, pooling, and classification.

Input to the classification system is provided by applying an edge operator to the original image and coding the resulting edges using 16 levels of quantization.

For the first processing layer, local features are learned from 64 pixel blocks using k-means clustering. Local features from four such patches within a 64x64 pixel region are combined to create the first level features.

A weight matrix from features to classes is then created using a support vector machine, and features are further merged based on covariance between these weights to create a second pooling layer of features.

A total of 1,024 features from the 64x64 patch, and 256 features from the neighboring 8 patches are fed through a discrete weight support vector machine for classification.

Finally, the results are smoothed with a spatio-temporal filter.

Different cores and neurons in the TrueNorth chip performs different functions, owing to its programmability.

16 TrueNorth: Benchmarking Public Datasets

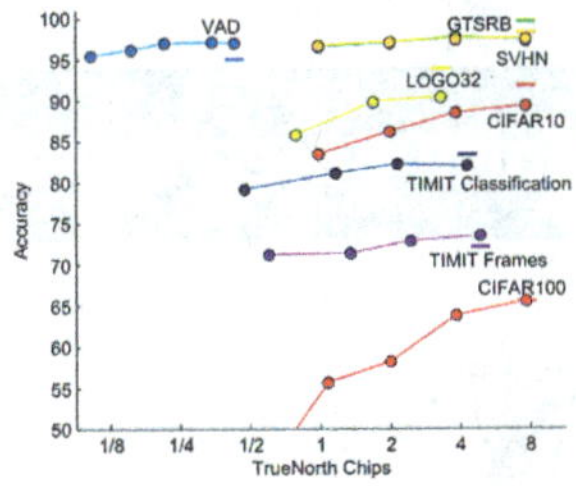

S. K. Esser et al., "Convolutional networks for fast, energy-efficient neuromorphic computing," **Proc. of the National Academy of Sciences (PNAS)**, 2016.

	State of the art		TrueNorth best accuracy		TrueNorth 1 chip				
Dataset	Approach	Accuracy	Accuracy	#cores	Accuracy	#cores	FPS	mW	FPS/W
CIFAR10	CNN (11)	91.73%	**89.32%**	31492	**83.41%**	4042	1249	204.4	6108.6
CIFAR100	CNN (34)	65.43%	**65.48%**	31492	**55.64%**	4042	1526	207.8	7343.7
SVHN	CNN (34)	98.08%	**97.46%**	31492	**96.66%**	4042	2526	256.5	9849.9
GTSRB	CNN (35)	99.46%	**97.21%**	31492	**96.50%**	4042	1615	200.6	8051.8
LOGO32	CNN	93.70%	**90.39%**	13606	**85.70%**	3236	1775	171.7	10335.5
VAD	MLP (36)	95.00%	**97.00%**	1758	**95.42%**	423	1539	26.1	59010.7
TIMIT Class.	HGMM (37)	83.30%	**82.18%**	8802	**79.16%**	1943	2610	142.6	18300.1
TIMIT Frames	BLSTM (38)	72.10%	**73.46%**	20038	**71.17%**	2476	2107	165.9	12698.0

The network for LOGO32 was an internal implementation. BLSTM, bidirectional long short-term memory; CNN, convolutional neural network; FPS, frames/second; FPS/W, fames/second per Watt; HGMM, hierarchical Gaussian mixture model; MLP, multilayer perceptron. Accuracy of TrueNorth networks is shown in bold.

16 TrueNorth: Benchmarking Public Datasets

By appropriately designing the structure, neurons, network input, and weights of convolutional neural networks (CNNs) during training, it is possible to efficiently map those networks to neuromorphic hardware. In the PNAS paper in 2016, IBM researchers explored mapping deep CNN and its variants to one or multiple TrueNorth chips.

One limitation is TrueNorth is that each neuron's fanout is limited to 256. For large-scale networks where the neuron fanout exceeds 256, multiple neurons on the same core are configured with identical synapses and parameters (and thus will have matching output), allowing distribution of the same data to multiple targets. If insufficient neurons are available on the same core, a feature can be "split" by connecting it to a core with multiple neurons configured to spike whenever they receive a spike from that feature. Neurons used in either duplication scheme are referred to here as copies.

In the top plot, accuracy of different sized networks running on one or more TrueNorth chips to perform inference on eight datasets for image classification, voice activity detection, and speech recognition.

For comparison, accuracy of state-of-the-art unconstrained approaches (as of 2016) are shown as bold horizontal.

The bottom table shows (left) the accuracy of state-of-the-art approaches for the eight datasets, (middle) TrueNorth's best accuracy for the datasets using multiple (or one) TrueNorth chips, and (right) TrueNorth's accuracy and measured power and classifications per energy (FPS per Watt) reported using a single TrueNorth chip (<4,096 cores). It is known that augmenting training data through manipulations such as mirroring can improve scores on test data, but this adds complexity to the overall training process. To maintain focus on the algorithm presented here, the training set is not augmented and the TrueNorth's results are compared to other works that also do not use data augmentation. The experiments show that for almost all of the benchmarks, a single-chip network is sufficient to come within a few percent of state-of-the-art accuracy. Increasing to up to eight chips improved accuracy by several percentage points.

17 TrueNorth: Scaling Up TrueNorth Based Systems

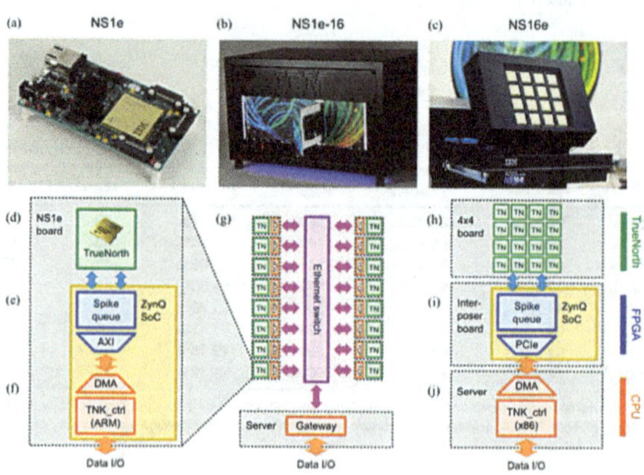

J. Sawada et al., "TrueNorth Ecosystem for Brain-Inspired Computing: Scalable Systems, Software, and Applications," **Proc. of the International Conference for High Performance Computing, Networking, Storage and Analysis (SC'16)**, 2016.

17 TrueNorth: Scaling Up TrueNorth Based Systems

The Neurosynaptic System, 1 million neuron evaluation platform (NS1e), shown in (a), supports embedded and mobile applications, and its compact form factor and modular nature make it a building block for scale-out systems of arbitrary size. The main processing element of the NS1e is a single TrueNorth chip (d). It is coupled on board with a Xilinx Zynq Z-7020 SoC (e)(f) containing FPGA programmable logic and two ARM Cortex-A9 cores connected to 1GB DDR3 SDRAM.

(b) shows the NS1e-16 system constructed using sixteen NS1e boards, with aggregate capacity of 16 million neurons and 4 billion synapses, interconnected via a 1Gig-Ethernet packet switched network (g). The system optionally includes a traditional server (3.4 GHz quad-core Xeon, 32GB RAM, 1 TB hard drive) to act as a host gateway. The NS1e-16 system is integrated into a single 6U rack, and consumes a total of approximately 68W not including the server (~56W for the NS1e node cards and ~12W for the network communication).

Using TrueNorth's native chip-to-chip asynchronous communication interfaces, (c) shows the NS16e (Neurosynaptic System 16 million neuron evaluation) platform, which seamlessly integrates 16 TrueNorth chips into a *scale-up* solution. The NS16e system consists of three boards assembled using vertical stacking/mating connectors: a custom 4x4 board (h), mated to a custom interposer board with an off-the-shelf Avnet 7Z045 Zynq SoC mini module (i).

18 Summary of TrueNorth Processor

- All-digital programmable neuron and binary SRAM synapse implementation
  - Synapses values are fixed during operation (offline training)
  - Does not feature on-chip learning
- Asynchronous operation with token controller
- Homogeneous design of 4,096 neurosynaptic cores in 28nm CMOS and network-on-chip
  - 1M neurons, 256M synapses
- Capability of scaling up towards larger systems with many TrueNorth chips

19 Tutorial Outline

- TrueNorth processor
  - IBM, 2014
  - Digital, no on-chip learning, 1M neurons, 256M synapses
- ROLLS processor
  - ETH Zurich, 2015
  - Analog/mixed-signal, long/short-term on-chip learning, 256 neurons, 128K synapses
- Loihi processor
  - Intel, 2018
  - Digital, on-chip learning, 128K neurons, 128M synapses
- RRAM-based computing
  - Efforts from academia/industry researchers
  - In-memory mixed-signal computing, small/moderate array demonstrations

Next, let's look into with the ROLLS processor developed by ETH Zurich.

20 ROLLS: Top-Level Block Diagram

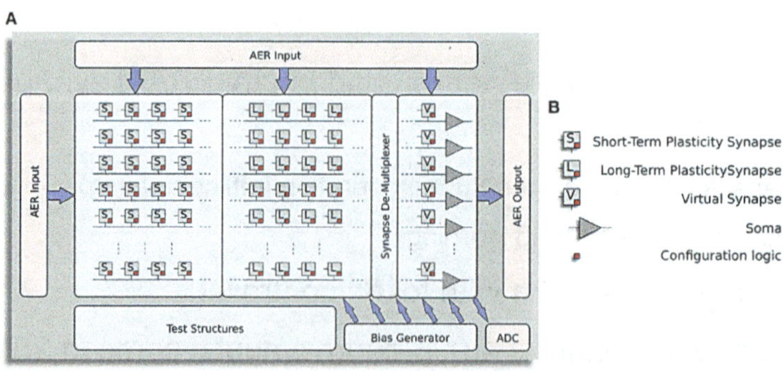

- ROLLS (Reconfigurable On-Line Learning Spiking) neuromorphic processor
  - 256 neurons, 64k long-term plasticity synapses, 64k short-term plasticity synapses
  - Online learning for image classification applications

N. Qiao et al., "A reconfigurable on-line learning spiking neuromorphic processor comprising 256 neurons and 128K synapses," **Frontiers of Neuroscience**, 2015.

20 ROLLS: Top-Level Block Diagram

This slide shows the architecture of the ROLLS neuromorphic processor. ROLLS processor integrated 256 neurons, 64k (256x256) long-term plasticity synapses, and 64k (256x256) short-term plasticity synapses.

Online learning with long-term and short-term plasticity for image classification applications has been demonstrated.

(A) shows the block diagram of the architecture, showing two distinct synapse arrays (short-term plasticity and long-term plasticity synapses), an additional row of synapses (virtual synapses) and a row of neurons (somas). A synapse de-multiplexer block is used to connect the rows from the synapse arrays to the neurons.

Peripheral circuits include asynchronous digital AER logic blocks, an analog-to-digital converter (ADC), and a programmable on-chip bias-generator.

(B) shows the block-diagram legend.

The architecture comprises also a "synapse de-multiplexer" static logic circuit, which allows the user to choose how many rows of plastic synapses should be connected to the neurons. It is a programmable switch-matrix that configures the connectivity between the synapse rows and the neuron columns. By default, each of the 256 rows of 1x512 synapses is connected to its corresponding neuron. By changing the circuit control bits it is possible to allocate multiple synapse rows to the neurons, thereby disconnecting and sacrificing the unused neurons. In the extreme case, all 256x512 synapses are assigned to a single neuron, and the remaining 255 neurons remain unused.

21 ROLLS: Analog Neuron Circuit

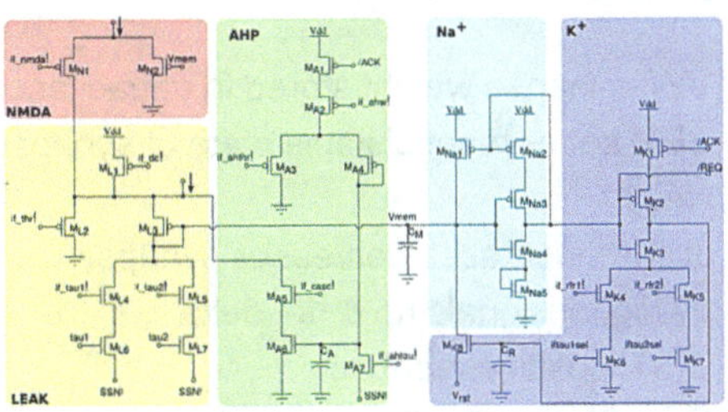

- LEAK: neuron's leak conductance
- AHP: spike-frequency adaption
- Na+: Sodium channels positive feedback → spike
- K+: Potassium conductance reset, refractory period

- Various bio-plausible neuron behaviors are achievable

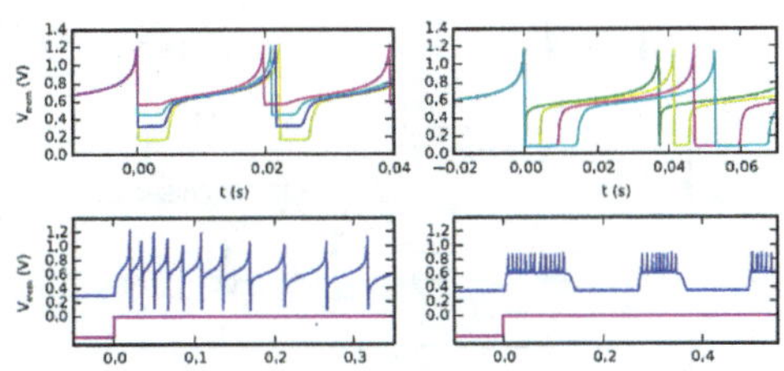

21 ROLLS: Analog Neuron Circuit

At the top figure, the schematics of the silicon neuron is shown.

It comprises an NMDA block ($M_{N1,N2}$), which implements the NMDA voltage gating function, a LEAK DPI circuit (M_{L1-L7}) which models the neuron's leak conductance, an AHP DPI circuit (M_{A1-A7}) in negative feedback mode, which implements a spike-frequency adaptation behavior, an Na+ positive feedback block($M_{Na1-Na5}$) which models the effect of Sodium activation and inactivation channels for producing the spike, and a K+ block (M_{K1-K7}) which models the effect of the Potassium conductance, resetting the neuron and implementing a refractory period mechanism. The negative feedback mechanism of the AHP block, and the tunable reset potential of the K+ block introduce two extra variables in the dynamic equation of the neuron that can endow it with a wide variety of dynamical behaviors (Izhikevich, 2003).

At the bottom figure, possible bio-plausible neuron behaviors that can be expressed by the silicon neuron are shown. The top-left quadrant shows measured data from the chip representing the neuron membrane potential in response to a constant current injection for different values of reset voltage. The top-right quadrant shows the neuron response to a constant current injection for different settings of its refractory period. The bottom-left quadrant demonstrates the spike-frequency adaptation behavior, obtained by appropriately tuning the relevant parameters in the AHP block of neuron schematic and stimulating the neuron with a constant injection current. By further increasing the gain of the AHP negative feedback block the neuron can produce bursting behavior (bottom-right quadrant).

22 ROLLS: Long-Term Plasticity Synapse Circuit

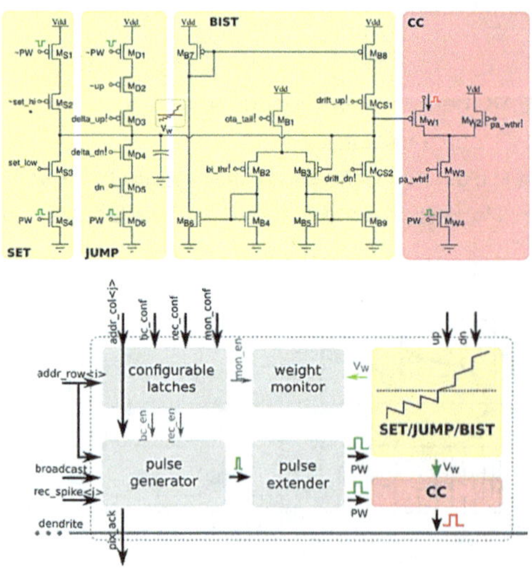

- VW: synapse weight stored in capacitance
- SET: set or reset bistable state of synapse weight
- JUMP: increases/decreases synapse weight by digital signals up & dn, delta
- BIST: positive feedback
- CC: VW thresholding → models EPSC (excitatory postsynaptic potential)

22 ROLLS: Long-Term Plasticity Synapse Circuit

Each of the 256×256 synapse circuits in the long-term plasticity array comprises event-based programmable logic circuits for configuring both synapse and network properties, as well as analog/digital circuits for implementing the learning algorithm.

The top figure shows the schematic of the analog/digital weight update circuits. These circuits are subdivided into four sub-blocks: the SET block can be used to set/reset the bistable state of the synaptic weight by sending an AER event with the matching address and properly asserting the configuration signals set_hi and set_low. The JUMP block increases or decreases the synaptic weight internal variable (i.e., the voltage *Vw*) depending on the digital signals up and dn, that are buffered copies of the ones generated in the silicon neuron stop-learning block. The heights of the up and down jumps can be set by changing the delta_up! and delta_dn! signals. The BIST block consists of a wide-range transconductance amplifier configured in positive feedback mode, to constantly compare the *Vw* node with the threshold bi_thr!: if *Vw* > bi_thr! then the amplifier slowly drives the *Vw* node, drifting toward the positive rail, otherwise it actively drives it toward the ground. The drift rates to the two states can be tuned by biases drift_up! and drift_dn!, respectively. The current converter (CC) block converts the *Vw* voltage into a thresholded EPSC with maximum amplitude set by pa_wht!.

The digital logic part, shown in the bottom figure, has an pulse generator circuit that manages the handshaking signals required by the AER protocol, and three one-bit configurable latches: one latch sets/resets the MON_EN signal, which enables/disable the synapse monitor circuit, which buffers the synapse weight *Vw* signal for off-chip reading. The remaining two latches are used to set the BC_EN and REC_EN signals, which control the activation modes of the synapse. There are three different activation modes can be configured: direct activation, broadcast activation and recurrent activation.

23 ROLLS: Spike-based Learning (Long-Term Plasticity)

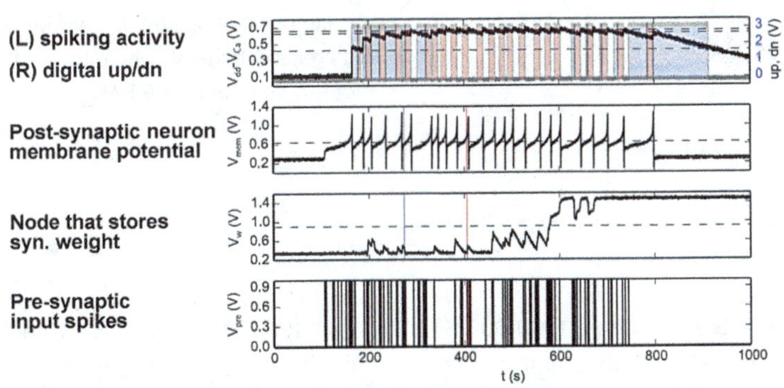

- Spike-based learning rule using pre-synaptic spike timing & state of post-synaptic neuron membrane potential (Brader, Senn, Fusi, Neural Comp. 2007)
- Blue line (t = 273 s): @ pre-synaptic spike, post-synaptic neuron Vmem is low, meaning spike occurred recently → no pre-to-post causality → VW decrease
- Red line (t = 405 s): @ pre-synaptic spike, post-synaptic neuron Vmem is high, meaning spike will likely occur soon → pre-to-post causality exists → VW increase

23 ROLLS: Spike-based Learning (Long-Term Plasticity)

This slide shows long-term plasticity experimental results that highlight the features of both spike-based neuron and synapse learning circuits in action: weight updates are triggered when the pre-synaptic spikes arrive, and when the post-synaptic neuron's Calcium concentration is in the appropriate range. Depending on the value of the Calcium concentration signal, the digital up and dn signal turn on or off. The weight internal variable is increased or decreased depending on where the membrane potential is with respect to the membrane threshold.

The bottom trace represents the pre-synaptic input spikes.

The second trace from the bottom represents the bi-stable internal variable (node Vw from the previous slide) for the synaptic weight.

The third trace represents the post-synaptic neuron's membrane potential.

The top trace shows both a voltage trace proportional to the neuron's integrated spiking activity as well as the digital control signals that determine whether to increase (red shaded area), decrease (blue shaded area) or leave Vw unchanged (no shaded area). The horizontal lines represent the thresholds used in the learning algorithm, while the vertical lines at t = 273 s (blue line) and t = 405 s (red line) are visual guides to show where the membrane potential is, with respect to its threshold, for down and up jumps in Vw respectively.

24 ROLLS: Short-Term Plasticity Synapse Circuit

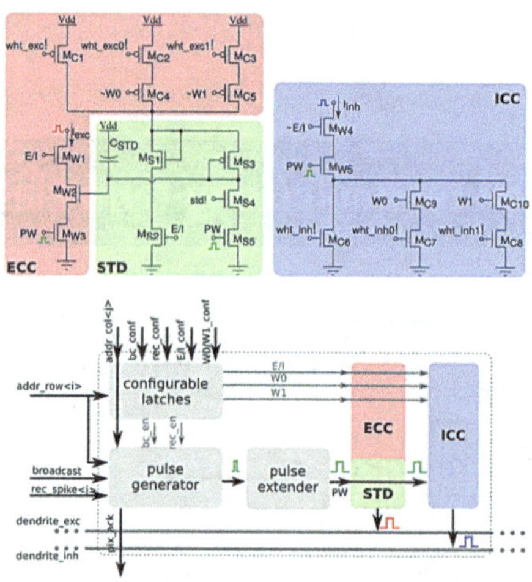

- Short-term depression: EPSC decreases with input spike and slowly recovers
- ECC: excitatory (E/I = high)
- ICC: inhibitory (E/I = low)
- STD: when excitatory spike arrives, PW lowers M_{W2} gate voltage
  → with burst of spikes, I_{exc} will be less
- M_{W2} gate voltage slowly recovers through diode-connected p-FET M_{S3}

The array of short-term plasticity (STP) synapses contains circuits that allow users to program the synaptic weights, rather than changing them with a fixed on-chip learning algorithm. Specifically, each synapse has a two-bit programmable latch that can be used to set one of four possible weight values.

The top figure shows the transistor-level schematic diagram of the excitatory and inhibitory pulse-to-current converters.

The left panel shows the excitatory CC and the STD circuit. The CC at the top generates a current that is proportional to the 2-bit weight. The proportionality constant is controlled through analog biases. This current charges up the C_{STD} capacitor through the diode connected p-FET M_{S3} so that at steady state, the gate voltages of M_{S1} and M_{W2} are equal. A pre-synaptic pulse on the PW port activates the I_{exc} current branch, and produces a current that initially

24 ROLLS: Short-Term Plasticity Synapse Circuit

is proportional to the 2-bit weight original current. At the same time, the *PW* pulse activates also the STD branch through transistor M_{S5} and an amount of positive charge that is controlled by the bias *STD* is removed from the capacitor *CSTD*. The gate voltage of M_{W2} is now momentarily lower than that of M_{S1}, and recovers slowly through the diode connected p-FET M_{S3}. Pulses that arrive before the capacitor voltage has recovered completely will generate a current that is smaller than the original one, and will further depress the effective synaptic weight through the STD branch. The excitatory block is only active if the *E/I* voltage is high. If *E/I* is low, the inhibitory current DAC in the right panel is active and generates a weight-proportional inhibitory current on *PW* pulses.

The bottom figure shows the block diagram of the synapse element.

In addition to the latches for setting the weight, there are two extra latches for configuring the synapse activation mode.

25 ROLLS: Effect of Short-Term Depression

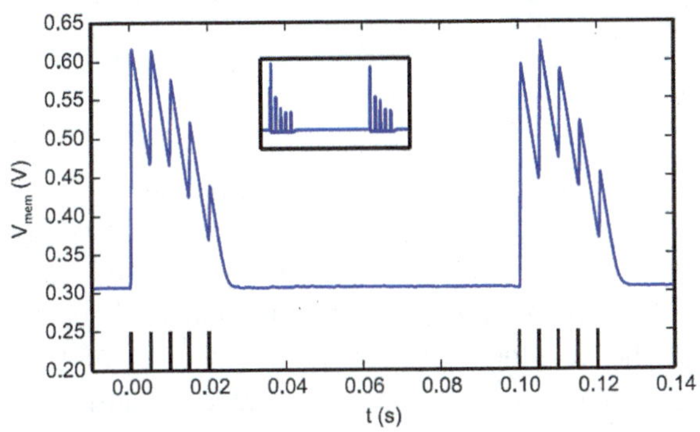

- With more burst spikes, the jump in V_{mem} gets smaller (synapse is temporally depressed)
- Inset: derivative of membrane potential = synaptic EPSC

The effect of short-term depression on EPSC magnitudes is illustrated in the figure.

Two bursts separated by 100 ms were sent to a programmable synapse. Each burst has 5 spikes with an inter-spike interval of 5 ms. Within a burst, the jumps in the neuron *Vmem* gradually get smaller as the synapse is depressed and the magnitude of the EPSCs it generates decreases. After the first burst, the synapse efficacy recovers as can be seen in the response to the second burst.

During a period of no stimulation the synapse recovered and responded with large Excitatory Post-Synaptic Potentials (EPSPs) to the initial part of the following burst, before depressing again.

The responses to the two bursts are not identical in the figure as the state of the neuron, synapse, and DPI circuits are not exactly the same at the onset of each burst.

The figure inset shows the derivative of the membrane potential which is equivalent to the synaptic EPSCs (minus the neuron leak).

26 ROLLS: 180nm Prototype Chip

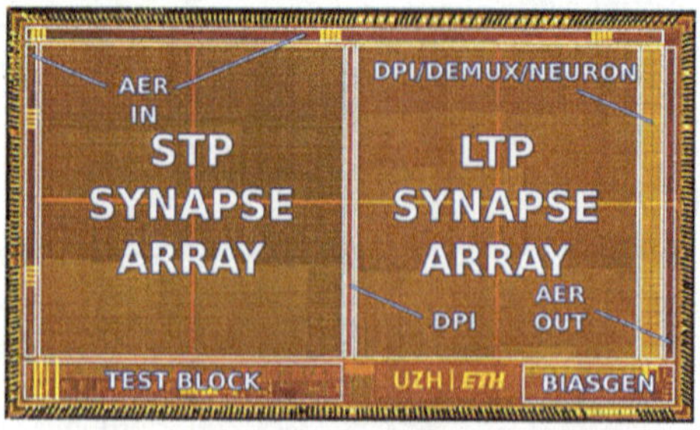

- 51.4 mm^2, 12.2 million transistors
- Synapses dominate the chip area

Circuit	Dimensions (μm x μm)	Number	Total area:	(mm^2)	(%)
Neuron	55.69×16.48	256		0.235	0.47
Post-synaptic learning	39.09×16.48	256		0.165	0.32
LTP synapse	15.3×16.48	64 k		16.147	31.41
STP synapse	16.24×16.48	64 k		17.129	33.32
Virtual synapse	35.6×16.48	512		0.300	0.58
Synapse de-mux	49.56×4389.4	1		0.218	0.42
AER in (columns)	8770×154	1		0.135	0.26
AER in (rows)	112×4357	1		0.488	0.95
AER out	166.2×4274.9	1		0.710	1.38
BiasGen	539.5×1973	1		1.064	2.07

Micro-photograph of the ROLLS neuromorphic processor is shown.

The chip was fabricated using a 180nm CMOS process and occupies an area of 51.4 mm^2, comprising 12.2 million transistors. Overall, the long-term and short-term synapses dominate the chip area.

The area distribution of main circuit blocks is shown in the right table. The silicon neurons contain circuits that implement a model of the adaptive exponential Integrate-and-Fire (I&F) neuron, post-synaptic learning circuits used to implement the spike-based weight-update/plasticity mechanism in the array of long-term plasticity synapses, and analog circuits that model homeostatic synaptic scaling mechanisms operating on very long time scales. The array of long-term plasticity synapses comprises pre-synaptic spike-based learning circuits with bi-stable synaptic weights, that can undergo either Long-Term Potentiation (LTP) or Long-Term Depression (LTD). The array of Short-Term Plasticity (STP) synapses comprises synapses with programmable weights and STP circuits that reproduce short-term adaptation dynamics. Both arrays contain analog integrator circuits that implement faithful models of synaptic temporal dynamics. Digital configuration logic in each of the synapse and neuron circuits allows the user to program the properties of the synapses, the topology of the network, and the properties of the neurons.

27 ROLLS: Image Classification Example

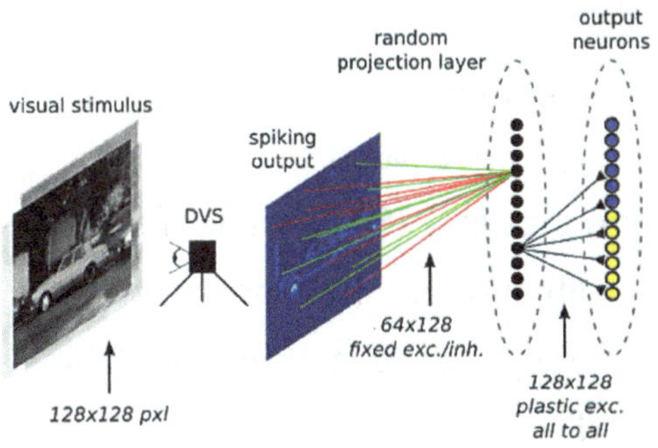

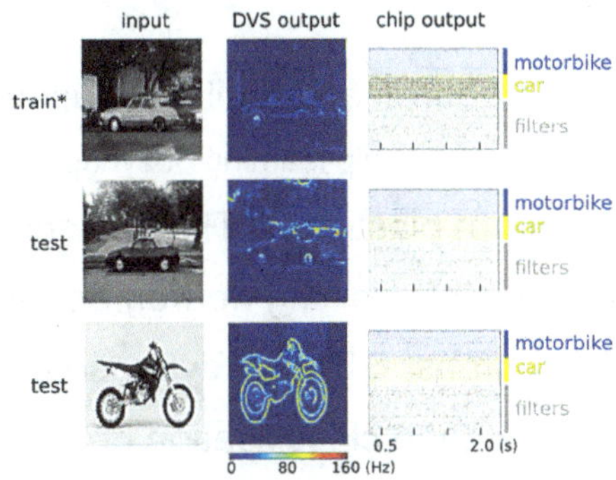

- DVS: dynamic vision sensor
- Sensor output: spike format (event-driven)

- Car image test:
  - Motorbike neuron: 6.0 Hz
  - Car neuron: **80.9** Hz
- Motorbike image test:
  - Motorbike neuron: **7.4** Hz
  - Car neuron: 4.9 Hz

A feasibility study in this slide demonstrates how the ROLLS neuromorphic processor can be used in conjunction with a spiking dynamic vision sensor (DVS) for learning to solve an image classification task. This experiment used a DVS as the front-end image sensor, interfaced to the ROLLS chip via a commercially available digital board. A two-layer spiking neural network is implemented, which processes the visual stimuli by extracting sparse random features in real- time. The network is composed of 128 hidden neurons and 128 output neurons on the ROLLS neuromorphic processor. 64 output neurons of the network are trained to become selective to one of two image classes, and the other 64 to become selective to the other class, via supervised learning protocol.

The left figure shows the neural network architecture and visual stimulus. Two different classes of images (motorbikes or cars) are displayed on a screen with a small jitter applied at 10 Hz. A random subset of the spikes emitted by the DVS are mapped to 128 hidden layer neurons. Specifically, each of the 128 neurons is connected to 64 randomly selected pixels with either positive or negative weights, also set at random. The output neurons in the last layer receive spikes from all the 128 hidden layer neurons, via plastic synapses. The output layer neurons are also driven by an external "teacher" signal which is correlated with one of the image classes.

The right figure shows the spiking activity of the hardware neurons during the training (upper panel) and testing phase (middle and lower panels). Left column shows examples of raw images from the Caltech 101 database. Middle column shows the heat map of the DVS spiking activity, where each pixel color represents the pixel's mean firing rate. Right column shows the raster plots of the ROLLS neuromorphic processor neurons. The star on the top panel label indicates that during the training phase an additional excitatory "teacher" signal was used to stimulate the "car" output neurons to induce plasticity. During testing no teacher signal is provided and only the excitatory currents from the input synapses drive the classifier activities. The average firing rates of the output layer for "motorbike" and "car" neuron pools are 6.0 Hz and 80.9 Hz during training. During testing with a "car" image they are 7.1 Hz and 11.1 Hz. During testing with a "motorbike" image they are 7.4 Hz and 4.9 Hz.

28 Summary of ROLLS Processor

- Analog neuron and digital synapse implementation
  - Support on-chip learning of synapses through plasticity rules
- Long-term plasticity (LTP) and short-term plasticity (STP) implemented
- Bio-plausible spiking neuron behaviors with sub-threshold operation
- 256 neurons, 64K LTP synapses, 64K STP synapses in 180nm testchip
- Simple image classification demonstrated

29 Tutorial Outline

- TrueNorth processor
  - IBM, 2014
  - Digital, no on-chip learning, 1M neurons, 256M synapses
- ROLLS processor
  - ETH Zurich, 2015
  - Analog/mixed-signal, long/short-term on-chip learning, 256 neurons, 128K synapses
- Loihi processor
  - Intel, 2018
  - Digital, on-chip learning, 128K neurons, 128M synapses
- RRAM-based computing
  - Efforts from academia/industry researchers
  - In-memory mixed-signal computing, small/moderate array demonstrations

Next, let's look into with the Loihi processor developed by Intel.

30 Loihi: Computation with Spikes and Parallelism

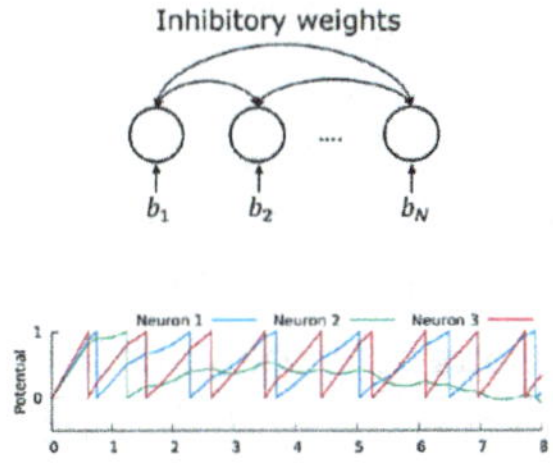

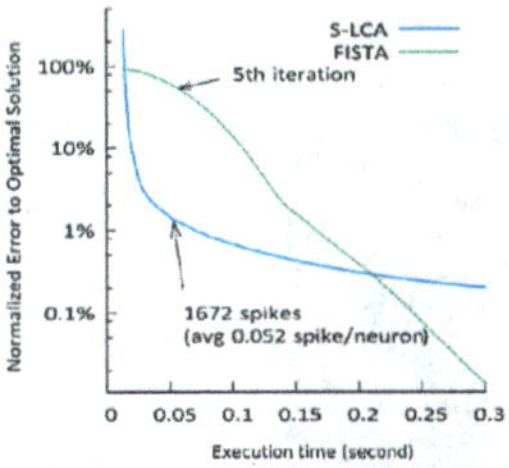

- (Top) network topology for solving LASSO. Objective is to determine sparse coefficients that best represents a given input as the linear combination of features from a dictionary.
- (Bottom) evolution of membrane potential in a three-neuron example; spike rates of the neurons stabilize to fixed values.
- An algorithmic efficiency comparison of a solution based on spiking network (S-LCA) and a conventional optimization method (FISTA).
- The efficiency-accuracy tradeoff shows that SNN solution (e.g. S-LCA) can be attractive for applications with solutions within 1% of the optimal solution

Computations in spiking neural networks (SNNs) are carried out through the interacting dynamics of neuron states. An instructive example is the $\ell 1$-minimizing sparse coding problem, also known as LASSO, which can be solved with the SNN using the Spiking Locally Competitive Algorithm (S-LCA). The objective of this problem is to determine a sparse set of coefficients that best represents a given input as the linear combination of features from a feature dictionary. The coefficients can be viewed as the activities of the spiking neurons in the left figure that are competing to form an accurate representation of the data. By properly configuring the network, it can be established that as the network dynamics evolve, the average spike rates of the neurons will converge to a fixed point, and this fixed point is identical to the solution of the optimization problem.

Such computation exhibits completely different characteristics from conventional linear algebra-based approaches. The right figure compares the computational efficiency of an SNN with the conventional solver FISTA3 by having them both solve a sparse coding problem on a single-threaded CPU. The SNN approach (labelled S-LCA) gives a rapid initial drop in error and obtains a good approximate solution faster than FISTA. After this, the S-LCA convergence speed significantly slows down, and FISTA instead finds a much more precise solution quicker. Hence, an interesting efficiency-accuracy tradeoff arises that makes the SNN solution particularly attractive for applications that do not require highly precise solutions, such as a solution that is 1 percent within the optimal solution.

31 Loihi: Neuromorphic Processor with On-Chip Learning

- Intel 14nm FinFET process, 60mm^2
- Mesh with 128 neuromorphic cores, 3 x86 cores, 2.07B transistors, 33MB SRAM
- Asynchronous network-on-chip
- Functional over 0.5 – 1.25V supply
- Features on-chip learning in spiking neural network (SNN) framework
- Includes computational primitives including:
  - Stochastic noise
  - Configurable and adaptable synaptic, axon, and refractory delays
  - Configurable dendritic tree processing

M. Davies et al., "Loihi: A Neuromorphic Manycore Processor with On-Chip Learning," **IEEE Micro**, 2018.

Learning in an SNN refers to adapting the synaptic weights and, hence, varying the SNN dynamics to a desired one. Similar to conventional machine learning, the intent is to express learning as the minimization of a particular loss function over many training samples. Learning in an SNN naturally proceeds in an online manner, and the SNN synaptic weight adaptation rules must satisfy a locality constraint where each weight can only be accessed and modified by the destination neuron.

When the learning rule satisfies the locality constraint, the inherent parallelism offered by SNNs will allow the adaptive network to be scaled up to large sizes in a way that can be computed efficiently. If the rule also minimizes a loss function, the system will have well-defined dynamics. To support the development of such scalable learning rules, Loihi offers a variety of local information to a programmable synaptic learning process.

Loihi was fabricated in Intel's 14nm FinFET process. The chip instantiates a total of 2.07 billion transistors and 33 MB of SRAM over its 128 neuromorphic cores and three x86 cores, with a die area of 60 mm2. The device is functional over a supply voltage range of 0.50 V to 1.25 V.

Loihi includes a total of 16 MB of synaptic memory. With its densest 1-bit synapse format, this provides a total of 2.1 million unique synaptic variables per mm^2, over three times higher than TrueNorth, the previously most dense SNN chip. This does not consider Loihi's hierarchical network support that can significantly boost its effective synaptic density. On the other hand, Loihi's maximum neuron density of 2,184 per mm^2 is marginally worse than TrueNorth's. Process normalized, this represents a 2X reduction in the design's neuron density, which may be interpreted as the cost of Loihi's greatly expanded feature set, an intentional design choice.

32 Loihi: Mesh Operation

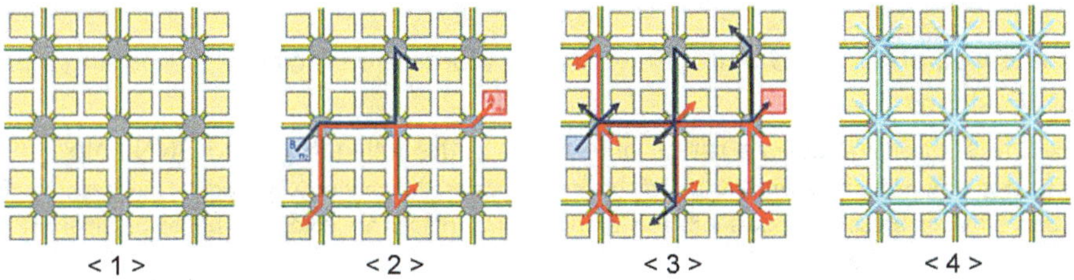

- < 1 > initial idle state for time-step ***t***
  - Each square represents a core in the mesh containing multiple neurons
- < 2 > neurons n1 and n2 in cores A and B fire and generate spike messages
- < 3 > spikes from all other neurons firing on time-step ***t*** in cores A and B are distributed to their destination cores
- < 4 > each core advances its algorithmic time-step to ***t+1*** as it handshakes with its neighbors through barrier synchronization messages

This slide shows the operation of the neuromorphic mesh as it executes an SNN model.

<1> All cores begin at algorithmic time-step t. Each core independently iterates over its set of neuron compartments, and any neurons that enter a firing state generate spike messages that the network-on-chip (NoC) distributes to all cores that contain their synaptic fan-outs.

<2> Spike distributions for two such example neurons $n1$ and $n2$ in cores A and B are illustrated in the second box.

<3> Additional spike distributions from other firing neurons adding to the NoC traffic in the third box. The NoC distributes spike (and all other) messages according to a dimension-order routing algorithm. The NoC itself only supports unicast distributions. To multicast spikes, the output process of each core iterates over a list of destination cores for a firing neuron's fan-out distribution and sends one spike per core.

<4> At the end of the time-step, a mechanism is needed to ensure that all spikes have been delivered and that it's safe for the cores to proceed to time-step $t + 1$. Rather than using a globally distributed time reference (clock) that must prepare for the worst-case chip-wide network activity, we use a barrier synchronization mechanism, illustrated in the fourth box. As each core finishes servicing its compartments for time-step t, it exchanges barrier messages with its neighboring cores. The barrier messages flush any spikes in flight and, in a second phase, propagate a time-step advance notification to all cores. As cores receive the second phase of barrier messages, they advance their time-step and proceed to update compartments for time $t + 1$.

33 Loihi: Flexible SNN Connectivity

- Flexible and well-provisioned SNN connectivity features are crucial for supporting a broad range of workloads
- To support this, Loihi includes a range of features in the fully integrated SNN chip:
  - ***Sparse network compression***: Besides a common dense matrix connectivity model, Loihi supports three sparse matrix compression models in which fan-out neuron indices are computed based on index state stored with each synapse's state variables.
  - ***Core-to-core multicast***: Any neuron may direct a single spike to any number of destination cores, as the network connectivity might require.
  - ***Variable synaptic formats:*** Loihi supports any weight precision between one and nine bits, signed or unsigned, and weight precisions may be mixed even within a single neuron's fan-out distribution.
  - ***Population-based hierarchical connectivity:*** As a generalized weight-sharing mechanism, such as to support convolutional neural network types, connectivity templates may be defined and mapped to specific population instances during operation. This feature can reduce a network's required connectivity resources by over an order of magnitude.

Flexible and well-provisioned SNN connectivity features are crucial for supporting a broad range of workloads. Some desirable networks might call for dense, all-to-all connectivity, while others might call for sparse connectivity; some might have uniform graph degree distributions, some might require high precision synaptic weights, such as to support learning, while others can make do with binary connections. As a rule, algorithmic performance scales with increasing network size, measured by not only neuron counts but also neuron-to-neuron fan-out degrees. We see this rule holding all the way to biological levels (1:10,000). Due to the $O(N2)$ scaling of connectivity state in the number of fan-outs, it becomes an enormous challenge to support networks with high connectivity using today's integrated-circuit technology.

To address this challenge, Loihi supports a range of features to relax the sometimes-severe constraints that other neuromorphic designs have imposed on the programmer:

- *Sparse network compression*. Besides a common dense matrix model, Loihi supports three sparse matrix compression models in which fan-out neuron indices are computed based on index state stored with each synapse's state variables.
- *Core-to-core multicast*. Any neuron may direct a single spike to any number of destination cores, as the network connectivity might require.
- *Variable synaptic formats*. Loihi supports any weight precision between one and nine bits, signed or unsigned, and weight precisions may be mixed (with scale normalization) even within a single neuron's fan-out distribution.
- *Population-based hierarchical connectivity*. As a generalized weight-sharing mechanism, connectivity templates may be defined and mapped to specific population instances during operation.

Loihi is claimed to be the first fully integrated SNN chip that supports the above features. Previous chips (e.g. TrueNorth) store their synapses in dense matrix form that significantly constrains the space of networks that may be efficiently supported.

34 Loihi: Core Top-Level Microarchitecture

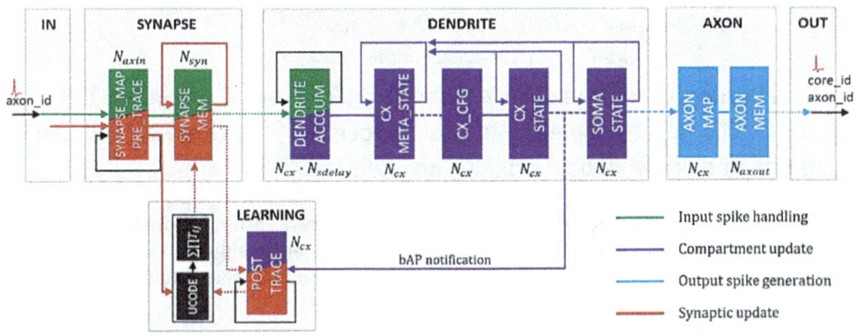

- SYNAPSE unit processes all incoming spikes and reads out the associated synaptic weights.
- DENDRITE unit updates the state variables u and v of all neurons in the core.
- AXON unit generates spike messages for all fanout cores of each firing neuron.
- LEARNING unit updates synaptic weights using the programmed learning rules.

This slide shows the core top-level microarchitecture of the Loihi neuromorphic chip. Colored blocks in this diagram represent the major memories that store the connectivity, configuration, and dynamic state of all neurons mapped to the core. The core's total SRAM capacity is 2 Mb, including ECC overhead. The coloring of memories and dataflow arcs illustrates the core's four primary operating modes: input spike handling (green), neuron compartment updates (purple), output spike generation (blue), and synaptic updates (red). Each of these modes operates independently with minimal synchronization at a variety of frequencies, based on the state and configuration of the core. The black structure marked UCODE represents the configurable learning engine.

The SYNAPSE unit processes all incoming spikes and reads out the associated synaptic weights from the memory.

The DENDRITE unit updates the state variables *u* and *v* of all neurons in the core.

The AXON unit generates spike messages for all fanout cores of each firing neuron.

The LEARNING unit updates synaptic weights using the programmed learning rules at epoch boundaries.

Varying degrees of parallelism and serialization are applied to sections of the core's pipeline to balance the throughput bottlenecks that typical workloads will encounter. Dataflow drawn with finely dotted arrows in the diagram indicate parts of the design where single events are expanded into a potentially large number of dependent events. In these areas, the hardware is generally parallelized.

35 Loihi: Learning Engine

- Supports baseline pairwise STDP (spike timing dependent plasticity)

$$\Delta w_{i,j} = \begin{cases} A_{-}\mathcal{F}(t - t_i^{post}), & \text{On presynaptic spike} \\ A_{+}\mathcal{F}(t - t_j^{pre}), & \text{On postsynaptic spike} \end{cases}$$

  and also aims to supports more advanced learning rules beyond pairwise STDP
- Learning rule functional form: on every learning epoch, a synapse will be updated whenever the appropriate pre- or post-synaptic conditions are satisfied

$$z := z + \underbrace{\sum_{i=1}^{N_P} S_i \prod_{j=1}^{n_i} \underbrace{(V_{i,j} + C_{i,j})}_{T_{i,j}}}_{P_i}$$

  z is the transformed synaptic variable
  ***V**$_{i,j}$* refers to input variable available to learning engine
  ***C**$_{i,j}$* and ***S**$_i$* are microcode-specified signed constants.

- Given a spike arrival sequence s[t] ∈ {0,1}, an ideal trace sequence x[t] over time is defined as:

$$x[t] = \alpha \cdot x[t-1] + \delta \cdot s[t]$$

  δ is an impulse amount added on every spike event
  α is a decay factor

  Loihi computes a low-precision (7-bit) approximation of this first-order filter (stochastic rounding)

Pairwise STDP is simple, event-driven, and highly amenable to hardware implementation. For a given synapse connecting presynaptic neuron *j* to postsynaptic neuron *i*, an implementation needs only maintain the most recent spike times for the two neurons. Given a spike arrival at time *t*, one local nonlinear computation needs to be evaluated to update the synaptic weight $\mathbf{w}_{i,j}$.

To support more advanced learning rules beyond pairwise STDP, a number of architectural challenges arise. First, the functional forms describing $\Delta\mathbf{w}_{i,j}$ become more complex and seemingly arbitrary. These rules are at the frontier of algorithm research and, therefore, require a high degree of configurability. Second, the rules involve multiple synaptic variables, not just weights. Finally, advanced learning rules rely on temporal correlations in spiking activity over a range of timescales, which means more than just the most recent spike times must be maintained. These challenges motivate the central features of Loihi's learning architecture.

In Loihi's learning rule, on every learning epoch, a synapse will be updated whenever the appropriate pre- or post-synaptic conditions are satisfied. A set of microcode operations associated with the synapse determines the functional form of one or more transformations to apply to the synapse's state variables. The rules are specified in sum-of-products form for *z* in the middle equation.

Given a spike arrival sequence $s[t] \in \{0,1\}$, an ideal trace sequence $x[t]$ over time is defined as the bottom equation. The Loihi hardware computes a low-precision (7-bit) approximation of this first-order filter using stochastic rounding. By setting δ to 1 (typically with relatively small α), $x[t]$ saturates on each spike, and its decay measures elapsed time since the most recent spike. Such trace configurations exactly implement the baseline STDP rules dependent only on the previously described nearest-neighbor pre/post spike time separations. On the other hand, setting δ to a value less than 1, specifically $1 - \alpha T_{min}$ (where T_{min} is the minimum spike period), causes sufficiently closely spaced spike impulses to accumulate over time, and $x[t]$ reflects the average spike rate over a timescale of $\tau = -1/\log\alpha$.

36

Loihi: Performance and Energy Results

This table provides a selection of energy and performance measurements from pre-silicon SDF and SPICE simulations, which are consistent with early post-silicon characterization.

- Pre-silicon performance and energy results:

Measured parameter	Value at 0.75 V
Cross-sectional spike bandwidth per tile	3.44 Gspike/s
Within-tile spike energy	1.7 pJ
Within-tile spike latency	2.1 ns
Energy per tile hop (E-W / N-S)	3.0 pJ / 4.0 pJ
Latency per tile hop (E-W / N-S)	4.1 ns / 6.5 ns
Energy per synaptic spike op (min)	23.6 pJ
Time per synaptic spike op (max)	3.5 ns
Energy per synaptic update (pairwise STDP)	120 pJ
Time per synaptic update (pairwise STDP)	6.1 ns
Energy per neuron update (active / inactive)	81 pJ / 52 pJ
Time per neuron update (active / inactive)	8.4 ns / 5.3 ns
Mesh-wide barrier sync time (1-32 tiles)	113-465 ns

37

Loihi: Comparison again Atom Processor

- Comparison of solving L1 minimization on Loihi and Atom:
  (Results are expressed as improvement ratios Atom/Loihi)

Number of Unknowns	400	1,700	32,256
Number of non-zeros in solutions	≈10	≈30	≈420
Energy	2.58x	8.08x	48.74x
Delay	0.27x	2.76x	118.18x
EDP	0.7x	22.33x	5760x

On an earlier iteration of the Loihi architecture, we quantitatively assessed the efficiency of Spiking LCA to solve LASSO. A 1.67-GHz Atom CPU was used to run both LARS and FISTA3 numerical solvers as a reference architecture for benchmarking. These solvers are among the best known for this problem. Both chips were fabricated in 14-nm technology, were evaluated at a 0.75-V supply voltage, and required similar active silicon areas (5 mm2).

The table in this slide shows the comparison in computational efficiency between the two architectures of Atom and Loihi, as measured by energy-delay-product (EDP).

Results are expressed as improvement ratios Atom/Loihi. The Atom numbers are chosen using the more efficient solver between LARS and FISTA.

The conventional LARS solver can handle problems of small sizes and very sparse solutions quite efficiently. On the other hand, the conventional solvers do not scale well for the large problem, and the Loihi predecessor achieves the target objective value with over 5,000 times lower EDP.

38 Summary of Loihi Processor

- Digital neuron and synapse implementation for on-chip learning
  - Support on-chip learning of STDP and more advanced learning rules
- 128 neuromorphic cores, 3 x86 cores
- 14nm Intel FinFET process, $60mm^2$
- Asynchronous network-on-chip
- 128K neurons, 128M synapses
- Large EDP gain against Atom processor for complex optimization tasks

39 Tutorial Outline

- TrueNorth processor
  - IBM, 2014
  - Digital, no on-chip learning, 1M neurons, 256M synapses
- ROLLS processor
  - ETH Zurich, 2015
  - Analog/mixed-signal, long/short-term on-chip learning, 256 neurons, 128K synapses
- Loihi processor
  - Intel, 2018
  - Digital, on-chip learning, 128K neurons, 128M synapses
- RRAM-based computing
  - Efforts from academia/industry researchers
  - In-memory mixed-signal computing, small/moderate array demonstrations

Finally, RRAM-based computing efforts from academia and industry researchers will be presented.

40 Emerging Non-Volatile Memory (eNVM)

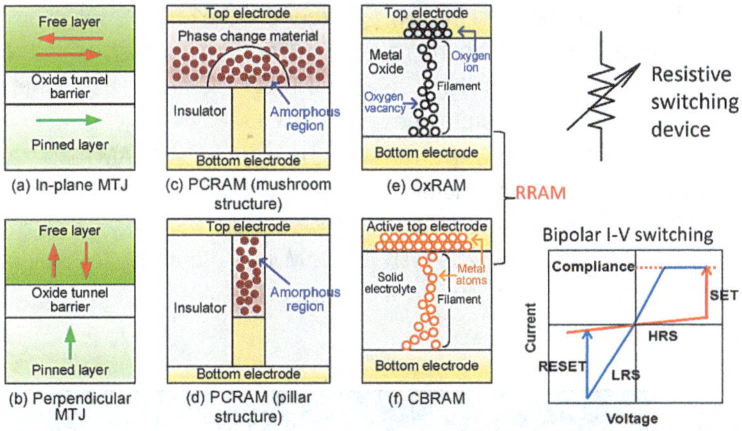

S. Yu, P.-Y. Chen, "Emerging memory technologies: recent trends and prospects," IEEE Solid State Circuits Magazine, vol. 8, no. 2, pp. 43-56, 2016.

To fill in the gap in the memory hierarchy, the research community have proposed different emerging NVMs, including the STT-MRAM based on the magnetic tunneling junction, and PCM based on the chalcogenide materials, as well as the resistive random access memory (RRAM). The physical mechanism of RRAM is based on the formation and rupture of conductive filaments in the oxide materials. In general, the eNVMs are based resistance switching in a two terminal device.

To make it simple, you can think the resistive switching device as a variable resistor. Here is a typical bipolar I-V switching characteristic of RRAM device. When applying a positive voltage beyond a threshold, it can SET from HRS to LRS. If reversing the voltage polarity to negative voltage, it can RESET from LRS to HRS. The LRS can represent the logic state "1" and the HRS can represent the logic state "0". Therefore, we can make it as a memory device.

41 Trends for eNVM Prototype Chips

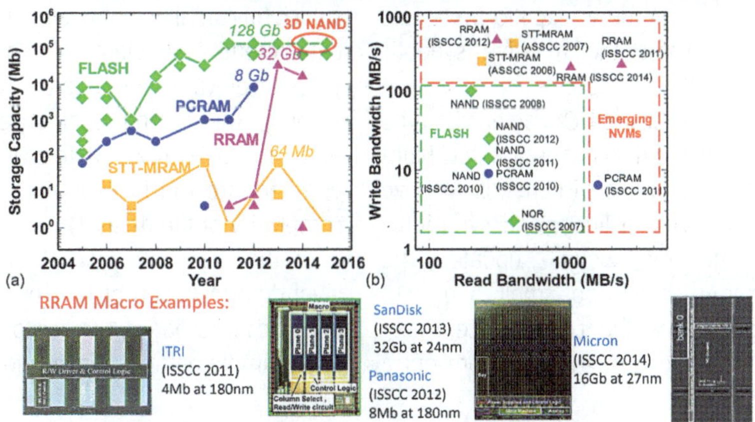

S. Yu, P.-Y. Chen, "Emerging memory technologies: recent trends and prospects," IEEE Solid State Circuits Magazine, vol. 8, no. 2, pp. 43-56, 2016.

41 Trends for eNVM Prototype Chips

In the past decade, there have been significant progresses of eNVM research and development, not only from academia but also from industry.

This chart summarizes the eNVM prototypes reported in the major conference like ISSCC. We can see the capacity of NAND Flash is still the largest, especially with 3D NAND available. However the PCM/RRAM is catching up quickly, up to 32Gb. STT-MRAM typically has tens of Mb capacity due to a larger cell area by the selection transistor. If we look at the performance, emerging NVMs outperform the NAND or NOR flash in both write and read bandwidth by 10X.

Here are a few RRAM macro examples by different companies, for example, ITRI, Panasonic, SanDisk, Micron, etc. These demonstrations show that RRAM is a promising technology from the industry perspective.

42 Current Status of Resistive Synaptic Devices

- Mostly focused on device-level engineering...

Performance metrics	Desired Targets
Device dimension	< 10 nm
Multilevel states number	>100* with a linear symmetric update
Energy consumption	<10 fJ/programming pulse
Dynamic range	>100*
Retention	>10 years* (offline)
Endurance	>10^9 updates* (online)

Note: * these numbers are application-dependent

- A few array-level demo with simple pattern classification, such as:
  - IBM's 500×661 1T1R array with PCM (IEDM 2014)
  - UCSB's 12*12 crossbar array with memristors (Nature 2015)
  - Umich's 32*32 crossbar array with memristors (Nature Nano 2017)
  - Tsinghua's 32*32 1T1R array with multilevel RRAM (Nature Comm. 2017)

S. Yu, "Neuro-inspired computing with emerging non-volatile memory," **Proc. IEEE**, vol. 106, no. 2, pp. 260-285, 2018

Emerging non-volatile memory (e.g. RRAM) should be engineered towards the resistive synaptic devices for neuromorphic applications.

This table shows the desired characteristics. The device dimension should be nanoscale for high-density integration, and the device needs to have >100 levels of states with a linear and symmetric conductance change for weight update. The programming energy should be small, e.g. <10 fJ. Dynamic range means the conductance max/min ratio should be large, e.g. >100 to represent a sufficiently small weight value. If offline training, the data retention requirement may be as long as 10 years at elevated temperature, e.g. 85 oC. If online training, the cycling endurance requirement may be as much as 1E9 updates. Of course, all of these numbers are much application-dependent (i.e. the complexity of the dataset).

In the past few years, there are a few array-level demonstration of tens by tens or hundreds by hundreds crossbar or 1-transistor-1-resistor (1T1R) for some simple pattern classification tasks.

43 Crossbar Architecture for Accelerating Weighted Sum and Weight Update

- Weighted sum (inference): all cells are activated in parallel, summing up column current→perform vector-matrix multiplication
- Weight update (training): cell's conductance could be updated by applying programming voltage row by row or in parallel.

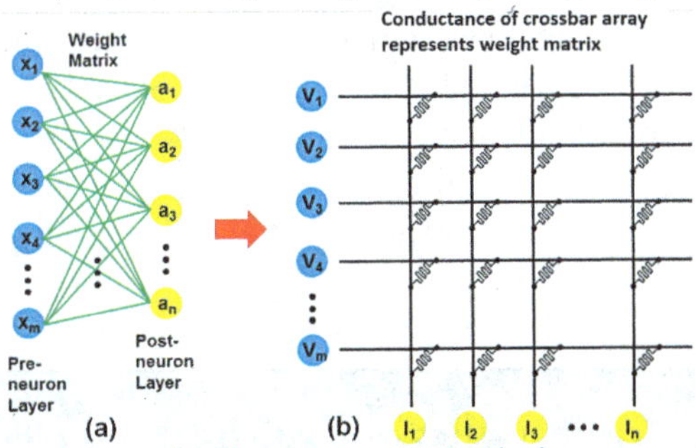

Task	Operations
$W \cdot X$	$I_i = \sum_j G_{ij} \cdot V_j$
W update	$\Delta G_{ij} = \eta \cdot V_i \cdot V_j$

(analog computation inside the array, may need ADC at edge of array)

Let us introduce how the crossbar array can accelerate the weighted sum and weight update processes in neural networks.

In typical learning algorithms, there are many operations like vector multiple by matrix, which is very time consuming.

Here we apply voltages representing vector to the row, and use the conductance of the resistive devices to represent the weight matrix. According to the Kirchhoff law, the column current is the weighted sum.

To update the weight, we can apply voltages from the two ends, row and column, the weight change delta_W is proportional to the values at the end of row and column, as in many delta-rule based learning algorithms. Or we could also pre-calculate the weight change delta_W by peripheral circuitry, and update the weight matrix row by row.

44 Selector and Pseudo-Crossbar Array with 1T1R

- If all the cells turned on (i.e., in the fully parallel read), there is no sneak path problem.
- If the cells are partially turned on (i.e., in the row-by-row write), it needs to suppress the sneak current in unselected cells → Need selector!

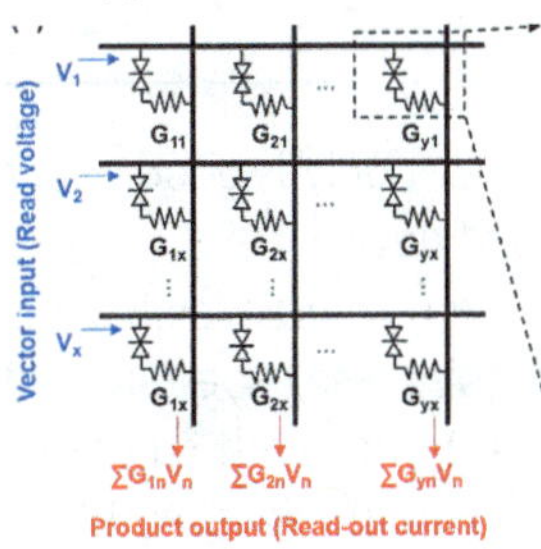

1S1R Array

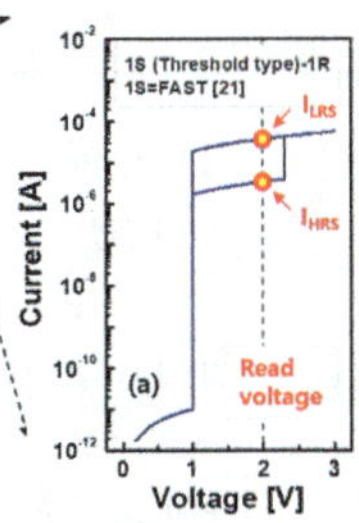

Phase transition in strongly correlated oxides and chalcogenides shows threshold switching.
Unfortunately, selector is not mature yet ☹

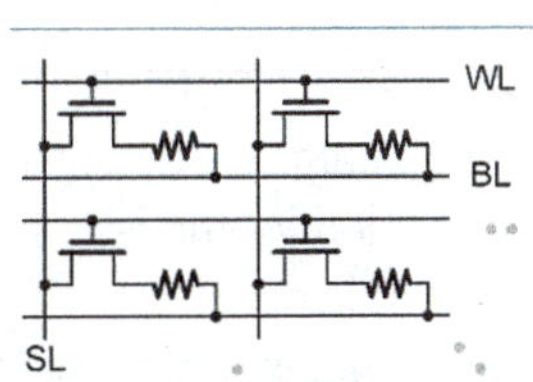

Pseudo-crossbar Array

44 Selector and Pseudo-Crossbar Array with 1T1R

We need to notice that in the fully parallel read out operation, all the cells are selected and participated in the analog computation, thus there is no sneak path problem.

In the partially selected row-by-row write operation, if not all cells are programmed in parallel, still the two-terminal selectors with strong I-V nonlinearity are needed to suppress the sneak current in the unselected cells.

There are many selector candidates reported in the literature based on different physical mechanisms, however, they are still in the premature stage in terms of large-scale manufacturing. Alternatively, the three-terminal transistors are preferred to be organized in the 1T1R with BL and SL perpendicular to form pseudo-crossbar, where the all the WLs could be turned on simultaneously for fully parallel-read out, and partially turned on for row-by-row write.

45 Resistive Synaptic Devices based on eNVM

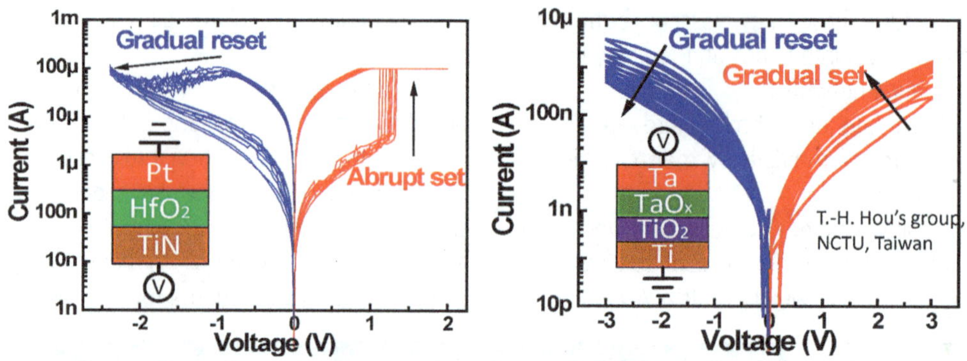

- **Offline training:** weights are pre-defined by software training, just need one-time loading to the array→ Conventional "binary" *filamentary* RRAM with gradual reset only is good enough
- **Online training:** weights are updated during run-time→ Special "analog" *interfacial* RRAM with both smooth set and reset is needed

Let us look at what I-V characteristics of resistive synaptic devices are needed for the learning.

There are two types of training:

Offline training, where weights are pre-defined by software training, then just we need one-time loading to the array for the following inference. In this case, conventional "binary" filamentary RRAM with gradual reset only is good enough, as shown in this HfOx based device.

Online training, where weights are updated during run-time. In this case, special "analog" interfacial switching RRAM with both smooth set and reset is needed, as shown in this TaOx/TiO2 based device.

46

Achieve Fast Convergence for Offline Training

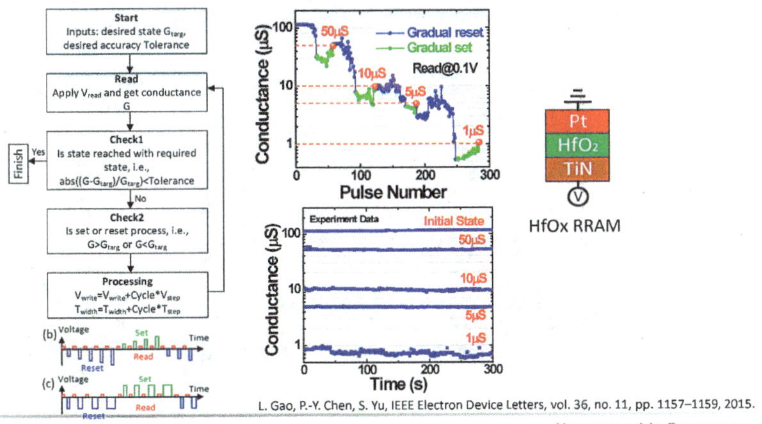

L. Gao, P.-Y. Chen, S. Yu, IEEE Electron Device Letters, vol. 36, no. 11, pp. 1157–1159, 2015.

Neuromorphic Processors

Let us first look at offline training, as we just need one time programming to tune the device conductance to the pre-defined weight (or target conductance).

We developed a testing protocol with iterative programming, by pulse train to tune the conductance with feedback control.

For example, here on HfOx RRAM, we set a few target weights (50 uS, 10 uS, 5 uS, 1 uS), then we can use pulse train to tune the conductance to the desired states, if it overshoots the target, a reversed voltage polarity pulses will be applied.

We can tune the conductance into very accurate level this way, of course, with the penalty of relatively long programming time (i.e. tens of cycles).

After the weight tuning, each conductance level should be stable over time (with read-out noise though).

47

Implementation of Convolution Kernel on Cross-Point Array

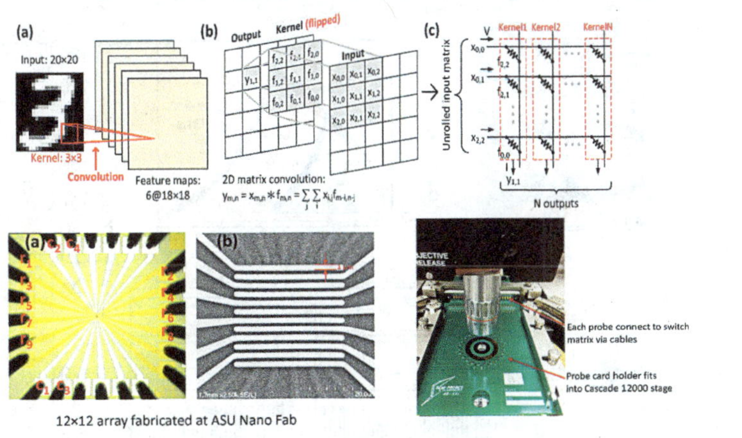

With the offline trained HfO2 RRAM devices, we demonstrated a convolution kernel mapping to a crossbar array. The 2D convolution kernel (e.g. 3x3) in convolutional neural network is unrolled into one column's conductance, while multiple columns could represent multiple kernels.

We set up the experimental platform under the probe card in a probe station, we pre-program the conductance to the desired weight. Then the read voltage representing the input image vector is applied to the rows, and the column current representing the weighted sum is measured by the SMU of the testing equipment.

48 Edge Detection By Prewitt Horizontal and Vertical Kernel Operation

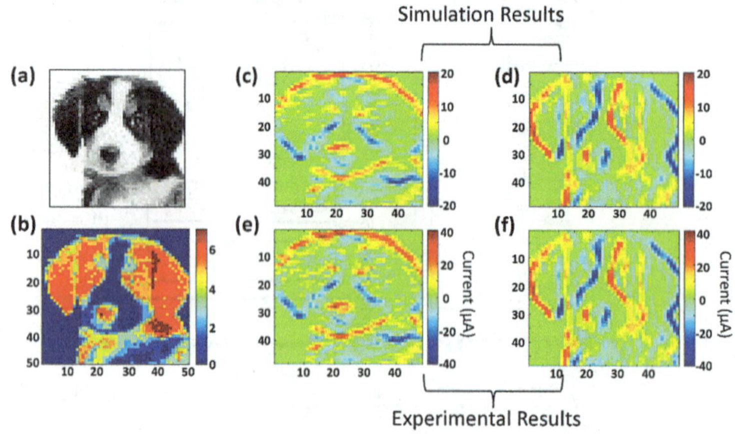

L. Gao, P.-Y. Chen, S. Yu, IEEE Electron Device Lett., vol. 37, no. 7, 870-873, 2016.

With this experimental set-up, we demonstrated the edge detection of the image. The horizontal and vertical edge detection kernel is preprogramed to two columns of the crossbar array, and then we scan the image window over the input image (an dog image in this case). The column current is measured as weighted sum, and it matches well with the simulated results.

49 Device and Array-level Demo for Online Training

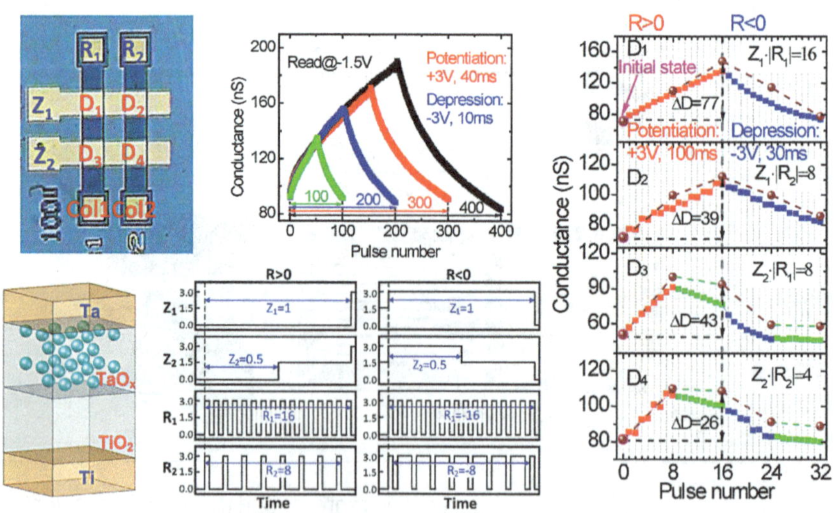

L. Gao, et al, Nanotechnology vol. 26, 455204, 2015, collaborated with T.-H. Hou's group, NCTU, Taiwan

49 Device and Array-level Demo for Online Training

For online training, the iterative programming is too slow, as we need to update the weights all the time during the training.

Then we need smooth weight update curve (both potentiation and depression) without the feedback control.

Therefore, special RRAM structure is needed. For example, here we have the bilayer TaOx/TiO2 device. It shows very smooth weight update under identical programming pulses, and more than 200 levels are achievable.

With this device, we further developed a scheme to program the entire crossbar array in parallel, and experimentally demonstrated the scheme on a 2x2 array as proof-of-concept.

According to the delta-rule in back-propagation, the weight change delta_W is proportional to the error and the gradient (from the post-layer) and the input (from the pre-layer). Therefore, we encode the Z value from the pre-layer as the duty cycle of the programming time-window, and encode the R value from the post-layer as the number of pulses in the time-window. When the R pulses overlap with Z, the effectively voltage drop on the device will program the device accordingly.

For example, here if Z1*R1 is 16, then we need to update 16 levels of the conductance of Cell D1; if Z1*R2 is 8, then we update 8 levels of Cell D2; if Z2*R1 is 8, then we update 8 levels of Cell D3; if Z2*R2 is 4, then we update 4 levels of Cell D4. In principle, we could update all the cells in parallel, however, when the array size scales up, there may be a limit on how much peak power could be delivered to program all the cells.

50 ADC Neuron Circuits: Integrate-and-Fire Model

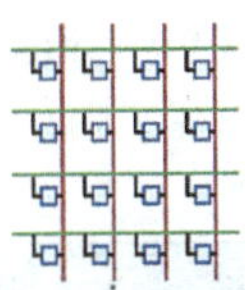

- **Analog current-to-digital output converter (ADC)**, operating as the Integrate-and-Fire neuron model
- Neuron circuit is much larger than the column pitch of crossbar array→column sharing→reduced parallelism

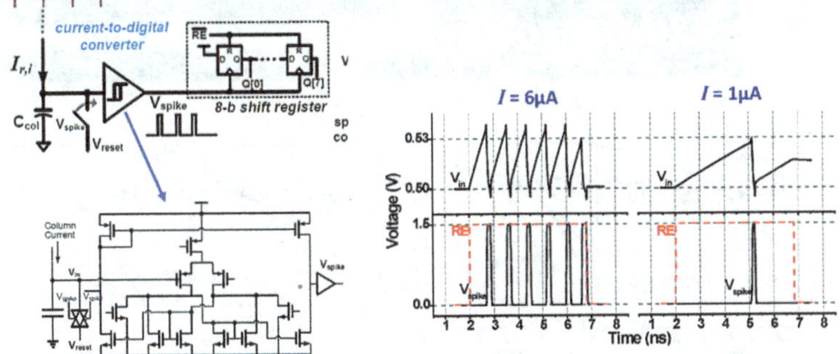

D. Kadetotad, et al. IEEE JETCAS, vol. 5, no. 2, pp. 194-204, 2015.

At the edge of the crossbar array, there requires the neuron circuits to convert the column analog current to the digital output, essentially functioning as the analog to digital (ADC) converter.

One example of the customized CMOS ADC circuitry is to employ the integrate-and-fire neuron model. The weighted sum current is integrated at the capacitor at the end of the column, and the charging and discharging frequency is proportional to the current, which will be converted to the number of spike at the output, which could be further encoded to binary data by shift register.

51 Oscillation Neuron with Threshold Switching Device

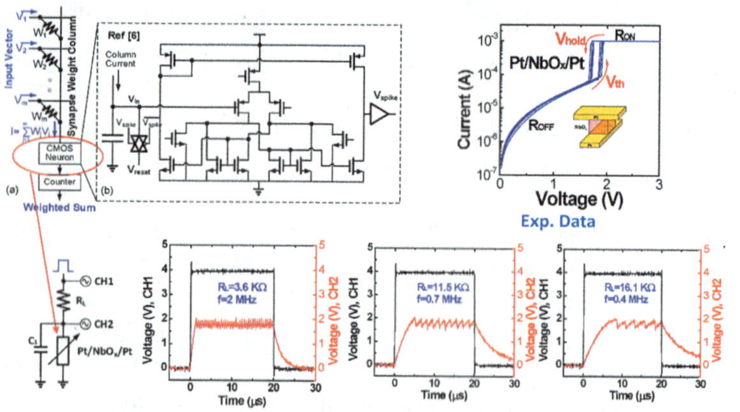

P.-Y. Chen, et al. ICCAD 2016 and L. Gao, et al. Applied Physics Letters, 111, 103503, 2017

One problem with the CMOS ADC circuitry is the relatively large area, which could be much larger than the column pitch of the crossbar array, thus multiple columns have to share one neuron node. As a result, time-multiplexing has to be used to read out all the columns sequentially, thereby reducing the parallelism of the crossbar array.

One possible solution to this problem is to employ the oscillation neuron node based on threshold switching device. For example, the I-V characteristics of the NbOx based device shows an abrupt turn-on above Vth and an abrupt turn-off below Vhold. If such device is collected to the end of the column, then the node voltage between the column equivalent resistance RL and the device may oscillate between Vth and Vhold under the condition that RL is in between on and off resistances of the threshold switching device.

In our initial experiments, we varied the RL externally, and show that the oscillation frequency is proportional to the conductance of the RL (thus the weighted sum from the column). With an following CMOS inverter, the oscillation waveform could be digitalized. The oscillation frequency could be improved to GHz range if the parasitic capacitance is limited to wire capacitance of the column only.

52 Exp. Data of Analog Synapses for Online Training

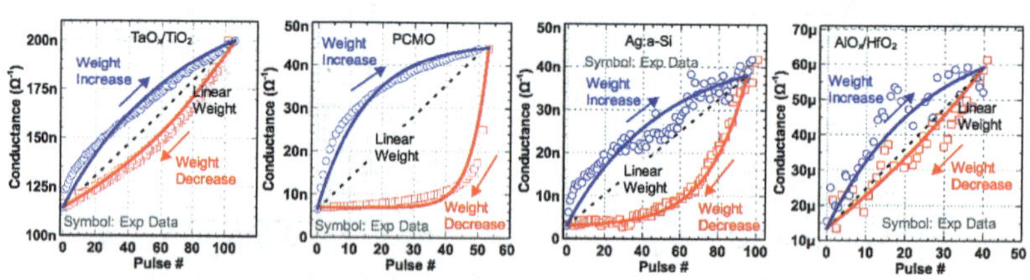

Non-ideal device properties:

- Limited weight precision
- Finite ON/OFF ratio
- Weight update nonlinearity and asymmetry
- Device variation

Ref: (a) L. Gao et al., Nanotechnology, 2015. (b) S. Park et al., IEDM, 2013. (c) S. H. Jo, et al., Nano letters, 2010. (d) J. Woo et al., EDL, 2016

52 Exp. Data of Analog Synapses for Online Training

There have been various resistive synaptic devices reported in the literature for online training. Here are a few representative candidates. The weight update curve (conductance vs # identical programming pulse) generally shows the following non-ideal device effects, including limited weight precision (number of levels), finite ON/OFF conductance ratio, weight update nonlinearity and asymmetry (weight increase with blue trajectory, while weight decrease with red trajectory), and the device variation from device to device and from cycle to cycle. Then the question is how would these non-ideal device properties impact the system-level performance.

53 NeuroSim+: A Simulator from Device to Algorithm

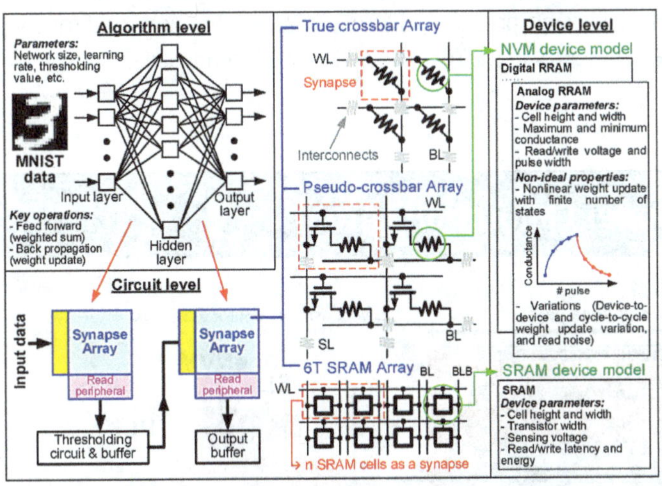

Input:

- Network structure,
- Array type and technology node
- Device type and non-ideal factors
- Training/testing traces

Output:

- Area,
- Latency,
- Energy,
- Accuracy

Algorithms supported: multilayer perceptron, convolutional neural network (on-going)

The 1st version available online and downloadable https://github.com/neurosim/MLP_NeuroSim

To quantify the impact of the non-ideal device properties, we developed NeuroSim framework.

NeuroSim+ is an integrated simulation framework for benchmarking synaptic devices and array architectures in terms of the system-level learning accuracy and hardware performance metrics. It has a hierarchical organization from the device level (transistor technology and memory cell models) to the circuit level (synaptic array architectures and neuron periphery) and then to the algorithm level (neural network topologies). The source code of NeuroSim+ version 1.0 is publicly available at https://github.com/neurosim/MLP_NeuroSim

54 Model Calibration (Latency, Energy, Leakage)

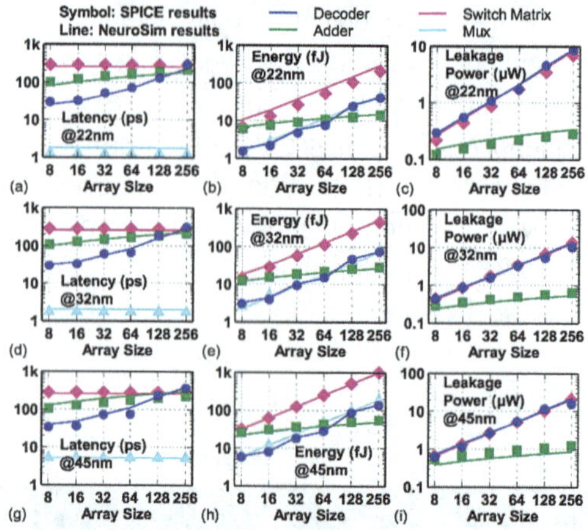

Latency (Horowitz equation)

$$= \tau_f \sqrt{ln(v_s)^2 + \frac{2}{rampInput \times \tau_f}\beta(1-v_s)}$$

Dynamic energy=NCV_{DD}^2

Static energy=$GV_W^2NT_{PULSE}$

Leakage=$NI_{off}V_{DD}$

Supported tech node from 130nm to 7nm with calibration of PTM model

P.-Y. Chen, X. Peng, S. Yu, "NeuroSim+: An integrated device-to-algorithm framework for benchmarking synaptic devices and array architectures," IEEE International Electron Devices Meeting (IEDM) 2017

To validate NeuroSim+ model based on the analytical equations for latency, energy and leakage power, we have performed the performance metrics in main circuit modules using SPICE with PTM model at 45nm, 32nm and 22nm. The predicted model prediction and the SPICE simulation results is close.

55 Model Calibration (Area)

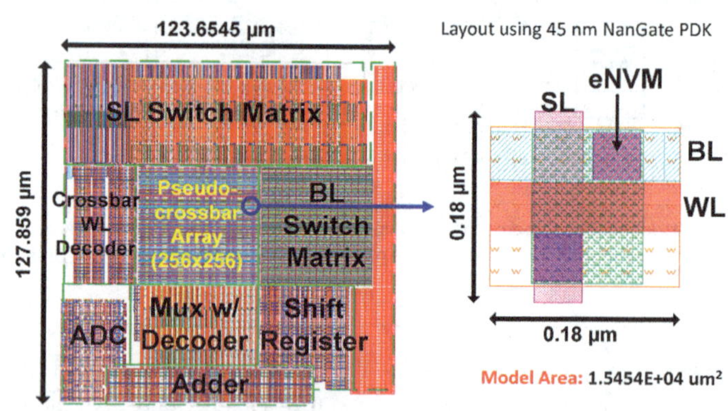

Layout Area: 1.5810E+04 um²

Model Area: 1.5454E+04 um²

P.-Y. Chen, et al. IEDM 2017

To validate NeuroSim+ model, we have performed the area calibration of a synaptic core (256x256 1T1R) using layout at FreePDK 45nm. The predicted layout area and the actual layout area is close.

56

A Case Study of Multilayer Perceptron (MLP)

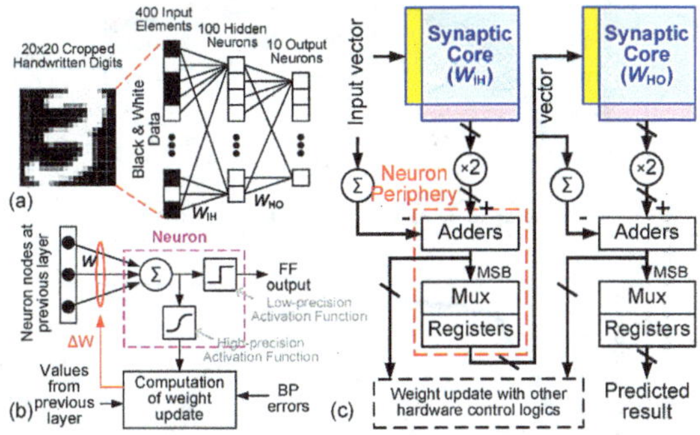

- A multilayer perceptron (MLP) 400-100-10 network is used for analog RRAM synapses benchmarking.

P.-Y. Chen, et al. IEDM 2017

NeuroSim+ can support the integration of neural network algorithm with the circuit-level model to form a complete simulation framework. In this 2-layer multilayer perceptron (MLP) case study, we have a fully connected network with topology 400 input neurons, 100 hidden neurons and 10 output neurons. The MNIST handwritten digit images are cropped and converted to black and white (1-bit) data to reduce the complexity of input encoding. For design simplicity, each neuron will truncate the weighted sum to 1-bit output value for the input of next neuron node. In this way, inference which is purely feed forward (FF) can be realized with low-precision. However, the computation on the back propagation (BP) of weight update still needs high precision for training.

57

Impact of Weight Precision and Weight Update Nonlinearity in Analog Synapses

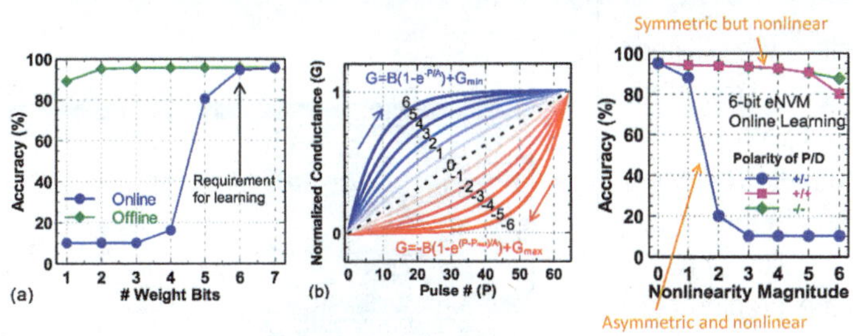

- At least 6-bit is required for MNIST dataset online learning, while 1-bit may work for offline classification.
- Nonlinearity significantly degrades accuracy for online learning if using analog synapses.

P.-Y. Chen, et al. IEDM 2017

To quantify the impact of non-ideal device properties, we performed sensitivity analyses in the online learning and offline classification or inference modes. The results suggest that the weight precision should be at least 6 bits for online learning to achieve high accuracy, while 1 bit seems sufficient for offline classification.

The major factor that kills the learning accuracy is the asymmetry in the weight update. We labeled the weight increase from 0 to 6 degrees in blue color and weight decrease from 0 to -6 degrees in red color. The learning accuracy is relatively resilient to the nonlinearity magnitude if both the potentiation (P) and depression (D) have the same polarity. However, if they have different polarities, for example, P is positive and D is negative as commonly observed in today's analog eNVM devices, the accuracy could quickly drop with increase of the nonlinearity magnitude.

58 Impact of Weight Update Variations

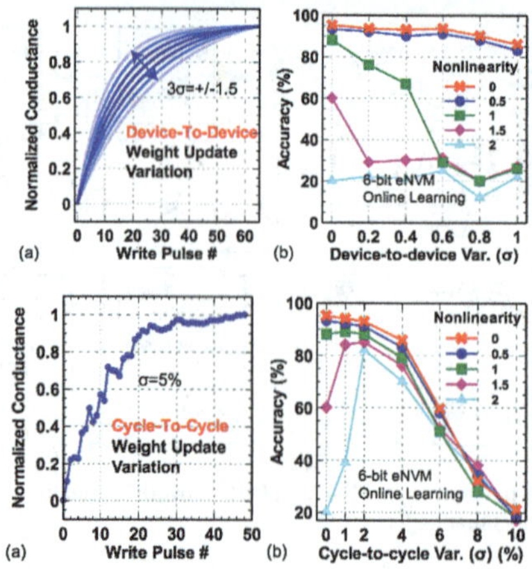

- The neural network has reasonably good resiliency to the device-to-device variation.
- A small cycle-to-cycle is beneficial to overcome the cycle-to-cycle variation.

P.-Y. Chen, et al. IEDM 2017

Regarding the impact of weight update variations, the neural network is generally resilient to the device-to-device variation, but its impact becomes prominent at high nonlinearity. A small cycle-to-cycle variation can alleviate the degradation of learning accuracy by high nonlinearity, as it helps the network to jump out of the local minima during the optimization. However, too large variation overwhelms the deterministic update amount defined by BP thus is harmful to the accuracy.

59 Specs and Online Learning Accuracy of Reported and Desired eNVMs

Benchmark for training for 1M MNIST images

	Reported eNVMs for learning				Desired eNVMs for learning	
Analog eNVM type	TaO_x/TiO_2	PCMO	Ag:a-Si	AlO_x/HfO_2	Targeted eNVM	Ideal eNVM
# of conductance states	102	50	97	40	64 (6 bits)	64 (6 bits)
Nonlinearity (weight increase/decrease)	0.66/-0.69	3.68/-6.76	2.4/-4.88	1.94/-0.61	1.0/-1.0	0/0
R_{ON}	5 MΩ	23 MΩ	26 MΩ	16.9 kΩ	200 kΩ	200 kΩ
ON/OFF ratio	2	6.84	12.5	4.43	50	50
Weight increase pulse	3V/40ms	-2V/1ms	3.2V/300μs	0.9V/100μs	2V/100ns	2V/10ns
Weight decrease pulse	-3V/10ms	2V/1ms	-2.8V/300μs	-1V/100μs	2V/100ns	2V/10ns
Weight update cycle-to-cycle variation (σ)	<1%	<1%	3.5%	5%	2%	0%
Accuracy for online learning	~10%	~10%	~73%	~41%	90%	94.8%
Area	1071.3 μm²	1071.3 μm²	1072.0 μm²	3657.2 μm²	1247.3 μm²	1247.3 μm²
Latency for online learning (1M images)	3.57E10 s	7.00E8 s	4.20E8 s	5.60E7 s	8.82E4 s	8.82E3 s
Energy for online learning (1M images)	65.86 mJ	29.4 mJ	87.94 mJ	150 mJ	29.80 mJ	29.80 mJ

Red: major causes for failure, green: good properties

- Today's analog eNVM suffers from large weight update nonlinearity, and small on/off ratio, making it challenging for achieving high accuracy for online learning.

Overall, we listed the reported devices with the extracted realistic device parameters in this table. Today's analog eNVM devices are problematic to be used in online learning. The red color highlights the killing factor to the accuracy. Also the red color labels on the latency of online training with 1 millions images, the training latency is more than 10 years due to the slow programming speed of these reported analog eNVM devices.

Therefore, we propose to set up the targeted and ideal specifications for eNVMs that is able to achieve a comparable accuracy to the MLP software baseline. The key improvements are linearity and symmetry of the weight update, higher on-state resistance (~100 kOhm), larger on/off ratio (>50) and faster programming pulse width (10ns~100ns).

60 Benchmark SRAM vs. eNVM-based System for Offline Inference

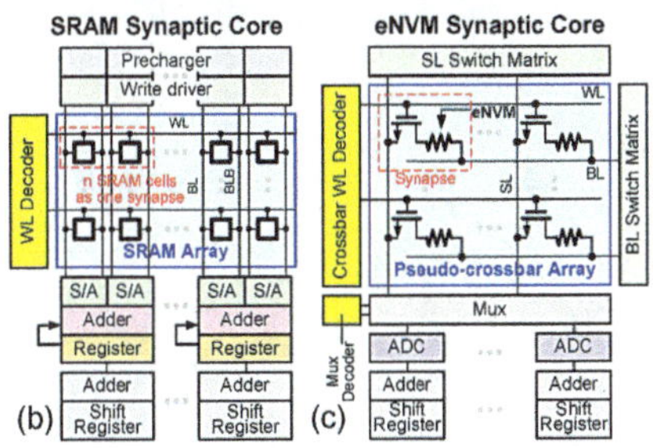

P.-Y. Chen, et al. IEDM 2017

Offline classification results (eNVM outperforms in all aspects)

	2-bit SRAM	2-bit eNVM
Area	4321 μm²	1561 μm²
Latency	135.48 ms	1.88 ms
Energy	99.37 μJ	7.54 μJ
Leakage Power	468.33 μW	26.65 μW

Regarding the offline inference or classification, eNVM based synaptic core actually could potentially outperform the SRAM based synaptic core at the same technology node and weight precision (low-precision such as 2-bit) in all the circuit-level performance metrics. It is always possible to tune the eNVM conductance to the target weight level by iterative one-time programming.

61 Summary of RRAM-based Computing

- Today's resistive memory devices can be tuned to multilevel (possibly by iterative programming), and offline inference is most suitable application.
- For online training, "analog" synapses with continuous weights need further device engineering to overcome challenges such as nonlinear and asymmetric weight update, and improve on/off ratio and programming speed.
- CMOS neuron node is complex. It is worth of exploring more compact oscillation neuron node with threshold switching devices.
- Co-design devices, circuits, architectures and algorithms is helpful. NeuroSim+ framework is the first trial in the co-design, and will be extended to deeper network and larger dataset.

About the Editors

Minkyu Je

Minkyu Je received the M.S. and Ph.D. degrees, both in Electrical Engineering and Computer Science, from Korea Advanced Institute of Science and Technology (KAIST), Daejeon, Korea, in 1998 and 2003, respectively. In 2003, he joined Samsung Electronics, Giheung, Korea, as a Senior Engineer. From 2006 to 2013, he was with Institute of Microelectronics (IME), Agency for Science, Technology, and Research (A*STAR), Singapore. He worked as a Senior Research Engineer from 2006 to 2007, a Member of Technical Staff from 2008 to 2011, a Senior Scientist in 2012, and a Deputy Director in 2013. From 2011 to 2013, he led the Integrated Circuits and Systems Laboratory at IME as a Department Head. He was also a Program Director of NeuroDevices Program under A*STAR Science and Engineering Research Council (SERC) from 2011 to 2013, and an Adjunct Assistant Professor in the Department of Electrical and Computer Engineering at National University of Singapore (NUS) from 2010 to 2013. He was an Associate Professor in the Department of Information and Communication Engineering at Daegu Gyenogbuk Institute of Science and Technology (DGIST), Korea from 2014 to 2015. Since 2016, he has been an Associate Professor in the School of Electrical Engineering at Korea Advanced Institute of Science and Technology (KAIST), Korea.

His research areas are advanced IC platform development including smart sensor interface ICs and ultra-low-power wireless communication ICs, as well as microsystem integration leveraging the advanced IC platform for emerging applications such as intelligent miniature biomedical devices, ubiquitous wireless sensor nodes, and future mobile devices. He is an author of 5 book chapters and has more than 290 peer-reviewed international conference and journal publications in the areas of sensor interface IC, wireless IC, biomedical microsystem, 3D IC, device modeling and nanoelectronics. He also has more than 50 patents issued or filed. He has served on the Technical Program Committee and Organizing Committee for various international conferences, symposiums, and workshops including IEEE International Solid-State Circuits Conference (ISSCC), IEEE Asian Solid-State Circuits Conference (A-SSCC) and IEEE Symposium on VLSI Circuits (SOVC).

Myung Hoon Sunwoo

Myung Hoon Sunwoo received B.S. degree from Sogang University, M.S. degree from Korea Advanced Institute of Science and Technology (KAIST), and Ph.D. degree from the University of Texas at Austin in Electrical and Computer Engineering. He worked for Electronics and Telecommunications Research Institute (ETRI) in Daejeon, Korea from 1982 to 1985, and for the Digital Signal Processor Operations, Motorola, in Austin, Texas, U.S.A. from 1990 to 1992. Since 1992, he has been with the School of Electrical and Computer Engineering, Ajou University in Suwon, Korea, where he is currently a Professor.

He has authored over 400 papers and holds over 90 patents. He received 50 awards including IEEE Circuits and Systems Society (CASS) Chapter of the Year Award-World in 2013 and CASS Chapter of the Year-R10 in 2016 and the Best Paper Awards from various conferences and societies. He has been a Technical Committee member for numerous conferences and societies. He was an Associate Editor for the IEEE Transactions on Very Large Scale Integration (VLSI) Systems (2002–2003), Guest Editors for the Journal of VLSI Signal Processing (Kluwer, 2005) and for the Journal of Signal Processing Systems (Springer-Verlag, 2012).

Currently, he is the Director of the micro Diagnostic Smart Devices (*u*DSD) Information and Telecommunication Research Center (ITRC) sponsored by the Ministry of Science and ICT of Korea, which consists of universities, hospitals, and companies to cover emerging interdisciplinary technical areas. His research interests include low power algorithms and architectures, medical devices, deep learning, and artificial intelligence circuits and systems (AICAS).

He served as the General Chair of International Symposium on Circuits and Systems (ISCAS) 2012, the successful event held in Seoul, Korea and will serve again as the General Co-chair of ISCAS 2021, Daegu,

Korea. He was a member of ISCAS steering committee (SC) from 2010 to 2015 and is a member of APCCAS SC. He also served as the Honorary Chair of Asian Pacific Conference on Circuits and Systems (APCCAS) 2016. He has been involved in various IEEE activities over three decades including a member of IEEE CASS BoG (Board of Governors) elected twice from 2011 to 2016 and elected a CASS Vice President (VP)–Conferences in 2018. He was a Distinguished Lecturer of the IEEE CASS from 2009 to 2010.

As the IEEE CASS VP-Conferences, he initiated the first International Conferences on AICAS in 2019, a successful and timely event in Hsinchu, Taiwan. He was the President of the IEIE Semiconductor Society from 2012 to 2013. He was an honorary ambassador of Korean Tourism Organization. He was a chair of IEEE CASS, Seoul Chapter from 2004 to 2018, is currently the IEEE CASS VP-Conferences and an IEEE Fellow.

About the Authors

Maysam Ghovanloo

Maysam Ghovanloo received the B.S. degree in electrical engineering from the University of Tehran, Tehran, Iran, in 1994, the M.S. degree in biomedical engineering from the Amirkabir University of Technology, Tehran, Iran, in 1997, and the M.S. and Ph.D. degrees in electrical engineering from the University of Michigan, Ann Arbor, in 2003 and 2004, respectively. From 2004 to 2007, he was a faculty member at the Department of Electrical and Computer Engineering, NC-State University, Raleigh, NC. From 2007 to 2019 he was a faculty member at the School of Electrical and Computer Engineering, Georgia Institute of Technology, Atlanta, GA. He is currently the founder and CTO of Bionic Sciences Inc., Atlanta, GA. He has authored or coauthored more than 250 peer-reviewed conference and journal publications on implantable microelectronic devices, integrated circuits and microsystems for medical applications, and modern assistive/rehabilitation technologies. He also holds 10 issued patents.

He is a Fellow of IEEE and a recipient of the National Science Foundation CAREER Award, the Tommy Nobis Barrier Breaker Award for Innovation, and Distinguished Young Scholar Award from the Association of Professors and Scholars of Iranian Heritage. He is an Associate Editor of the IEEE Transactions on Biomedical Engineering, and serves on the Senior Editorial Board of the IEEE Journal on Emerging and Selected Topics in Circuits and Systems (JETCAS). He served as an Associate Editor of IEEE Transactions on Biomedical Engineering, and Transactions on Circuits and Systems, Part II, as well as a Guest Editor for the IEEE Journal of Solid-State Circuits and IEEE Transactions on Neural Systems and Rehabilitation Engineering. He chaired the IEEE Biomedical Circuits and Systems (BioCAS 2015) in Atlanta, GA, and co-chaired the technical program committee for BioCAS 2014, 2016, 2018, and 2019. He has also served on the technical subcommittees of the IEEE International Solid-State Circuits Conference (ISSCC) and Custom Integrated Circuits Conference (CICC).

Michael Haas

Michael Haas received the BSc and the MSc degree in electrical engineering from the University of Ulm, Ulm, Germany, in 2011 and 2013, respectively and the Dr.-Ing. degree from the University of Ulm, Germany, in 2019.

From 2013-2014 he joined Rohde & Schwarz, working as an antenna test range engineer. In 2014, he began his work at the Institute of Microelectronics, University of Ulm, in the field of integrated circuits for multi-channel, bidirectional neural interfaces. In 2019 he joined the Laboratory Equipment Solutions team at Boehringer Ingelheim Pharma GmbH & Co. KG, Biberach as a development engineer.

Dr. Haas received the VDE award for his Master studies in electrical engineering from the University of Ulm in 2014.

Nick Van Helleputte

Nick Van Helleputte received the MS degree in electrical engineering in 2004 from the Katholieke Universiteit Leuven, Belgium. He received his Ph.D. degree from the same institute in 2009 (MICAS research group). His PhD research focused on low-power ultra-wide-band analog front-end receivers for ranging applications. He joined imec in 2009 as an Analog R&D Design Engineer. He is currently R&D manager of the Connected Health Solutions group. His research focus is on ultra-low-power circuits for biomedical applications. He has been involved in analog and mixed-signal ASIC design for wearable and implantable healthcare applications. Dr. Van Helleputte has developed ultra-low-power custom ICs for multi-modal vital signs sensing. His research focused on complete system-on-chip solutions covering all aspects including analog amplification and filtering, analog-to-digital conversion, digital signal and processing power management. He also worked on neural interfaces in the form of active high-density neural probes for the central and peripheral nervous system. In addition to IC design, his research group has a strong focus on highly miniaturized and ultra-low-power systems based on both COTS as well as their custom ASICs. His research collaborations included early pathway research (TRL 1-5) as well as bilateral collaborations with industrial partners towards novel product developments (TRL 5-8). Nick is an IEEE and SSCS member (SSCS Distinguished Lecturer '17-'18) and served on the technical program committee of VLSI circuits symposium and ISSCC.

Kwonjoon Lee

Kwonjoon Lee received B.S. degree in School of Electrical Engineering from the Sung Kyun Kwan University (SKKU), Suwon, Korea, in 2011, and M.S degree in the School of Electrical Engineering from the Korea Advanced Institute of Science and Technology (KAIST), Daejeon, Korea, in 2014. From 2014 to 2015, he was with Next Generation Product Development, System LSI, Samsung Electronics for healthcare IC design. From 2015 to 2018, he was with Healthrian, which is a start-up company for mobile healthcare solution as a healthcare system & IC designer. Since 2017, he has been working toward Ph. D. degree in School of Electrical Engineering from Korea Advanced Institute of Science and Technology (KAIST), Daejeon, Korea.

His research areas are system design for wearable healthcare including application searching, determination of system specification, and bio-medical SoC design. He is also interested in bio-signal processing algorithm based on physiology, signal processing, and machine learning techniques.

Maurits Ortmanns

Maurits Ortmanns received Dr.-Ing. degree from the University of Freiburg, Germany, in 2004.

From 2004 to 2005, he has been with Sci-Worx GmbH, Hannover, Germany, working in the field of mixed- signal circuits for biomedical implants. In 2006, he joined the Integrated Interface Circuits, University of Freiburg, as an Assistant Professor. Since 2008, he has been a Full Professor with the University of Ulm, Ulm, Germany, where he is currently the Head of the Institute of Microelectronics. He holds several patents. He has authored the book Continuous-Time Sigma–Delta A/D Conversion, published several other book chapters, and over 250 IEEE journal and conference papers. His current research interests include mixed-signal integrated circuit design, self-correcting and reconfigurable analog circuits, with special emphasis on data converters and biomedical applications.

Dr. Ortmanns received the VDI and the VDE Award from the Saarland University in 1999, the ITG Publication Award 2015, and the Best Student Paper Awards at MWSCAS 2009 and at SampTA 2011, and the best demo awards at ICECS 2016 and Sensors 2017. He also received the faculty's teaching award 2012 and 2015. He served as a Program Committee Member for ESSCIRC, DATE, ECCTD. He was an Associate Editor of the IEEE Transactions of Circuits and Systems I and II and the Guest Editor of the IEEE Journal Solid-State Circuits. He was a Technical Program Committee Member of the IEEE ISSCC from 2012 to 2016, an Executive Committee Member of ISSCC from 2013 to 2016, and the European TPC Chair of ISSCC from 2015 to 2016.

Jae-sun Seo

Jae-sun Seo received the B.S. degree from Seoul National University in 2001, and the M.S. and Ph.D. degree from the University of Michigan in 2006 and 2010, respectively, all in electrical engineering. He spent graduate research internships at Intel circuit research lab in 2006 and Sun Microsystems VLSI research group in 2008. From January 2010 to December 2013, he was with IBM T. J. Watson Research Center, where he worked on cognitive computing chips under the DARPA SyNAPSE project and energy-efficient integrated circuits for high-performance processors. In January 2014, he joined ASU as an assistant professor in the School of ECEE. During the summer of 2015, he was a visiting faculty at Intel Circuits Research Lab. His research interests include efficient hardware design of machine learning / neuromorphic algorithms and integrated power management. Mr. Seo was a recipient of Samsung Scholarship (2004-2009), IBM Outstanding Technical Achievement Award (2012), and NSF CAREER Award (2017). He is a IEEE Senior Member, and has served on the technical program committee for ISLPED (2013-2019), DAC (2018-2020), ICCAD (2018-2019), on the review committee member for ISCAS (2017-2019), and on the organizing committee for ICCD (2015-2017).

Hoi-Jun Yoo

Hoi-Jun Yoo is an ICT Chair professor in the School of Electrical Engineering and the director of the System Design Innovation and Application Research Center (SDIA) at Korea Advanced Institute of Science and Technology (KAIST). He was the TPC Chair of the ISSCC 2015, a Plenary Speaker of the ISSCC 2019 entitled *"Intelligence on Silicon: From Deep Neural Network Accelerators to Brain-Mimicking AI-SoCs"*, the Chair of Technology Direction (TD) subcommittee of the ISSCC 2013. He is an IEEE fellow since 2008, an Executive Committee member of the Symposium on VLSI, and a Steering Committee member of the A-SSCC, of which he was nominated as the Steering Committee Chair from 2020 to 2025.

He was the VCSEL pioneer in Bell Communications Research at Red Bank, NJ. USA and the manager of the DRAM design group at Hyundai Electronics during the era of 1M DRAM up to 256M SDRAM. From 2003 to 2005, he served as the full time Advisor to the Minister of Korean Ministry of Information and Communication for SoC and Next Generation Computing.

His current research interest includes Bio-Inspired Artificial Intelligence (AI) chip design, Multicore AI SoC design including DNN accelerators, Wearable Healthcare Systems, Network-On-Chip, and High-Speed Low-Power Memory. He has published more than 400 papers, and wrote or edited 5 books: *"DRAM Design"* (1997, Hongneung), *"High Performance DRAM"* (1999 Hongneung), *"Low Power NoC for High Performance SoC Design"* (2008, CRC), *"Mobile 3D Graphics SoC"* (2010, Wiley), and *"Bio-Medical CMOS ICs"* (Co-editing with Chris Van Hoof, 2010, Springer), and co-written chapters in numerous books.

He received the Order of Service Merit from the Korean government in 2011 for his contributions to the Korean memory industry, the Scientist/Engineer of the month Award from the Ministry of Education, Science and Technology of Korea in 2010, the Kyung-Am Scholar

Award in 2014. He also received the Electronic Industrial Association of Korea Award for his contributions to the DRAM technology in 1994, the Hynix Development Award in 1995, the Korea Semiconductor Industry Association Award in 2002, the Best Research of KAIST Award in 2007, the Excellent Scholar of KAIST Award in 2011, and the Best Scholar of KAIST Award in 2019. In addition, he was a co-recipient of the ASP-DAC Design Award in 2001, the A-SSCC Outstanding Design Awards in 2005, 2006, 2007, 2010, 2011, 2014, the ISSCC/DAC Student Design Contest Awards in 2007, 2008, 2010, 2011, and the ISSCC Demonstration Session Recognition in 2016, 2017, 2019 and the Best Paper Award of the IEEE AI-CAS in 2019.

He has served as an Executive Committee member of the ISSCC, the IEEE SSCS Distinguished Lecturer ('10-'11), and the TPC Chair of the International Symposium on Wearable Computers (ISWC) 2010 and the A-SSCC 2008. He was the Editor-In-Chief (EIC) of the Journal of Semiconductor Technology and Science (JSTS) published by the Korean Institute of Electronics and Information Engineers from 2015 to 2019, a guest editor of the IEEE JSSC and the T-BioCAS, and is currently an associate editor of the IEEE JSSC and the SSCL.

Shimeng Yu

Shimeng Yu is an associate professor of electrical and computer engineering at the Georgia Institute of Technology in Atlanta, Georgia. He received the B.S. degree in microelectronics from Peking University, Beijing, China in 2009, and the M.S. degree and Ph.D. degree in electrical engineering from Stanford University, Stanford, California, in 2011 and in 2013, respectively. From 2013 to 2018, he was an assistant professor of electrical and computer engineering at Arizona State University, Tempe, Arizona. Prof. Yu's research interests are nanoelectronic devices and circuits for energy-efficient computing systems. His expertise is on the emerging non-volatile memories (e.g., RRAM) for different applications, such as machine/deep learning accelerator, neuromorphic computing, monolithic 3D integration, and hardware security. Among Prof. Yu's honors, he was a recipient of the NSF Faculty Early CAREER Award in 2016, the IEEE Electron Devices Society (EDS) Early Career Award in 2017, the ACM Special Interests Group on Design Automation (SIGDA) Outstanding New Faculty Award in 2018, the Semiconductor Research Corporation (SRC) Young Faculty Award in 2019, etc. He is a senior member of the IEEE.